BIOLOGICAL AND MEDICAL PHYSICS,
BIOMEDICAL ENGINEERING

BIOLOGICAL AND MEDICAL PHYSICS, BIOMEDICAL ENGINEERING

The fields of biological and medical physics and biomedical engineering are broad, multidisciplinary and dynamic. They lie at the crossroads of frontier research in physics, biology, chemistry, and medicine. The Biological and Medical Physics, Biomedical Engineering Series is intended to be comprehensive, covering a broad range of topics important to the study of the physical, chemical and biological sciences. Its goal is to provide scientists and engineers with textbooks, monographs, and reference works to address the growing need for information.

Books in the series emphasize established and emergent areas of science including molecular, membrane, and mathematical biophysics; photosynthetic energy harvesting and conversion; information processing; physical principles of genetics; sensory communications; automata networks, neural networks, and cellular automata. Equally important will be coverage of applied aspects of biological and medical physics and biomedical engineering such as molecular electronic components and devices, biosensors, medicine, imaging, physical principles of renewable energy production, advanced prostheses, and environmental control and engineering.

Peter Lenz (Ed.)

Cell Motility

 Springer

Peter Lenz
Fachbereich Physik
Philipps-Universität Marburg
Renthof 5
35032 Marburg
Germany

ISBN: 978-0-387-73049-3   e-ISBN: 978-0-387-73050-9

Library of Congress Control Number: 2007938098

Printed on acid-free paper.

9 8 7 6 5 4 3 2 1

springer.com

Preface

Cell motility is a fascinating example of cell behavior that is fundamentally important to a number of biological and pathological processes. Motility of eukaryotic cells is based on a complex, self-organized, mechano-chemical machine consisting of cytoskeletal filaments and molecular motors. This network is highly dynamic, but able to show precise spatial and temporal organization. The machine is regulated by a complex network of biochemical reactions coupled to force and movement-generating processes.

In general, the cytoskeleton is responsible for the movement of the entire cell and for movements within the cell. There are (roughly) two ways by which cells can move: swimming (i.e., movement through liquid) and crawling (i.e., movement across a rigid surface). Swimming cells experience viscous forces that are orders of magnitude greater than inertial forces. Some cells move by undergoing a non-symmetric (i.e., non-reciprocal) sequence of shape changes. Others use the whipping of flagella or the coordinated beating of cilia to propel themselves through the surrounding liquid.

The movement of cells across rigid surfaces is even more complex. One has to distinguish between crawling and gliding. Crawling of a cell (attached to a rigid substrate) requires the coordinated activity of the cytoskeleton, membrane, and adhesion system. Growing actin filaments (typically organized into cross-linked filament networks) exert forces on the membranes creating projections (filopodia, lamellipodia, and pseudopodia) at the leading edge. These attach then to the substrate, converting protrusion into movement along the substrate. Contraction of the cell then leads to forward motion that continues as a tread-milling cycle of front protrusion and rear retraction. Gliding cells slide across a rigid substrate by various mechanisms. Examples include a twitching motion based on the retraction of pili or propulsion by slime extrusion. The bacterium *Myxococcus xanthus* has engines for both modes of motion, and switches between them depending on whether cells move individually or collectively. Gliding of eukaryotic cells is less common, but some parasites move this way by using an actomyosin-based machinery.

Many cellular processes involve transport and movements within the cell boundaries. For example, the replicated chromosomes have to be brought into one of the two daughter cells during mitosis. This process requires many coordinated movements and major structural changes in the cytoskeleton. Also, for many other large molecules and vesicles, the most direct way to specific locations within the cell is along cytoskeletal filaments. This direct transport (which is driven by molecular motors) is much more precise and quicker than diffusional motion.

Molecular motors are essential for many processes of cellular motion. There is a whole variety of different motor proteins. The most important classes include linear motors (such as myosin, kinesin, and dynein), rotatory motors (such as ATPsynthase and bacterial flagella), and nucleic acid motors (such as helicases and topoisomerases). The linear motors use ATP to move along filaments. They are much more than simple transporters. Two-headed motors attach to adjacent filaments leading to sliding of oppositely oriented filaments which is responsible for, e.g., muscle contraction. In collections of motors these induced interactions give rise to a complex cooperative behavior allowing cells to actively deform their shape. On the other hand, single motors can exhibit quite complex shape changes. For example, ATPsynthase (the motor that produces ATP) performs a rotational motion. Other rotatory motors enable bacteria to swim, such as the rotating flagellum of *E.coli*. The latter motors are brought to rotation by a proton or ion flux.

One of the key challenges in cell motility is to develop a complete physical understanding of how and why cells move. Considering the wealth of processes taking place in these systems, and the large number of different cellular components involved, it becomes clear that this can only be a long-term goal. Nevertheless, our understanding of cell motility has increased greatly during the last few years. Progress has been mainly driven by combined theoretical and experimental efforts in studying model systems, by the availability of detailed mechanical, biochemical, and structural data on key components, and by novel theoretical approaches in modeling motile cells. In this book, these new trends are illustrated by several reviews on specific examples of some of the fundamental phenomena and concepts of cell motility.

The first three chapters deal with the best-studied model systems for actin-based cellular motion, namely *Listeria*, keratocytes and *Dictyostelium*. Many cellular movements are based on the dynamics of the actin network. Actin is often associated with myosin to produce the forces required for motion. However, this is not always necessary: the intracellular bacterium *Listeria monocytogenes* uses the host cell's own actin machinery to propel itself. Its motion simply relies on actin polymerization and cross-linking of the network. In this system, theoretical modeling has been very successful partially because actin-based propulsion can be investigated experimentally in reconstituted systems with purified proteins where beads replace the bacterium. The first chapter by Prost et al. summarizes the theoretical approaches and the newest experimental developments.

In whole-cell motility, the interplay between biochemical and mechanical processes is more complicated. Typically, a moving cell needs to coordinate the action of a large number of individual molecular building blocks. While the molecular basis of these dynamical processes are beginning to be understood, only little is known about the large-scale mechanisms the cell uses to achieve coherent cell movement. A summary of the current efforts to bridge this gap is given by Keren and Theriot in the second chapter. They describe the analysis of the interplay of biophysical aspects of actin-based cell motility with the underlying biochemical processes in fish epithelial keratocytes. Keratocytes appear to be the simplest available model system for this purpose because of their simple overall geometry and persistent motion. The role of the main mechanical modules involved in keratocyte motility (namely the actin-myosin cytoskeleton, the cell membrane, and the cytoplasmic fluid) is summarized, and their interplay with biochemical processes in the large-scale coordination of cell motility is discussed.

The amoeba *Dictyostelium discoideum*, also a eukaryotic microorganism, moves by protrusion of pseudopods. In contrast to keratocytes and *Listeria*, however, this motion is directed. In response to starvation, the placement of pseudopods is controlled by the external concentration of a small molecule, cyclic AMP. Surface proteins sense this chemical information, which is then processed by a signal transduction network controlling actin dynamics. In the third chapter Levine and Rappel review theoretical efforts in modeling both the behavior of single cells and multicellular aggregation phenomena. They report on current successes but also demonstrate that realistic models of *Dictyostelium* motility will ultimately require the combination of a microscopic description of the gradient-sensing system with the macroscopic mechanics of force generation, shape transformation, and cell translocation.

Chapters 4 and 5 deal with the mechanical and structural properties of two very important components of the eukaryotic cytoskeleton. Dogterom et al. review the properties of microtubules in Chapter 4. These semi-flexible polymers form networks that fulfill various tasks in cells ranging from providing tracks for motor proteins to dividing the duplicated chromosomes in replicating cells. The focus of this contribution is the force-generation by growing microtubules. Important experimental concepts as well as the theoretical framework required to explain the underlying conversion from chemical to mechanical energy are introduced.

Molecular motors are the topic of the fifth chapter. Sarah Rice presents detailed structural data on kinesin, dynein, and myosin. It is shown how conclusions can be drawn from this data about the conformational changes the motors undergo. The implications for our understanding of the mechanisms of motility and cargo-binding are analyzed.

The sixth chapter is the only one concerned with motion based on the bacterial motility machinery. Makoto Miyata summarizes recent progress in our understanding of the gliding motility of *Mycoplasma*. These bacteria use membrane protrusions at a cell pole for motion. Their gliding machinery is composed of several proteins that can undergo rather complex shape changes. The mysteries on how the bacterium is actually being pulled over the surface by this machinery are just being resolved. It becomes clear from Miyata's review, however, that the involved proteins provide a cytoskeletal structure allowing bacterial motion.

Recently, a new theoretical description of the cytoskeleton has been developed (independently by several groups) that is based on generalized hydrodynamic theories. Such approaches have been highly successful in describing a large variety of properties of complex fluids such as liquid crystals, polymers, and gels. Their extension to active systems (in which energy is continuously supplied by internal or external sources and internally consumed) is a very promising new approach to describe the dynamical properties of mixtures of cytoskeletal filaments and motor proteins. The contribution of Liverpool and Marchetti in Chapter 7 gives an overview on these current theoretical efforts.

Collective effects in living matter are also the topic of the eighth chapter, in which phenomena arising from hydrodynamic interactions in arrays of cilia and rotational motors are discussed. It is shown how these interactions can lead to coordination of ciliar beating and emergence of wave-like structures. Collective ciliar motion is not only important for feeding and swimming of cells, but also seems to play an essential role in the establishment of left and right in developing vertebrates. Despite considerable experimental progress in investigating the generation of symmetry breaking fluid flow by ciliar beating, some fundamental questions remain. Theoretical modeling might be very helpful in clarifying some of these issues. The open questions are introduced together with the current theoretical efforts.

This book contains a few examples of biological problems that illustrate the wealth of novel phenomena one encounters in cell motility. The contributions demonstrate that a truly interdisciplinary approach is required to successfully investigate these systems. A key role is here played by physical concepts. Quantitative studies of moving cells and direct comparison between theory and experiment have only become possible by the application of ideas and models of hydrodynamics, elasticity theory, and statistical physics to living matter.

In such a rapidly growing field, of course, many interesting developments and discoveries are still to come. These reviews, however, already give us an indication in which direction future research has to go. As will be demonstrated by several authors, complex interrelationships between different modular components are characteristic of biological systems. New experimental techniques have to be developed to extract information about the interactions between these components. On the theoretical side, models are needed that bridge the gap between molecular and macroscopic length scales coupling thus biochemical reactions with the mechanics of cellular motion.

Finally, from the point of view of a theoretical physicist, it is highly exciting to observe that physical concepts and theoretical modeling are becoming increasingly important and accepted in biology, and in cell motility in particular. On the other hand, I think these developments are also of interest to physics itself because these living systems give us the opportunity to investigate such fundamental phenomena as transport, dynamical phase transitions, and emergence of order and pattern formation far from thermodynamic equilibrium in a completely novel context.

Marburg, March 2007   *Peter Lenz*

Contents

List of Contributors

Marileen Dogterom
FOM Institute for Atomic and
Molecular Physics (AMOLF),
Kruislaan 407, 1098 SJ Amsterdam,
The Netherlands

Julien Husson
FOM Institute for Atomic and
Molecular Physics (AMOLF),
Kruislaan 407, 1098 SJ Amsterdam,
The Netherlands

Jean-François Joanny
Institut Curie, UMR 168, 26 rue
d'Ulm, F-75248 Paris Cédex 05,
France

Kinneret Keren
Department of Biochemistry, Stan-
ford University, Stanford, CA 94305,
USA

Liedewij Laan
FOM Institute for Atomic and
Molecular Physics (AMOLF),
Kruislaan 407, 1098 SJ Amsterdam,
The Netherlands

Peter Lenz
Fachbereich Physik, Philipps-
Universität Marburg, D-35032
Marburg, Germany

Herbert Levine
Center for Theoretical Biological
Physics, University of California San
Diego, 9500 Gilman Drive, La Jolla,
CA 92093, USA

Tanniemola B. Liverpool
Department of Mathematics, Uni-
versity of Bristol, University Walk,
Bristol BS8 1TW, UK

M. Cristina Marchetti
Physics Department, Syracuse
University, Syracuse, NY 13244,
USA

Makoto Miyata
Department of Biology, Graduate
School of Science, Osaka City
University Sumiyoshi-ku, Osaka
558-8585, JAPAN

Laura Munteanu
FOM Institute for Atomic and
Molecular Physics (AMOLF),
Kruislaan 407, 1098 SJ Amsterdam,
The Netherlands

Jacques Prost
ESPCI, 10 rue Vauquelin, F-75231
Paris Cédex 05, France

Wouter-Jan Rappel
Center for Theoretical Biological
Physics, University of California San
Diego, 9500 Gilman Drive, La Jolla,
CA 92093, USA

Sarah Rice
Northwestern University, Department of Cell and Molecular Biology,
303 E. Chicago Ave., Ward 8-007,
Chicago, IL 60611, USA

Cécile Sykes
Institut Curie, UMR 168, 26 rue
d'Ulm, F-75248 Paris Cédex 05,
France

Julie A. Theriot
Department of Biochemistry and
Department of Microbiology and
Immunology, Stanford University,
Stanford, CA 94305, USA

Christian Tischer
FOM Institute for Atomic and
Molecular Physics (AMOLF),
Kruislaan 407, 1098 SJ Amsterdam,
The Netherlands

The Physics Of *Listeria* Propulsion

Jacques Prost[1,2], Jean-François Joanny[1,3], Peter Lenz[4], and Cécile Sykes[1,5]

[1] Institut Curie, UMR 168, 26 rue d'Ulm, F-75248 Paris Cédex 05, France
[2] ESPCI, 10 rue Vauquelin, F-75231 PARIS Cédex 05, France
   jacques.prost@espci.fr
[3] jean-francois.joanny@curie.fr
[4] Fachbereich Physik, Philipps-Universität Marburg, D-35032 Marburg, Germany
   peter.lenz@physik.uni-marburg.de
[5] cecile.sykes@curie.fr

1.1 Introduction

Listeria is a pathogenic bacterium, which can be dangerous for immune deficient individuals. It can be found almost everywhere, in particular in food such as soft cheese and smoked salmon. After ingestion, it is able to penetrate into the cellular system where it moves from cell to cell and divides on average every twenty minutes. After a few hours of growth, host-cell actin filaments begin to form a dense cloud on the surface of the bacterium. Polarization of the cloud leads to the formation of a comet tail consisting of an oriented, cross-linked network of filaments which pushes the bacterium forward (see Figure 1.1). This movement can be very rapid (up to 1μm/s [1]). Finally, the infection spreads as the bacteria push their way through their plasma membrane to invade the neighboring cells. Because it is inside the cellular system, it is hard to be detected by the immune system.

To understand the *Listeria* propulsion mechanism, a particularly intense scientific activity has been developed over these last years [2, 3, 4, 5, 6]. Why should one be particularly interested in this problem? The reason is that *Listeria* motility motion is due to the polymerization and cross-linking of an actin gel (i.e., the comet) just like eukaryotic cell motility is due to the polymerization of actin in the cell lamellipodium. It is then believed that learning something about *Listeria* is useful for understanding eukaryotic cells as well. Of course, studying *Listeria* does not avoid studying eukaryotic cells because there are many more aspects to eukaryotic motility than to *Listeria* motility [7] (like adhesion, molecular motors, etc.). Yet this allows us to select one aspect, namely actin polymerization and cross-linking, in geometrical conditions that are much simpler than those of eukaryotic cells because the process is exterior to the bacteria. One can use cell extracts in particular to perform *in vitro* experiments in reconstituted media. Yet, a priori simpler than eukaryotic cell motility, *Listeria* motility has its mysteries: in particular, a mutant has been observed to move by a succession of jumps followed by periods during which the bacteria is essentially immobile [8], see Figure 1.2. During the waiting period, the gel grows around the whole bacteria producing some kind of sheath. Eventually, the bacterium gets expelled from the sheath, hence the jerky motion. In the following, we give a few guidelines for thinking about the physics of the propulsion mechanism, and show that it is essentially a continuum mechanics problem with very unusual boundary conditions.

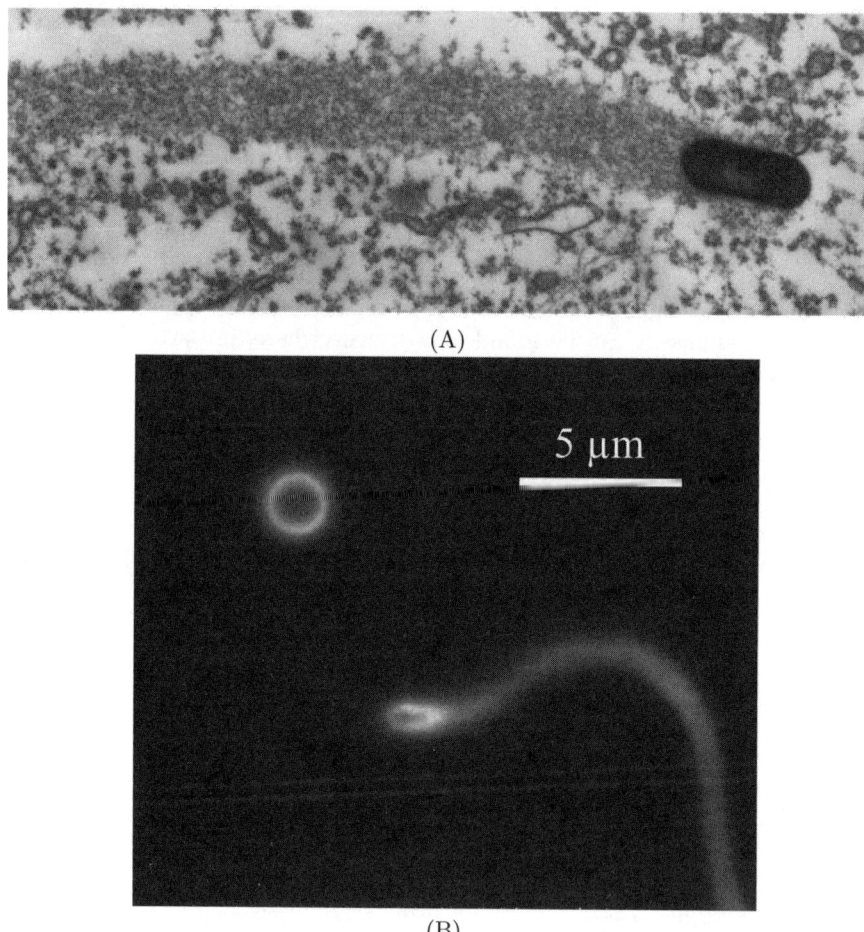

(A)

(B)

Figure 1.1. Example of wild type *Listeria* and its homogeneous comet as seen by electron microscopy (A) and confocal microscopy (B). Note the actin layer in front of the bacterium is missing in (A) while the fluorescence image (B) clearly shows that there is a gel. Figure (A) is reprinted from [9]. ©(1992), with permission from Elsevier, Figure (B) courtesy of Vincent Noireaux. (See color insert.)

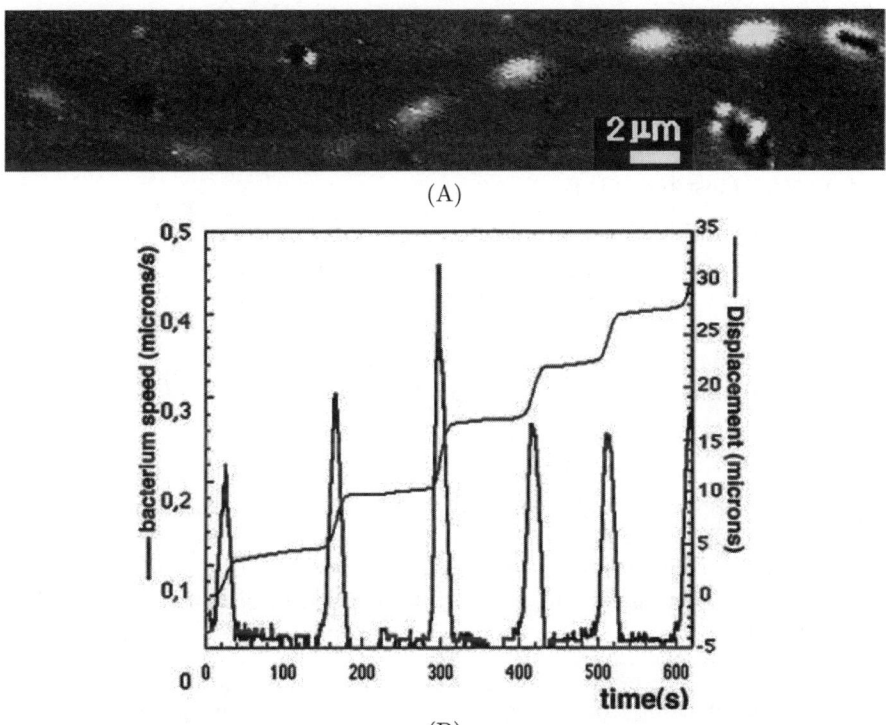

Figure 1.2. $ActA_{\Delta 21-97}$ *Listeria* mutant (courtesy Lasa et al. [8]): (A) snapshots of the motion of the mutant seen at the same time by phase-contrast and fluorescent microscopy. (B) Displacement and velocity as function of time. Figure is reprinted from [10] with permission from The Biophysical Society.

1.2 A Genuine Gel

1.2.1 A Little Chemistry

Before getting to the characterization of the gel, it is necessary to give a few tips concerning the biochemistry of the polymerization process. First, it is now well-established that to observe the formation of a comet, a particular enzyme called ActA must be present on the surface of the bacterium. *In vitro* experiments can be made by placing bacteria in cell extracts or in reconstituted extracts. The presence of ActA is necessary but not sufficient for getting a polymerization process comparable to the one observed *in vivo*. Actin filaments are polar: they have two fundamentally different extremities. One, called barbed (or plus) end, can polymerize while the other, called pointed (or minus) end, can simultaneously depolymerize. Of course this can only happen if energy is constantly fed into the system and this is done by hydrolyzing ATP into ADP. Phosphorylated monomers polymerize at the barbed end, while dephosphorylated monomers depolymerize at the pointed end. In the absence of any other protein, the polymerization rate of actin is about two orders of magnitude slower than *in vivo*. To obtain the right values, three other types of proteins must be added:

a) cofactors, which speed up depolymerization at the pointed end and thus speeds up turn over and polymerization at the barbed end,
b) capping proteins, which cap free barbed ends and localize polymerization strictly at the bacterium surface or its immediate neighborhood,
c) a protein complex called Arp2/3 which provides branching to the network [11, 12], and also speeds up the gel formation.

The identification of the minimum number of constituents necessary to reproduce *in vitro*, the bacterium motion was an important step towards a quantitative understanding of the process [13]. Another important step is the demonstration that it is possible to replace the bacteria itself by inert beads such as polystyrene beads, on which the enzyme ActA is grafted or adsorbed [14, 15]. More details on this aspect will be given in Sections 1.4 and 1.5 of this article.

1.2.2 Elastic Behavior

In the introduction of this article we wrote without further justification that the comet-like structure of Figure 1.1 was a gel, and that polymerization was taking place at the bacterium surface. That polymerization is indeed taking place at the bacterium surface was shown by using fluorescently labeled actin [16, 17]. This still does not tell us that the actin gel is a real gel in the sense that it has the mechanical properties of an elastic body. This can be done by cutting pieces of the comet using laser surgery techniques, and measuring the bending modulus of the comet [18, 19]. These experiments show that the comet does have elastic behavior over time scales of minutes. Elastic moduli are found to be in the kilo-Pascal range with a large total spread of two orders of magnitude. This spread is not related to a slow depolymerization process known to exist in the comet and that can be studied independently. These experiments show that the gel elastic properties must be explicitly taken into account in the physics of the motion.

Next, one wants to know about the connection between the bacterium surface and the gel. By using laser tweezers or better electric fields one can exert piconewton forces between bacterium and comet during typically a minute: no relative motion at a micron resolution can be detected [19]. This shows that the bacterium is firmly connected to the gel. More quantitatively, if one describes the bacterium-gel lateral interaction by a friction coefficient, such experiments put a lower limit to the friction coefficient, four orders of magnitudes larger than the hydrodynamic friction coefficient of the bacterium on the surrounding fluid!

1.3 Hydrodynamics and Mechanics

1.3.1 Motion in the Laboratory Frame

One can split the problem of *Listeria* motion into two parts. First, an external and simple part is deciding which of the bacterium or the comet moves with respect to the surrounding fluid, and second is an internal part describing the motion of the bacterium relative to the comet. The Reynolds numbers in this problem are extremely low (i.e., of the order of 10^{-7}) which means that only friction forces should be retained. Under such conditions, the external dynamics reads

$$f^{ext} = \zeta_b v_b + \zeta_c v_c, \tag{1.1}$$

where f^{ext} is an external force acting on the bacterium/comet system, ζ_b and ζ_c are the friction coefficients on the surrounding fluid, and v_b and v_c are the velocities of the bacterium and the comet, respectively. The polymerization process itself is responsible for the existence of a relative velocity between bacterium and comet: $v_b - v_c = v$. Note that v is not related simply to the polymerization rate v_p, as illustrated in the following. A combination of the two equations allows us to extract the velocity of the bacterium with respect to the surrounding fluid

$$v_b = \frac{\zeta_c v + f^{ext}}{\zeta_b + \zeta_c}. \tag{1.2}$$

Only in the absence of external force, and when $\zeta_c \gg \zeta_b$, does one have $v_b \approx v$.

This limit is obtained as soon as the comet length is larger than the bacterium size, and does not depend on the surrounding fluid viscosity because both friction coefficients are proportional to it.

1.3.2 Propulsion and Steady Velocity Regimes

Because experiments show that the comet is indeed a gel in the continuum mechanics sense, this implies that one has to understand what kind of stresses are generated by the polymerization-gelation process. If polymerization was taking place only at the rear part of *Listeria* then life would be simple. The velocity v would be simply the polymerization velocity v_p. One could write, for instance, $v \simeq v_p = a(k_+^b c_i - k_-^b)$, in which k_+^b, k_-^b, c_i, a are, respectively, the polymerization and depolymerization rates, the actin monomer concentration at the bacterium surface, and the actin monomer size. Most microscopic theories are concerned with the calculation of v_p [20, 21, 22, 23, 24].

Let us first consider the case of a flat disk of radius R, replacing the bacterium, and show that it corresponds to this simple limit. One side is treated with a suitable actin nucleator, the other side is not. As we will see in Section 1.5.3, approaching conditions have been experimentally obtained [25]. Actin polymerizes on the treated side only, a comet develops, and the disk moves, see Figure 1.3. The argument that we used to describe the bacterium motion is exactly the same provided we replace v_b by v_d the disk velocity. The conservation of the polymerized actin flux tells us that v is the polymerization rate. There are slight corrections due to a possible strain relaxation of the gel, discussed in [10]. With this proviso, one can write: $v \simeq v_p = a(k_+^b c_i - k_-^b)$.

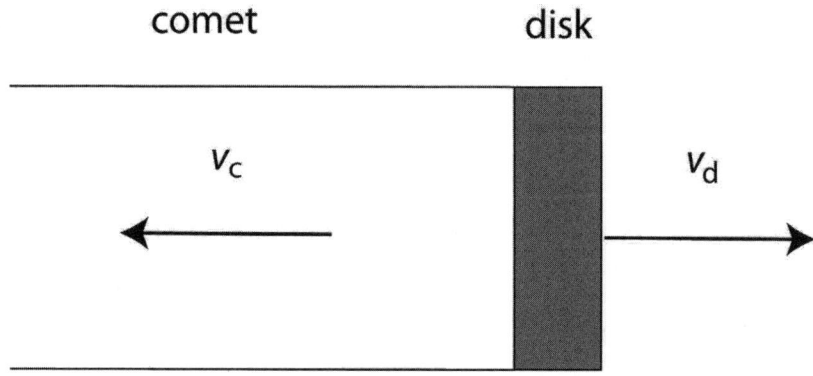

Figure 1.3. In the absence of external force, a flat disk moves with the polymerization velocity $v = v_d - v_c \simeq v_p$. Note, that motion also occurs if the disk is larger than the comet tail.

We will describe the stress dependence of the polymerization process in more detail, but we know experimentally that forces have to reach values of the order of piconewtons per filament to significantly influence the polymerization process. All we need to know here is that polymerizing filaments exert a force of $f = \sigma l^2$ on the nucleator, where σ is the stress normal to the surface and l^2 is the average area per polymerizing filament. Where does the stress come from in this example? In view of the small velocities we are considering here, the actin gel is to a very good approximation in mechanical equilibrium. Hence the divergence of the stress vanishes and the sum of all forces exerted by the gel on its external surrounding add up to zero. This implies that the force exerted by the gel on the disk is exactly equal and opposite to the force exerted by the fluid on the gel. The condition of a constant polymerization rate requires a constant stress, and in the absence of external force we obtain

$$\sigma = \frac{\zeta_c v_c}{\pi R^2} = \frac{\zeta_d \zeta_c v}{(\zeta_c + \zeta_d)\pi R^2} \simeq \frac{\zeta_d v}{\pi R^2}, \tag{1.3}$$

and

$$f \simeq \eta v l^2 / R. \tag{1.4}$$

With extract viscosities of the order of a few times that of water, a disk radius of a few microns, and a distance l of a few tens of nanometers one obtains a force of the order of 10^{-8}pN. This force is much smaller than typical forces required to have any influence on the polymerization process. The observed velocity corresponds to a stress-free polymerization rate. Note also that an increase of the viscosity by eight orders of magnitude would be required to influence v in a significant way via the developed stress. Such an experiment has been reported [26]. Yet, for most practical purposes, one can ignore the world external to *Listeria* for discussing its motion. This flat geometry is unique in that it allows for a direct comparison between observed velocities and microscopic theories.

Note further that we have ignored stresses arising from the fact that the addition of monomers at the disk-comet interface requires the removal of extract molecules, say water, from the surface to the external fluid through the gel. This is a permeation process. It is straightforward to show that the resulting force scales exactly as f, provided one replaces v by

$$v' \simeq va^2/l^2. \tag{1.5}$$

The permeation force is even weaker and the obtained results hold.

In the case of a real bacterium, however, the gel grows not only at the rear of the bacterium. In other bacteria, like *Shigella* [27], or in vesicles developing comets [28, 29], the comet is hollow which means that there is even no rear gel. To understand the propulsion mechanism, one has to understand that once a first layer has been polymerized and cross-linked, a subsequent polymerization can only occur if the first layer is stretched to a new position leaving space for it (see Figure 1.4). The stretching costs elastic energy and the release of this energy is the driving force for the motion. More precisely, if E is the shear modulus of the gel, the stored elastic energy per unit length reads

$$W_e \simeq E(e/r^2)2\pi reL, \tag{1.6}$$

where $e, r, L, (e/r)$ are the gel thickness, the bacterium radius, length and tensile strain, respectively (see Figure 1.5). Upon gel escape from the bacterium shell a fraction of this energy is released. If this fraction is of order one the propelling force is [18, 19]

$$f_p \simeq 2\pi Ee^3/r. \tag{1.7}$$

The gel-bacterium friction balances this force (remember we have shown that external hydrodynamic friction is, in most practical situations, entirely negligible). We will discuss in more details the notion of gel/bacterium friction, but for the sake of argument let us first describe the friction force by a friction coefficient ξ

$$f_f = \xi v 2\pi rL. \tag{1.8}$$

One immediately obtains an expression for the velocity: $v \simeq v_i e^3/(r^2L)$, in which $v_i = \left(\frac{E}{\xi}\right)$ is an intrinsic velocity scale, related only indirectly to the polymerization process. Two limits merit discussion.

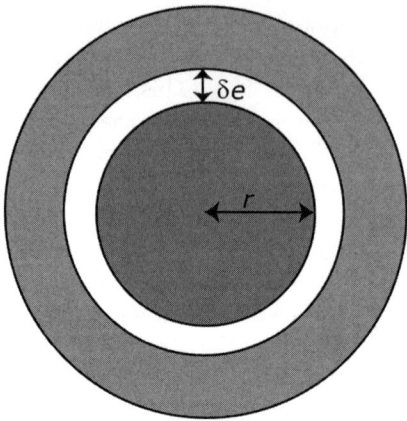

Figure 1.4. The grey gel layer, initially with inner radius r must stretch to a new radius $r + \delta e$ to allow a new gel layer to form.

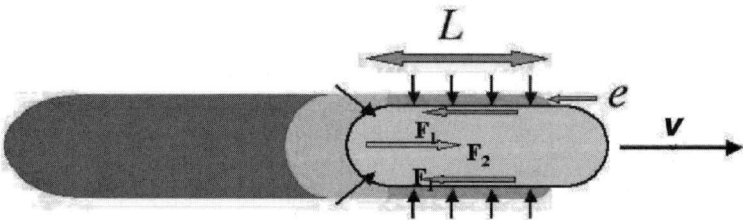

Figure 1.5. The elastic force gives rise to a propulsion force F_1 along the bacterium axis. The surface friction equilibrates F_1 with F_2.

In the first, neither the developed stress nor the actin monomer depletion are large enough to influence the polymerization process. Then, in steady state

$$e = \left(\frac{v_p}{v}\right) L, \qquad (1.9)$$

because the time e/v_p needed to generate a thickness e must equal the time L/v to advance one length L. Then

$$v \simeq v_i^{1/4} v_p^{3/4} \left(\frac{L}{r}\right)^{1/2}. \qquad (1.10)$$

Note that in this regime, the bacterium velocity relative to the gel v can be larger than the polymerization rate v_p, and that it is not proportional to it. Note also that it depends only weakly on the comet gel properties v_i: a two orders of magnitude change of the gel elastic modulus (everything else being kept constant) results in a factor of three change of the velocity v only. This explains why the experimentally observed velocity spread is by no means comparable to the one found for the gel modulus.

In the second regime, the thickness saturates to a value e^* controlled either by the developed stresses or by the actin monomer diffusion process. Then,

$$v = v_i \frac{(e^*)^3}{r^2 L}. \tag{1.11}$$

We will show that under appropriate circumstances $e^* = e_2 \log(v_p^0/v_{dp}^0)$, in which e_2 is a length proportional to the bacterium radius r, and v_p^0, v_{dp}^0 are the polymerization and depolymerization rates in the absence of stress, respectively. In this case, the dependence of the bacterium velocity on the polymerization rate is extremely weak!

1.3.3 Gel/Bacterium Friction and Saltatory Behavior

In the above discussion, we have used the notion of surface friction without further justification. The physical nature of this friction may be understood the following way: during the polymerization process, the actin filaments spend some time τ_c connected to the bacterium surface, and some other time τ_d disconnected to it. When the gel moves with respect to the bacterium, the connected filaments gradually distort until they detach. A force results from the distortion, as first understood by Tawada and Sekimoto [30]. The average force per unit area reads $F_f = n_c \phi$, where n_c is the average number of connected filaments and ϕ is a typical force per filament. In steady state: $n_c = n \frac{\tau_c}{\tau_c + \tau_d}$, where n is the number of enzymes per unit area on the bacterium surface. The force ϕ is simply given in terms of the product of a filament elastic modulus K multiplied by a typical displacement $v\tau_c$.

A first regime of small velocities is easy to discuss. Both τ_c and τ_d have their intrinsic thermodynamic value, τ_c^0 and τ_d^0, and the notion of a friction coefficient with a velocity independent value emerges as anticipated:

$$\xi = n \frac{(\tau_c^0)^2}{\tau_c^0 + \tau_d^0} nK. \tag{1.12}$$

If one estimates the gel elastic modulus on dimensional grounds by $E \simeq kT\lambda_p/\lambda^4$, and the filament surface modulus by $K \simeq kT\lambda_p/\lambda^3$, in which $\lambda_p \simeq 10\mu$m is the actin filament persistence length and λ the average distance between cross-links, then the intrinsic velocity takes the very simple form: $v_i \simeq \frac{(\tau_c^0 + \tau_d^0)}{\lambda n (\tau_c^0)^2}$, which further simplifies to

$$v_i \simeq \frac{\lambda(\tau_c^0 + \tau_d^0)}{(\tau_c^0)^2}. \tag{1.13}$$

Although it is possible to have reasonable values of λ, nothing is known on the connected and disconnected times.

The second regime is that of high velocities. The connections are broken in times much shorter than the thermodynamic connection time τ_c^0, such that the work done on the connection with the enzyme is of the order of the potential barrier w_b hindering the escape of the filament from its bound state. This condition requires $Kv\tau_c a_b \simeq w_b$, in which a_b is a length of order a. The essential result is that now the connected time is inversely proportional to the velocity and the friction force becomes also inversely proportional to it: $F_f \simeq \frac{nw_b^2}{\tau_c K v}$. The friction force due to this phenomenon tends to zero, simply because all bonds break. The total friction does not vanish though, because there is always a conventional hydrodynamic friction. The total curve $F_f(v)$ plotted in Figure 1.6 exhibits the typical shape of a solid-on-solid friction with stick/slip behavior. Under such circumstances, the saltatory mutant is easy to understand. A conventional steady state smooth motion is obtained when the curve characterizing the elastic force intersects the friction curve once. The saltatory behavior is obtained when the friction oscillates between the high and low friction regimes with the oscillation frequency depending on the size of the bead [31]. Typically, the bacterium starts to accumulate a thick gel layer until the elastic force reaches a value such that the unstable regime is reached, that is until surface bonds break. In this phase, the bacterium velocity is very small. Then the gel layer is quickly expelled, which gives a velocity burst and leaves the bacterium surface in its initial state; the cycle can start again. A more elaborate description can be found in [10, 18]. As we will see in Section 1.5.1, diffusion and convection of the actin polymerization activators on the surface of the beads (or cell membrane) is another source of saltatory behavior.

It is interesting to remark that all known phenotypes can be assembled in a single dynamical state diagram, provided one treats the side gel and the rear gel at the same level [10, 32]. This work shows in particular that the rear part of the gel does not in general participate positively in the propulsive force. Only close to stall force does the rear part contribute positively. As a result, the force velocity relation was predicted to exhibit two regimes [10] (as meanwhile also observed experimentally, see Section 1.5.1). This analysis further shows that what is called a saltatory mutant is in fact nothing but the crossing of a state boundary due to the mutation. The crossing, however, might result from many other causes. We illustrate this remark in the following section. Note eventually that although microscopic models are interesting in their own right, a comprehensive analysis cannot ignore the elastic level of interpretation, which naturally provides a correct distinction between internal and external forces.

1.4 Bio-Mimetic Approach

1.4.1 A Spherical *Listeria*

If it is true that *Listeria* needs only to display the enzyme ActA on its surface and for the rest it steals all the needed compounds from the surrounding cell,

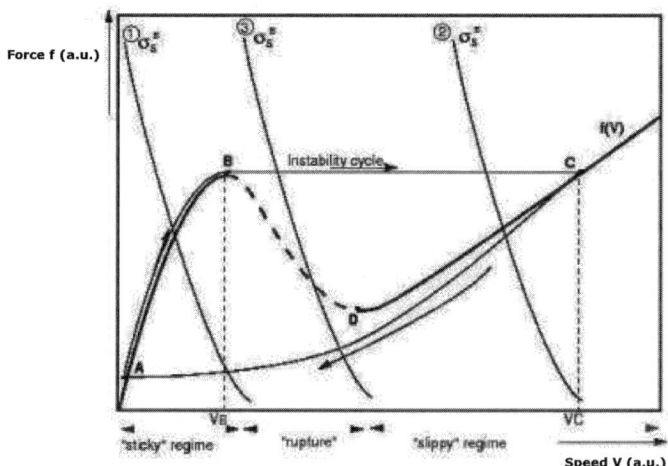

Figure 1.6. Typical curve giving the friction as a function of the bacterium/gel differential velocity. Figure is from [6].

it should be possible to replace the bacterium by an inert bead on which this enzyme is grafted. Then, placing the bead in a cell extract or in a reconstituted extract containing all relevant proteins and energy sources, one should be able to observe actin polymerization and hopefully a comet formation. This has been done in several laboratories [14, 15], proving that indeed only ActA was needed at the surface. It was further shown, that human actin polymerization enzymes could give rise to similar observations [33, 34]. Can one learn more with these *in vitro* bio-mimetic assays?

1.4.2 Spherical Symmetry

In many cases, the beads that are used are spherical, and unless a symmetry-breaking process develops, the produced actin gel respects the beads' symmetry. One observes that gel growth stops after a given thickness is reached. It is possible to prove that it corresponds to a steady state [35]: polymerization is still going on at the bead surface, while depolymerization takes place at the outer one. The observed thickness is always a fraction of the bead radius, and is orders of magnitude smaller than the comet length. Why is it so?

The point is that both the polymerization rate at the inner surface, and the depolymerization rate at the outer surface depend on the stresses that develop as we have already explained in the preceding paragraphs. The polymerization rate at the inner surface decreases under the action of the compressive normal stress, while the depolymerization rate at the outer surface increases under the action of the tensile stress. When both take on the same value, a steady state is reached.

It is possible to be more quantitative by writing standard chemical rate equations

$$\frac{dv_i}{dt} = k_+^b c_i - k_-^b \tag{1.14}$$

$$\frac{dv_e}{dt} = -k_-^p \tag{1.15}$$

$$\frac{de}{dt} = a\left(\frac{dv_i}{dt} + \frac{dv_e}{dt}\right). \tag{1.16}$$

Here, $\frac{dv_i}{dt}$ and $\frac{dv_e}{dt}$ are the average number of added monomers per unit time per filament at the inner and the outer surfaces, k_+^b is the second order rate constant for the addition of a monomer at the inner surface where the monomer density is c_i, and k_-^b, k_-^p are the first order rate constants for removing a monomer from the filament at the inner and outer surfaces, respectively. The monomer addition events at the outer surface are rare enough that the corresponding term can be safely neglected in all cases. The superscripts b, p, stand for barbed and pointed and indicate that the polymerization takes place at the barbed end while the depolymerization takes place at the pointed end. As in the preceding paragraph, e is the gel thickness and a the typical size of a monomer.

The stress dependence of the rates results from the fact that the potential barrier for adding or suppressing a monomer is shifted from its value at zero stress, from a quantity equal to the work given by the force that a given filament exerts on the link of interest. It is thus of the form $k = k_0 \exp\left(\frac{-fa}{kT}\right)$. The forces are deduced simply from the stresses. We have already noted that the tensile strain was $\frac{e}{r}$ in the geometry of *Listeria* and it is still the case in spherical geometry. Thus, the tensile stress at the outer surface is $\sigma_t \simeq E\frac{e}{r}$ and the force per filament is $f_t = \sigma_t l^2$, where $l = n^{-1/2}$ is the average distance between filaments. The normal stress obeys Laplace's law: $\sigma_r = \frac{2T}{r}$ where $T \simeq r\sigma_t$ is the total tension across the gel layer. We thus get $\sigma_n \simeq 2E\frac{e^2}{r^2}$. With all these remarks, we can write

$$k_+^b = k_+^{b0} \exp\left(-\frac{e^2}{e_0^2}\right) \tag{1.17}$$

$$k_-^b = k_-^{b0} \exp\left(\frac{e^2}{e_1^2}\right) \tag{1.18}$$

$$k_-^p = k_-^{p0} \exp\left(\frac{e}{e_2}\right). \tag{1.19}$$

With $e_i = r\left(\frac{kT}{a_i l^2 E}\right)^j$ and $j = 1/2$ for $i = 0, 1$, and $j = 1$ for $i = 2$. In all cases a_i is a length of order a. Note that all e_i scale like r.

1.4.3 Steady State

To discuss the conditions for steady state, one still needs to express the monomer concentration at the inner surface c_i as a function of its concentration at infinity c_∞. In a first approximation, it is reasonable to assume that the actin monomer concentration obeys a standard diffusion law. Thus, in steady state, and for $e \ll r$, the flux $j = -D\frac{\partial c}{\partial r}$ =const. in the gel and $c = c_\infty$ outside. Monomer conservation further imposes: $l^2 D\frac{\partial c}{\partial r} = \frac{dn_i}{dt} = \frac{dn_e}{dt}$. These conditions specify entirely the problem. One finds two regimes connected by a smooth crossover. For small radii, the inner concentration is essentially c_∞, and the steady state thickness is solution of the equation

$$\frac{c_\infty k_+^{b0}}{k_-^{p0}} = \frac{k_-^{b0}}{k_-^{p0}} \exp\left[(e^*)^2 \left(e_0^{-2} + e_1^{-2}\right)\right] + \exp\left(e^* e_2^{-1} + (e^*)^2 e_0^{-2}\right). \quad (1.20)$$

Because all the a_i scale like r, the steady state thickness e^* also scales like r. Actually, with numbers relevant to experimental situations one expects

$$e^* \simeq e/10. \quad (1.21)$$

This is what is observed experimentally for radii smaller than 10 microns [15]. Such numbers imply that the normal stress exerted by the gel on the bead is of the order of one atmosphere.

Note that if the leading term is provided by the depolymerization at the barbed end, one finds exactly the expression announced in Section 1.3.2 that is (with transparent notations), $e^* = e_2 \log\left(\frac{v_p^0}{v_{dp}^0}\right)$.

The other limit corresponds to what happens on a flat surface. Then no stress develops, but the thickness is still limited by the monomer depletion due to the need for the monomers to diffuse in from outside. Now, the steady state condition reads simply

$$l^2 D\left(\frac{c_\infty - c_i}{e^*}\right) = c_i k_+^b - k_-^b = k_-^p. \quad (1.22)$$

For all practical purposes, k_-^b can be neglected in this stress-free situation. It is then easy to infer

$$e^* = \frac{l^2 D c_\infty}{k_-^p}. \quad (1.23)$$

For large enough beads, this regime is always obtained. The crossover radius between the two regimes is given by $r_e \simeq \frac{a^{1/2} D l^3 c_\infty E^{1/2}}{(kT)^{1/2} k_-^p}$. Plugging the value of the diffusion constant as measured in solution, we estimate the crossover radius to be in the millimeter range. It turns out that one can clearly observe the two regimes, which implies that the monomer diffusion constant is about the same as in cells [36]. There may be many reasons for this large difference (temporary fixation sites on the gel, steric hindrance, other objects like the Arp2/3 complex diffusing slowly and limiting the polymerization rate).

What is more important is that the measured value ($D \simeq .02\mu m^2 s^{-1}$) is such that diffusion processes cannot be neglected in cells. The crossover length corresponds precisely to typical cell lengths.

1.4.4 Growth with Spherical Symmetry

The arguments developed above allow us to write the gel thickness growth as

$$\frac{de}{dt} = a \left[k_+^{b0} \exp \left(-\frac{e^2}{e_0^2} \right) - k_-^{p0} \exp \left(-\frac{e^2}{e_0^2} \right) \right], \tag{1.24}$$

where we have omitted the k_-^b term for the sake of simplicity. The solution of this equation, with initial condition $e = 0$, is a continuous monotonic function with essentially two regimes

- at short times, the growth is predicted to be linear: $\frac{de}{dt} \simeq a(k_+^{b0} - k_-^{p0})$, that is with obvious notations, $e = v_p^0 t$. Such a relation is probably too naive because it ignores the problems of the nucleation of filaments and of their multiplication with Arp2/3, etc.
- at long times, the gel thickness is close to its steady state value e^* and the dynamical equation can be linearized as a function of $\delta e = e - e^*$. The solution is then

$$\delta e(t) = \delta e(t_1) \exp \left(-\frac{t - t_1}{\tau} \right) \tag{1.25}$$

$$\tau^{-1} = k_+^{b0} \left(2 \frac{e^* a}{e_0^2} + \frac{a}{e_2} \right) \exp \left(-\frac{(e^*)^2}{e_0^2} \right) = \left(\frac{2e^*}{e_0^2} + \frac{1}{e_2} \right) v_p. \tag{1.26}$$

The thickness approaches its steady state value exponentially with a time constant growing linearly with the sphere radius r.

1.4.5 Symmetry Breaking

Next, we consider the transition from a spherical gel to the symmetry-broken state. In the following, we keep the arguments as simple as possible, and deal only with the thin shell regime. A more elaborate version can be found in [37]. As understood first by K. Sekimoto, the important point is that mechanical equilibrium requires that the total integrated tension across the gel thickness must be constant everywhere. The tensile stress must be larger in regions of smaller gel thickness because the integrated tension is the same. Now, because the depolymerization rate at the exterior surface depends exponentially on the tensile stress, this means that the depolymerization is faster where the gel is

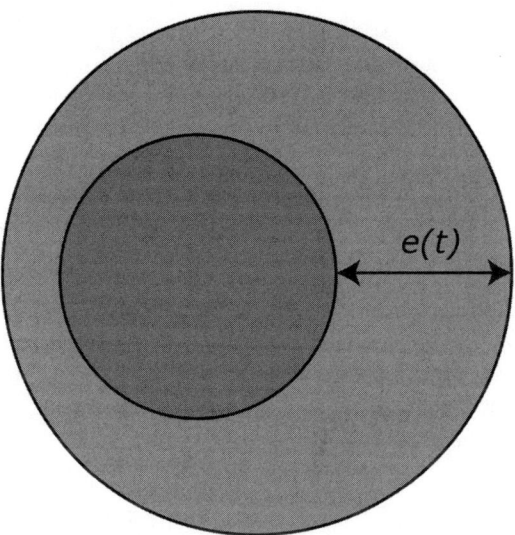

Figure 1.7. Example of a symmetry-breaking perturbation.

thinner, which amplifies the thinning process. This is a clear sign of instability, which we investigate in the following.

Assume that at some point the gel thickness has picked up a sinusoidal variation of the sort shown in Figure 1.7

$$e(t) = e_i(t) + \varepsilon(t) \cos \theta. \tag{1.27}$$

In this equations θ is the polar angle in spherical coordinates, and e_i and ε are the isotropic and anisotropic parts of the gel thickness. We consider the system in the late time regime because the stresses play essentially no role in the earlier regime. Thus we can consider $e_i - e^*$ and ε as small quantities that we need to treat at first order only for the discussion of stability. The tension being constant, it cannot depend on θ, and can only depend on e_i and ε^2, thus to lowest order

$$T = E \frac{e_i^2}{r} + \mathcal{O}(\varepsilon^2) \tag{1.28}$$

$$\sigma_r = \frac{T}{r} = \text{const.} \tag{1.29}$$

On the contrary,

$$\sigma_\theta = \frac{T}{e(\theta, t)} \simeq \frac{T}{e_i} \left(1 - \frac{\varepsilon}{e_i} \cos \theta \right) + \mathcal{O}(\varepsilon^2). \tag{1.30}$$

As announced, the tensile stress is maximal where the thickness is minimal.

If we extend the use of the dynamical equation to this non-homogeneous case, we can write by linearizing Equation (1.24)

$$\frac{d\delta e_i}{dt} = -\frac{\delta e_i}{\tau} \tag{1.31}$$

$$\frac{d\varepsilon}{dt} = \frac{\varepsilon}{\tau_0}. \tag{1.32}$$

As expected, we find that the isotropic part is stable and converges exponentially toward the steady state value e^*, and that the symmetry-breaking part is always unstable and grows exponentially fast with a time scale $\tau_0 = \tau\left(1 + 2\frac{e_2 e^*}{e_0^2}\right) = \frac{e_2}{v_p}$. Note that τ_0 grows linearly with the bead radius, just like τ, and is of the same order of magnitude.

Now we know that symmetry-breaking fluctuations are amplified, but where do they come from? Could thermal fluctuations be sufficient?

The typical fluctuation amplitude can be expressed as

$$\varepsilon_b \simeq \left(\frac{kTe^*}{Er^2}\right)^{1/2}. \tag{1.33}$$

For a 100nm bead such as that used in Reference [14], the corresponding amplitude is in the nanometer range. It might be sufficient to trigger the instability because of its exponential amplification. However, other factors such as enzyme heterogeneous distribution or deviation from pure spherical shape may be more important (similar effects can also arise from polymerization fluctuations [38]). For instance, if there is a total number N of enzymes at the bead surface, there will always be an imbalance of the order of $N^{1/2}$ between the two sides of the bead. The integrated tension T is still a constant and so is the normal stress σ_r at the bead surface. However, the force acting on filaments now depends on position through the enzyme density dependence. With $l^2(\theta) = l_i^2 + 2l_a l_i \cos\theta$, and l_i, l_a being the isotropic and anisotropic parts of the average distance between nucleation enzymes, one obtains a force per filament, $l^2\sigma_r$ that also depends on the angle. The equation for the anisotropic part of the gel thickness then becomes

$$\frac{d\varepsilon}{dt} = \frac{\varepsilon}{\tau_0} + v_b, \tag{1.34}$$

where $v_b = \frac{2l_a}{l_i}v_p \simeq \frac{2}{\sqrt{N}}v_p$. v_p is the polymerization velocity as defined in the first part of this article. The solution reads

$$\varepsilon(t) = v_b \tau_0 \left(\exp\left(\frac{t}{\tau_0}\right) - 1\right). \tag{1.35}$$

To assess which of thermal fluctuations or enzyme heterogeneity is the leading term in the symmetry-breaking source, one should then compare $v_b\tau_0$, and ε_b. Their bead radius dependence is such that at r large enough, the enzyme disorder should always win. Plugging reasonable orders of magnitudes suggests that even for beads of a few nanometers size the enzyme disorder is already more important.

Another symmetry-breaking source is the lack of sphericity of the beads. Arguments very similar to the one we have used for enzymes heterogeneity can be made. One has essentially to replace l_a/l_i by r_a/r_i, in which the subscripts refer to the beads' radius of curvature variation, with obvious meaning. It is difficult to put numbers on this term, because it depends on the preparation chemistry of the beads. It seems that beads smaller than a micron, have excellent sphericity, controlled by surface tension. In that range, one expects enzyme heterogeneity to provide the main symmetry-breaking term. It seems more difficult to obtain beads larger than a few microns, in which case spherical aberrations might provide the main symmetry-breaking term. It would be interesting to design carefully controlled experiments, for quantitatively checking these predictions. Clever microscopic models have been imagined in order to obtain symmetry-breaking conditions [23, 38]. However, they do not take into account the very nature of the actin gel.

Finally, degradation of the gel could also occur through rupture (instead of depolymerization). In both cases, degradation starts at the outer surface. In one case, however, symmetry is broken by a smooth perturbation of the thickness (depolymerization), and in the other by a sharp (localized) opening of the gel (rupture) [37]. Indeed, symmetry-breaking of an elastic actin gel driven by fracture was observed in Reference [39] (see Figure 1.8). Saltatory motion can also occur in this case, if time scales are such that the ruptured gel can regrow before the bead moves.

1.4.6 Limitations of the Approach and Possible Improvements

In the discussion developed above, we have kept only diagonal stresses. However, as soon as the gel thickness is inhomogeneous, some of the elastic energy is released by shear, and a complete analysis should contain the corresponding terms. One can show, however, that the exposed results are not changed in any significant way [37]. Furthermore, whenever the anisotropy changes with time, the gel redistribution causes friction at the bead/gel interface, a phenomenon that is not included in the present analysis. It is possible in fact to show that it does not modify the structure of the equations but simply renormalizes the onset time τ_0 of the anisotropy.

Indeed, under such circumstances, the tension has an angular dependent part, which must be proportional to the friction coefficient, the velocity of the gel relative to the bead, and have the right dimensions

$$T(\theta) = T_i + \kappa \xi r \frac{\mathrm{d}\varepsilon}{\mathrm{d}t} \cos\theta, \tag{1.36}$$

where κ is a dimensionless number and ξ is the bead/gel friction coefficient already discussed. The equations are formally unchanged and only the onset time of the modulation is modified to a new value

$$\tau_0^* = \tau_0 + \frac{r^2}{v_i}\left(\frac{e_2}{e_0^2} + \frac{1}{e^*}\right). \tag{1.37}$$

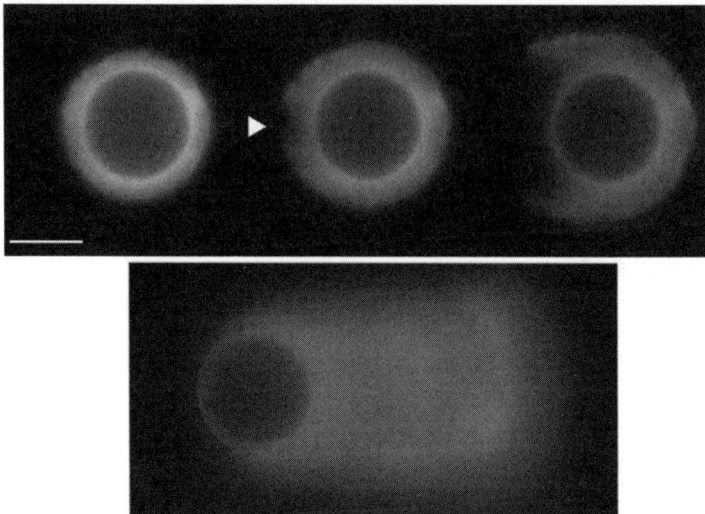

Figure 1.8. Timelapse of a symmetry-breaking event (white arrow head) preceding actin-based movement of a bead. Actin is marked with a fluorophore and the samples are observed in epifluorescence. The first three images were taken 21, 24, and 40 minutes after start of incubation. The last image shows the developed comet. Bar, 10μm. Images are taken from [39]. (Copyright 2005 National Academy of Sciences, U.S.A.).

The main conclusions are similar to the one obtained before. For instance the symmetry-breaking onset time is again proportional to the sphere radius, since e_0, e_2 and e^* are. This expectation is indeed born out by experiment [40]. Furthermore, if the friction is very high, the spherical steady state can be reached much before symmetry is actually broken. This will happen in particular if the gel is dense, an expectation also born out by experiment.

There are several other implicit simplifications in the above presentation: we have considered only one elastic modulus, without specifying whether it corresponds to compression or shear or a combination of them. A proper description is possible by the use of a covariant description of the gel [37]. Actually, the very geometry of the polymerization/cross-linking process implies that the gel should be anisotropic as well, but keeping this feature adds in complexity without bringing further understanding to the question.

A more important limitation comes from the hidden assumption that the gel density is constant throughout the gel and that all depolymerization is located at its external surface. In fact, it is known that the gel density decreases exponentially in a *Listeria* comet, over length scales comparable to the comet total length (i.e., several tens of microns [19]). One could be tempted to argue that this length scale is much larger than the one we discuss here and forget about this slow bulk depolymerization. It would, however, be a wrong argument, since the depolymerization mechanism is certainly stress dependent. In fact, the dynamical equation for the gel density should read

$$\frac{\partial \rho}{\partial t} + \boldsymbol{v} \cdot \nabla \rho = S - k_d \rho, \tag{1.38}$$

where $\boldsymbol{v}$ is the gel velocity relative to the bead or the bacterium, S is the gel density source localized at or close to the surface, and k_d is the depolymerization probability per monomer connected to the gel, per unit time. In general, one expects stress dependence similar to the one already described, that is, $k_d = k_d^0 \exp\left(\frac{\sigma_t \phi^2 a_3}{kT}\right)$, in which $\phi^2 = 1/(\rho a)$ is the average area spanned by a filament, and a_3 is a model-dependent length. Far from the surface, in an essentially unstressed comet tail, one does obtain an exponential decrease of density over a length $L = \frac{v}{k_d^0}$. Knowing v and the comet length, one deduces k_d^0 easily. The physical mechanism behind this depolymerization is not obvious: it could be that actin filaments can spontaneously break anywhere, or that reticulation points stabilize the structure and provide the rate-limiting step in the depolymerization, or that there is a one-to-one mapping of the pointed end density on the connected monomer density. In all cases, a constant average number of monomers should leave the gel for each event. Note that a depolymerization from the pointed ends of the filaments cannot generally be represented by such a mathematical structure. For instance, if the filaments were parallel on average, all starting from the surface at the barbed end and with a length distribution, the term would read $-k_-^p a \boldsymbol{p} \cdot \nabla \rho$, in which $\boldsymbol{p}$ is the unit vector in the filaments direction. Indeed, under such circumstances, $a \boldsymbol{p} \cdot \nabla \rho$ is a measure of the pointed end density.

Because of the exponential dependence of the depolymerization coefficient on stress, the length over which the density significantly decreases may become very short in the presence of such a stress and this mechanism could provide an alternative interpretation of the steady state in spherical symmetry. Under such circumstances, the density decrease occurs essentially over a length such that $\sigma_t = \frac{kT}{\phi^2 a_3}$, or with the scaling laws derived in spherical geometry, $e \simeq \frac{kT\rho}{E a_3} r$. The sharpness of the density decrease is controlled by the ρ dependence of E. In all reasonable cases, it is quite pronounced. For instance, if E is proportional to ρ, which is the case whenever the cross-link angular elasticity determines the elastic modulus, then

$$\rho = \rho_0 \exp\left[\frac{r_0}{\alpha L}\left(1 - \exp\left\{\frac{\alpha(r - r_0)}{r_0}\right\}\right)\right]. \tag{1.39}$$

In this equation, r_0, ρ_0 denote the radius and density at the bead surface, and α is a dimensionless number of the order of ten. There is a sharp cutoff for $r - r_0 \simeq \frac{r_0}{\alpha}$. Also considering that below a threshold density the gel integrity is totally lost, it is clear that any thickness measurement will give a value very close to r_0/α, essentially equivalent to the one derived in Section 1.4. The drawback of this type of presentation is the added complexity. Its merit is the connection of the comet's slow density decrease with the fast one under stress. The same type of result holds for other ρ dependences of E, such as the one used in Section 1.2.

An observation that is not accounted for by the present analysis is the strong density increase of the comet in the saltatory mutant. Simulations, taking into account the detailed chemistry of the polymerization process but ignoring the role of the polymerization enzyme, show a strong density dependence on external stress [21]. A simple phenomenological way of taking this effect into account consists in writing the gel density source

$$S = v\rho_- \exp\left(\frac{l_s}{v\tau}\right)\delta_s, \tag{1.40}$$

in which δ_s is the delta function at the surface. $\rho_- = \frac{1}{l^2 a}$ is determined by the enzyme density. The exponential factor expresses the exponential growth due to Arp2/3, l_s the length over which the branching phenomenon can occur, and τ a typical capping time. The gel density ρ_+ just outside the proximal domain reads: $\rho_+ = \rho_- \exp\left(\frac{l_s}{v\tau}\right)$. The expression of the growth velocity v is model dependent. If one assumes that a fraction of order unity of all filaments contributes to the stress, then the natural generalization of the polymerization law under stress reads

$$v = v_p^0 \exp\left(-\frac{\sigma_n a_0}{kTa\rho_+}\right). \tag{1.41}$$

Solving for this set of equations reproduces the results of the simulation fairly well. Indeed, in the large-force, slow-velocity limit, one finds

$$v \simeq \frac{l_s}{\tau \log\left(\frac{\sigma_n}{kT\rho_-}\right)}, \tag{1.42}$$

and

$$\rho = \frac{\sigma_n}{kT \log\left(\frac{v_p^0}{v}\right)}, \tag{1.43}$$

in which we have assumed $a \simeq a_0$ for simplicity. In this regime, the velocity is essentially independent of stress and determined by the branching and capping processes. Conversely, the density is practically proportional to the stress. Both features show up clearly in the simulation. For "large" velocities, a more conventional regime in which $\rho_+ \simeq \rho_-$ and $v \simeq v_p^0 \exp\left(-\frac{\sigma_n}{kT\rho_-}\right)$ may exist. Such a formulation should be exploited further.

1.5 New Experimental Developments

During the last few years, several novel experimental techniques have been developed. With these novel methods it is now possible to obtain detailed information about the forces involved in actin-based propulsion, to analyze in detail the properties of the comet tail, and to test the theoretical predictions on the influence of bead curvature. Here, we give a short summary of these new developments.

1.5.1 Direct Measurement of Actin-Generated Forces and Velocities

The forces generated in the process of actin-based propulsion can be directly probed in novel micromanipulation experiments [41]. Here, motility of polystyrene spheres attached to a "flexible handle" are probed with their growing comets held with micropipettes. In this way, external pulling and pushing forces of a few nanonewtons can be directly applied.

By using this setup, forces in the range from -1.7 to 4.3 nN have been applied to beads (where pushing forces are positive). An example is shown in Figure 1.9(a) and (b) where a force of 1.7 nN was imposed. Thus, the comet lengthens with constant velocity. By combining such measurements at different f^{ext}, the force velocity relation shown in Figure 1.9(c) is obtained.

As can be seen from the experimental data in Figure 1.9, the force velocity relation is essentially linear for pulling forces, while for pushing forces the decrease in v with increasing f^{ext} is somewhat slower. The stall force cannot be reached experimentally due to comet buckling at large forces f^{ext}. The gel thickness e increases with decreasing velocity, see Figure 1.9(d). For small velocities, e tends to a saturated value.

Within the framework provided above, the force velocity relation and the velocity dependence of the gel thickness can be understood quantitatively. By using Equations (1.7) and (1.8) for spherical beads with radius R, the relation between external force and velocity is given by

$$f^{ext} + E\frac{e^3}{R} \simeq A\xi v, \qquad (1.44)$$

where $A = 4\pi R^2$ is the contact area between bead and gel. The dependence of the velocity v on the gel thickness e is well-described by the empirical formula

$$e(v) = e^* \left[1 - \exp\left(-\frac{v_p^0 R}{ve^*} \right) \right]. \qquad (1.45)$$

The last equation simply interpolates between the steady state value $e = e^*$ for small velocities and the value of e for large velocities given by Equation (1.9).

It is also possible to demonstrate actin-based propulsion with softer objects than polystyrene beads. For example, (VCA-covered) oil droplets [42] or (ActA-covered) synthetic vesicles [43, 44] move once put into cell extract. Their surfaces are deformed by the elastic stresses exerted by the actin tail comet. Such squeezing effects have also been observed in endosomes [29].

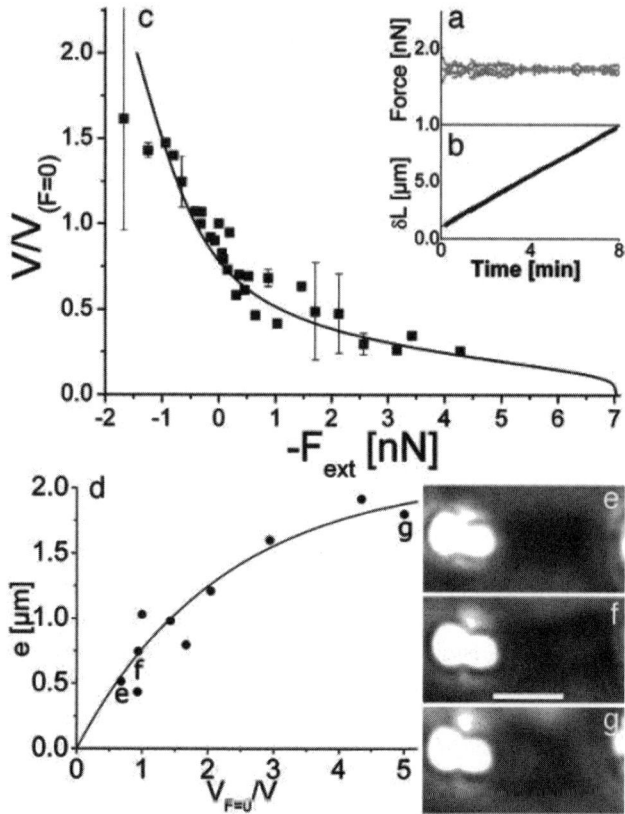

Figure 1.9. Force velocity diagrams as obtained from micromanipulation experiments [41]. Force (a) and comet length δL (b) as a function of time. Force velocity relation (c) where v is obtained from the slope of (b). The velocity is normalized by the velocity for $f^{ext} = 0$. The full curve is the fit by Equation (1.44), where $E \simeq 10^3$Pa and $\xi \simeq 3 \cdot 10^{10}$Pa · s/m. Velocity dependence of gel thickness e (d). Experimental data is fitted by Equation (1.45) for $e^* = 2.0\mu$m and $v_p^0 = 1.6\mu$m/min. Images of comets for different values of velocity (e-g) corresponding to data points in (d). Scale bar, 5μm. Figure is from [41]. (Copyright 2004 National Academy of Sciences, U.S.A.).

For liquid drops, the elastic stress distribution on the drop surface can be calculated from

$$\frac{2\gamma_0}{R} + \delta P(h) = \gamma \left(\frac{\cos\theta(h)}{h} + \frac{\mathrm{d}\cos\theta(h)}{\mathrm{d}h} \right) - \sigma_n(h), \qquad (1.46)$$

where h is the local thickness and θ the local angle of the tangent. Furthermore, δP is the difference between internal pressure and Laplace pressure, γ is the (spatially varying) surface tension, and (as above) σ_n is the normal stress at the surface of the comet. Numerically, the last equation together with Equation (1.41) can be solved numerically to calculate the droplet shape. By fitting experimentally determined profiles, one obtains for the parameters $\sigma_0 = kT\rho_+ \simeq 32\text{nN}/\mu\text{m}^2$ and $v_p^0 \simeq 1.4\text{nm/sec}$ (see Figure 1.10).

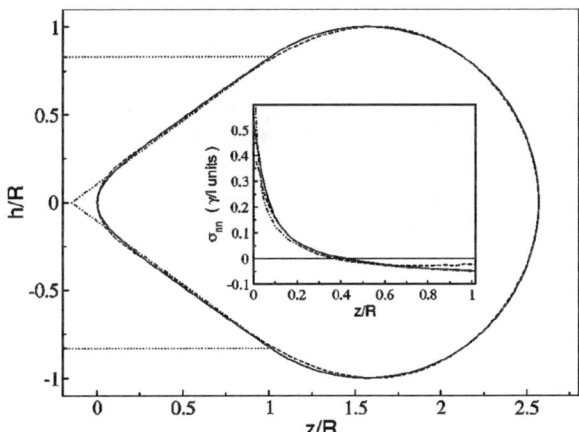

Figure 1.10. Calculated drop profile (full line) compared with the experimentally determined profile (dashed line). The inset shows the normal stress distribution along the drop surface. Figure is reprinted from [42]. (Copyright (2004), with permission from The American Physical Society).

As can be seen from the inset of Figure 1.10, the normal stress is positive at the back of the drop thus pulling the liquid droplet. Recent experiments confirm this [45]. This finding is in accordance with the theoretical prediction made in [10] that the actin gel could pull at the rear of *Listeria*.

Experimentally, it was also found that saltatory motion of soft beads (droplets) is different from that of hard beads [45]. Because on liquid droplets the actin activator (VCA) detached from actin filaments is free to move over the surface, the step size of the droplet motion depends on the diffusion time τ_d of VCA over the sphere's surface. Because τ_d increases with bead size, the step size does as well. The opposite is true for hard beads: here, larger beads have smaller step sizes (see Figure 1.11).

1.5.2 Properties of Comet Tails

Actin-based movement of beads is governed by the mechanical properties of the actin network. Experimentally, these can be fine-tuned by changing the concentrations of proteins involved in actin dynamics and assembly.

For example, the coating of polystyrene beads can be varied [46]. It was found that the degree of filament alignment in the comet tails depends on

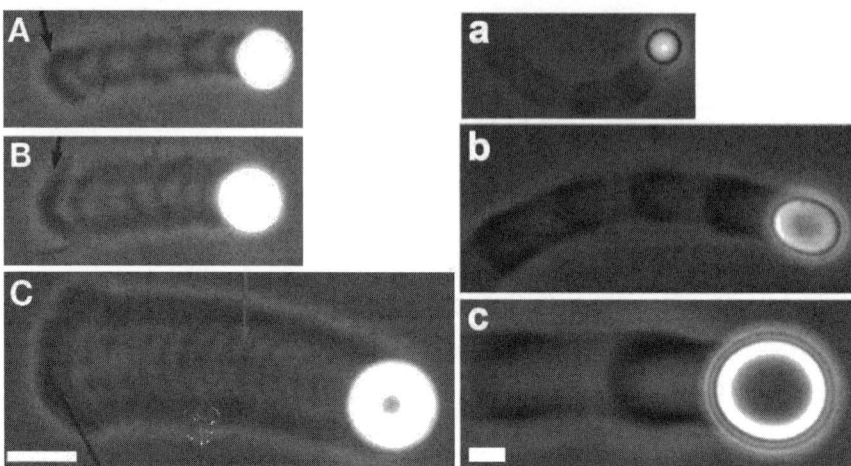

Figure 1.11. Saltatory motion of hard and soft beads observed by phase contrast microscopy. Note that the step size decreases with bead size for hard beads (A, B, C), whereas it increases with bead size for soft beads (droplets) (a,b,c). This effect is due to diffusion and convection of the actin polymerization activators on the fluid surface that is impeded in the case of hard polystyrene spheres. A, B, C: polystyrene beads of diameter 4.5μm to 9.1μm diameter covered with the VCA fragment and placed in the purified protein motility medium. The black arrow indicates the actin gel corresponding to the symmetry-breaking event. Bar, 10μm. Taken from [31]. (Copyright (2005), with permission from The Biophysical Society). a,b,c soft beads coated with VCA and PRO (which recruits VASP), placed in HeLa cell extracts. Bar, 2μm. Taken from [45]. (Copyright (2007), with permission from The Biophysical Society).

the surface ratio of VASP (which decreases the actin-branching frequencies) to Arp2/3 (which is involved in the formation of actin branches). For high ratios, the actin filaments align parallel to the direction of bead movement, accompanied by an increase in velocity.

Similar results were obtained *in vivo*. Brieher et al. found that tail elongation of *Listeria* in the presence of an Arp2/3 inhibitor generates hollow cylindrical comet tails of parallel bundles that attach along the sides of the bacterium [47]. In this case, the speed of *Listeria* was also enhanced significantly.

These effects were most systematically investigated for actin-propelled polystyrene spheres [48]. By varying the amount of gelsolin (which caps the actin filaments) various dynamical states could be observed.

One has to distinguish the following regimes:

- **Elastic polymerization-limited regime.** As shown in Section 1.3.2 for sufficiently small elastic modulus E, the polymerization rate is unaffected by the stress and the velocity v is given by Equation (1.10).

- **Elastic stress-limited regime.** For larger E actin polymerization at the surface might be limited by the accumulated elastic stress. In this case, the gel thickness saturates at a value close to the stress-limited, steady-state thickness of the gel, leading to a velocity [48]

$$v \sim \left(\frac{\Delta\mu}{l^2 a}\right)^{3/2} \frac{1}{\xi E^{1/2}}, \tag{1.47}$$

where $\Delta\mu$ is the energy of the polymerization reaction, and l and ξ denote (as above) the mesh-size of the actin network and the friction coefficient, respectively.
- **Diffusion-limited regime.** Here, one must take into account that monomers have to diffuse to the bead surface. An analysis similar to the one given in Section 1.4.2 yields [48]

$$c_i \simeq \frac{c_\infty D}{k_+^b R/l^2 + D}. \tag{1.48}$$

Thus, the polymerization velocity $v_p = v$ and $v_p = k_+^b c_i a$, implying

$$v \simeq \frac{c_\infty D l^2 a}{R}. \tag{1.49}$$

The state of the system is determined by two dimensionless variables [48] a dimensionless modulus $e = E\xi v_p^0(l^2 a/\Delta\mu)^2$ and a dimensionless diffusion coefficient $d = (Dc_\infty l^3 a^{3/2}/R)(\xi/v_p^0 \Delta\mu)^{1/2}$. The corresponding state diagram is shown in Figure 1.12.

Experimentally, the different morphologies of the comet tails in the dynamic regimes can be characterized by photobleaching of lines perpendicular to the comet axis. In the diffusion-limited regime, bleached vertical lines deform and newly grown actin layers open up and reduce their curvature. In the elasticity-dominated regime newly grown gel layers are first stretched. Further away from the bead surface, the layer reduces its area by reducing its curvature.

Finally, the surface concentration of Wiskott-Aldrich Syndrome protein (WASP) also has a direct influence on the motion of polystyrene beads. Similar to *Listeria*, actin-driven beads can also exhibit different kinds of motion. The different regimes are characterized by, e.g., velocity of motion and frequencies of periodic motion. In [31] a dynamic state diagram has been experimentally derived and it has been shown that bead motion generally depends on both bead diameter and protein surface density.

1.5.3 Actin-Based Propulsion of Disks

Recently, Schwartz et al. have experimentally confirmed the theoretical prediction (see Reference [19] and Section 1.3.2) that actin-polymerization can

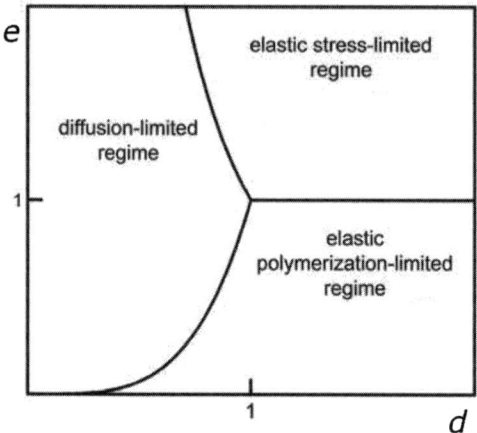

Figure 1.12. State diagram for bead motion as function of the dimensionless modulus e and the dimensionless diffusion coefficient d.

also provide the propulsive force to push non-curved objects [25]. They have made polystyrene disks (diameter $\simeq 5\mu$m and thickness $\simeq 0.8\mu$m) by compressing polystyrene beads. These disks were assayed similarly to the way described above: disks were coated with ActA and put into cellular extract containing enzymes and ATP.

It was found that the disks move in a manner similar to *Listeria* and beads. Typically, the entire disks were surrounded by actin clouds but actin tails predominantly formed on the flat surfaces. The most common motile configuration had a tail emanating from each face of the disk. Disks with perimeter tails and single face tails, however, could also be observed (see Figure 1.13).

By analyzing the direction of motion of the moving disks, Schwartz et al. concluded that the largest pushing forces occur along the flat face of the disks. For face tails, the motion is steady, while disks with perimeter tails exhibit a hopping motion. In contrast to the conclusions the authors draw from these results, all experimental findings are in complete agreement with the theoretical picture developed above. For example, the observation that flat disks move faster than the polystyrene spheres they are made from can be simply explained by the reduced polymerization rate on a curved surface due to higher elastic stresses. Similarly, the hopping motion of disks with perimeter tails can be explained within our theoretical framework.

1.6 Conclusion

In this review, we have shown that if one wants to understand the physics of the propulsion mechanism of the bacterium *Listeria*, one must analyze the

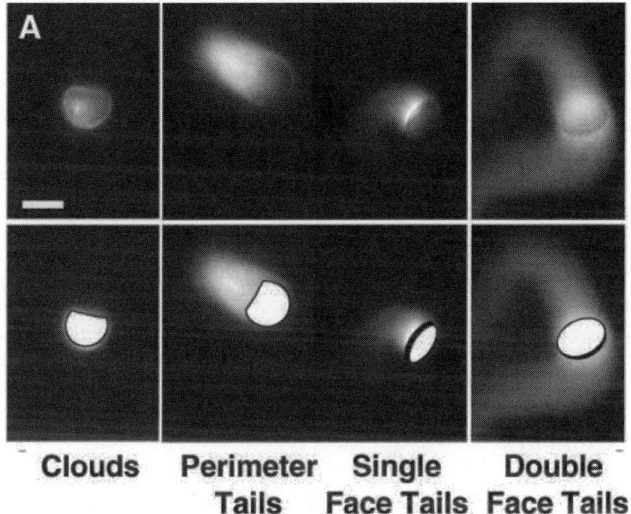

Clouds   Perimeter   Single   Double
Tails   Face Tails Face Tails

Figure 1.13. ActA coated polystyrene disks in motility media can have clouds (typically in the non-motile nucleation stage), perimeter tails, single-face tails or double-face tails. Figure is reprinted from [25]. (Copyright (2004), with permission from Elsevier).

mesoscopic stress distribution in the actin gel, and the solid-on-solid friction that the gel exerts on the bacterium surface. One should keep in mind, that only the polymerization process misses (so to speak) half of the problem. In particular, recognizing the importance of the nonlinear friction allows us to understand the saltatory mutant as a system working in a "stick-slip" regime familiar in solid friction. In general, the system gets into the saltatory regime at a well-defined threshold called a Hopf bifurcation. In its vicinity, measurable quantities such as velocity and comet density modulations are sinusodal functions (of time and space, respectively). Away from the bifurcation, variations are more abrupt, like in Figure 1.2. It is important to understand that the role of mutation is simply to drive the system in a region of phase space where the behavior is saltatory, but that many other external perturbations could have the same result without changing the genome. This point is well-illustrated in bio-mimetic experiments, where a simple bead diameter change drives the behavior from regular to saltatory. This example shows that the relationship between genotype and phenotype will not be easy to unravel. One will need to fully understand all the complex dynamical diagrams governing biological systems before being able to fully exploit the formidable genetic data that we are getting now.

It is important to stress that the mesoscopic approach is not antagonistic to microscopic ones, but rather complementary. Microscopic theories can be used as input for writing boundary conditions in mesoscopic theories. They should not only focus on the polymerization process, but on friction and de-polymerization as well. Another message is that, even if the *Listeria* propulsion mechanism can be, in some way, representative of the biochemistry involved in more complex Eukaryotic cells, the physics of it will be very different for two simple reasons. First, the stress distribution will be very different because the gel is now produced in the inside of a topologically spherical object, and second the tangential friction is very different because it is on a fluid membrane. Yet, what we learn here will be very useful for understanding cell motility. In particular, there are indications that mechanical instability of the acto-myosin cytoskeleton also play a major role in symmetry-breaking and polarization of cellular systems [49]. At last, what we find most rewarding, is the potential importance of the bio-mimetic approach for medical applications [50, 51].

Acknowledgements

This review is an updated version of [6].

References

1. Dabiri, G, Sanger, J, Portnoy, D, & Southwick, F. (1990) *Proc. Natl. Acad. Sci. USA* **87**, 6068–72.
2. Tilney, L & Portnoy, D. (1989) *J. Cell Biol.* **109**, 1597–608.
3. Southwick, F & Purich, D. (1998) *Trans. Am. Clin. Climatol. Assoc.* **109**, 160–72; discussion 172–3.
4. Cameron, L, Giardini, P, Soo, F, & Theriot, J. (2000) *Nat. Rev. Mol. Cell. Biol.* **1**, 110–9.
5. Pantaloni, D, Le Clainche, C, & Carlier, M. (2001) *Science* **292**, 1502–6.
6. Prost, J. (2002) *The Physics of Listeria Propulsion*, eds. Flyvberg, H, Julicher, F, Ormos, P, & David, F. (Springer, Berlin).
7. Bray, D. (2001) *Cell motility.* (Garland, New York).
8. Lasa, I, David, V, Gouin, E, Marchand, J, & Cossart, P. (1995) *Mol. Microbiol.* **18**, 425–36.
9. Kocks, C, Gouin, E, Tabouret, M, Berche, P, Ohayon, H, & Cossart, P. (1992) *Cell* **68**, 521–31.
10. Gerbal, F, Chaikin, P, Rabin, Y, & Prost, J. (2000) *Biophys. J.* **79**, 2259–75.
11. Svitkina, T & Borisy, G. (1999) *J. Cell Biol.* **145**, 1009–26.
12. Blanchoin, L, Amann, K, Higgs, H, Marchand, J, Kaiser, D, & Pollard, T. (2000) *Nature* **404**, 1007–11.
13. Loisel, T, Boujemaa, R, Pantaloni, D, & Carlier, M. (1999) *Nature* **401**, 613–6.
14. Cameron, L, Footer, M, van Oudenaarden, A, & Theriot, J. (1999) *Proc. Natl. Acad. Sci. USA* **96**, 4908–13.
15. Noireaux, V, Golsteyn, R, Friederich, E, Prost, J, Antony, C, Louvard, D, & Sykes, C. (2000) *Biophys. J.* **78**, 1643–54.
16. Theriot, J, Mitchison, T, Tilney, L, & Portnoy, D. (1992) *Nature* **357**, 257–60.
17. Kocks, C, Marchand, J, Gouin, E, d'Hauteville, H, Sansonetti, P, Carlier, M, & Cossart, P. (1995) *Mol. Microbiol.* **18**, 413–23.
18. Gerbal, F, Noireaux, V, Sykes, C, Julicher, F, Chaikin, P, Ott, A, Prost, J, Golsteyn, R. M, Friederich, E, Louvard, D, Laurent, V, & Carlier, M. F. (1999) *Pramana J. Phys.* **53**, 155–70.
19. Gerbal, F, Laurent, V, Ott, A, Carlier, M, Chaikin, P, & Prost, J. (2000) *Eur. Biophys. J.* **29**, 134–40.
20. Mogilner, A & Oster, G. (1996) *Biophys. J.* **71**, 3030–45.
21. Carlsson, A. (2001) *Biophys. J.* **81**, 1907–23.
22. Mogilner, A & Edelstein-Keshet, L. (2002) *Biophys. J.* **83**, 1237–58.
23. Mogilner, A & Oster, G. (2003) *Biophys. J.* **84**, 1591–605.
24. Gov, N & Gopinathan, A. (2006) *Biophys. J.* **90**, 454–69.
25. Schwartz, I, Ehrenberg, M, Bindschadler, M, & McGrath, J. (2004) *Curr. Biol.* **14**, 1094–8.
26. McGrath, J, Eungdamrong, N, Fisher, C, Peng, F, Mahadevan, L, Mitchison, T, & Kuo, S. (2003) *Curr. Biol.* **13**, 329–32.
27. Gouin, E, Gantelet, H, Egile, C, Lasa, I, Ohayon, H, Villiers, V, Gounon, P, Sansonetti, P, & Cossart, P. (1999) *J. Cell Sci.* **112**, 1697–708.
28. Merrifield, C, Moss, S, Ballestrem, C, Imhof, B, Giese, G, Wunderlich, I, & Almers, W. (1999) *Nat. Cell Biol.* **1**, 72–4.
29. Taunton, J, Rowning, B, Coughlin, M, Wu, M, Moon, R, Mitchison, T, & Larabell, C. (2000) *J. Cell Biol.* **148**, 519–30.

30. Tawada, K & Sekimoto, K. (1991) *J. Theor. Biol.* **150**, 193–200.
31. Bernheim-Groswasser, A, Prost, J, & Sykes, C. (2005) *Biophys. J.* **89**, 1411–9.
32. Gerbal, F. (1999) *Ph.D. Thesis.* (Université Paris VII).
33. Yarar, D, To, W, Abo, A, & Welch, M. (1999) *Curr. Biol.* **9**, 555–8.
34. Fradelizi, J, Noireaux, V, Plastino, J, Menichi, B, Louvard, D, Sykes, C, Golsteyn, R, & Friederich, E. (2001) *Nat. Cell Biol.* **3**, 699–707.
35. Noireaux, V. (2000) *Ph.D. Thesis.* (Université Paris XI).
36. Plastino, J, Lelidis, I, Prost, J, & Sykes, C. (2004) *Eur. Biophys. J.* **33**, 310–20.
37. Sekimoto, K, Prost, J, Julicher, F, Boukellal, H, & Bernheim-Groswasser, A. (2004) *Eur. Phys. J. E* **13**, 247–59.
38. van Oudenaarden, A & Theriot, J. (1999) *Nat. Cell Biol.* **1**, 493–9.
39. van der Gucht, J, Paluch, E, Plastino, J, & Sykes, C. (2005) *Proc. Natl. Acad. Sci. USA* **102**, 7847–52.
40. Bernheim-Groswasser, A, Wiesner, S, Golsteyn, R, Carlier, M, & Sykes, C. (2002) *Nature* **417**, 308–11.
41. Marcy, Y, Prost, J, Carlier, M, & Sykes, C. (2004) *Proc. Natl. Acad. Sci. USA* **101**, 5992–7.
42. Boukellal, H, Campas, O, Joanny, J.-F, Prost, J, & Sykes, C. (2004) *Phys. Rev. E.* **69**, 061906.
43. Upadhyaya, A, Chabot, J, Andreeva, A, Samadani, A, & van Oudenaarden, A. (2003) *Proc. Natl. Acad. Sci. USA* **100**, 4521–6.
44. Giardini, P, Fletcher, D, & Theriot, J. (2003) *Proc. Natl. Acad. Sci. USA* **100**, 6493–8.
45. Trichet, L, Campas, O, Sykes, C, & Plastino, J. (2007) *Biophys. J.* **92**, 1081–9.
46. Plastino, J, Olivier, S, & Sykes, C. (2004) *Curr. Biol.* **14**, 1766–71.
47. Brieher, W, Coughlin, M, & Mitchison, T. (2004) *J. Cell Biol.* **165**, 233–42.
48. Paluch, E, van der Gucht, J, Joanny, J.-F, & Sykes, C. (2006) *Biophys. J.* **91**, 3113–22.
49. Paluch, E, van der Gucht, J, & Sykes, C. (2006) *J. Cell Biol.* **175**, 687–92.
50. Friedrich, E, Golsteyn, R, Louvard, D, Noireaux, V, Prost, J, & Sykes, C. (2000) Institut Curie/CNRS, patent no. PCT/FR00/03469.
51. Fradelizi, J, Friedrich, E, Golsteyn, R, Louvard, D, Noireaux, V, & Sykes, C. (2001) Institut Curie/CNRS, patent no. PCT/FR01/00843.

2

Biophysical Aspects of Actin-Based Cell Motility in Fish Epithelial Keratocytes

Kinneret Keren[1] and Julie A. Theriot[1,2]

[1] Department of Biochemistry
[2] Department of Microbiology and Immunology, Stanford University, Stanford, CA 94305, USA

Summary. Cell motility is a fascinating dynamic process crucial for a wide variety of biological phenomena, including defense against injury or infection, embryogenesis, and cancer metastasis. A cell using actin-based motility to crawl across a substrate must coordinate the action of numerous individual molecular building blocks to achieve coherent cell movement. While the molecular basis of cell motility is beginning to be understood, relatively little is known about the large-scale mechanisms responsible for this remarkable self-organization that bridges many orders of magnitude in both space and time. In this chapter, we discuss the biophysical aspects of actin-based cell motility and the importance of their interplay with the underlying biochemical processes, focusing on fish epithelial keratocytes as a relatively simple model system. We review the current understanding regarding the mechanical and biochemical aspects of keratocyte motility, and at the same time highlight some gaps in our knowledge.

2.1 Introduction

2.1.1 Actin-Based Cell Motility

The directed motility of individual cells is important to a wide variety of basic biological processes. Free-living cells in liquid environments usually move by swimming, employing cilia or flagella. Cells that live on solid surfaces, however, move by crawling; this is the primary mode of active cell locomotion for motile animal cells, with the exception of sperm. Amoeboid cell crawling is used by a large assortment of cell types for different purposes; unicellular organisms such as amoebae and slime molds crawl through the soil in search of food, while in animals, similar crawling movements of neutrophils and macrophages in pursuit of microbial invaders are a key element in defense against infection. Directed motility of epithelial cells is essential for rapid closure of wounds, whereas programmed cell movements in developing embryos are responsible for rearranging tissue layers and wiring the nervous system. Cell motility also contributes to pathological processes such as cancer, where cell migration from the initial tumor site in a process called metastasis is responsible for spreading the malignancy to additional sites in the body.

In vivo, animal cells crawl within a tissue among other cells and extracellular matrix material in either a 2D or 3D environment. *In vitro*, cell motility is most often studied by plating cells on a 2D artificial substrate such as a glass coverslip. To crawl across a substrate, a cell must coordinate four mechanistically distinct processes in both space and time: protrusion of the leading edge, adhesion to the substrate, forward translocation of the cell body, and retraction of the rear [1, 2]. The migrating cell must be highly polarized so that these distinct processes can be localized to different regions of the cell. The majority of animal cells, as well as many unicellular organisms, move by actin-based crawling, in which protrusion at the leading edge is powered by polymerization of polar actin filaments that assemble into a dense dendritic network. Specific molecular components are required for the dynamic assembly of this cytoskeletal network, including factors that nucleate growth of new actin filaments, cross-link filaments, and catalyze filament disassembly. The actin cytoskeleton is anchored to the substrate through a complicated array of adhesion molecules (reviewed in [3, 4]), and this attachment enables the force generated by actin polymerization to be translated into cell protrusion. This general mechanism of actin-based motility is conserved from protozoa through vertebrates.

Research in the last few decades has focused on uncovering the biochemical basis of cell motility, and indeed numerous biochemical, structural, and genetic studies are now converging into an overall picture of the order of events and identities of the major molecular players involved in cell motility (reviewed in [3, 5, 6, 7]). With the characterization of the molecular basis of cell motility well underway, a new challenge has come to the fore: understanding how the cell coordinates the enormous number of molecules involved to achieve robust whole-cell motility that can be intrinsically persistent over time scales that are very long compared to the dynamics of individual protein-protein interactions, and yet remain responsive to changes in the mechanical and chemical features of its environment. A great deal of recent work in the field has focused on the cell-signaling mechanisms involved in the establishment and maintenance of large-scale cell polarity (reviewed in [8]). Less attention has been given to the equally important mechanical and physical aspects of the large-scale coordination of cell dynamics.

The importance of the interplay between molecular processes and mechanical and physical characteristics of the system are well illustrated by the actin-based motility of the intracellular bacterial pathogen *Listeria monocytogenes*. This pathogen hijacks the host cell's actin machinery and forms an actin "comet tail" which it uses to propel itself within the cell and through the membrane into neighboring cells, allowing efficient infection of a large number of cells without encountering the immune system. This form of motility depends solely on protrusion by actin polymerization, and unlike whole-cell motility, does not involve adhesion or contraction. *Listeria* motility was reconstituted using purified proteins [9] which is the "gold standard" of biochemical understanding. However, further work on the physics and mechanics of the actin comet tail [10, 11, 12, 13, 14, 15, 16, 17] was required to advance the understanding of actin-based propulsion and allow quantitative comparison between modeling and experiments. It is now important to extend the lessons learned from this simpler system to whole-cell motility, a more complicated system that is also fundamentally a coordinated molecular and mechanical process [1, 7].

2.1.2 Keratocytes as a Model System for Cell Motility

Uncovering the self-organization principles underlying whole-cell motility and understanding the complex relationship between biochemical and mechanical processes are exciting, yet highly demanding, goals. Confronted with such a challenge, it is useful to seek the simplest available model system, which for cell motility appears to be fish epidermal keratocytes (Figure 2.1; see Reference [18]). While the distribution and function of most of the major molecular players responsible for keratocyte motility are comparable to other well-characterized cells [19, 20], their simple overall geometry and persistent motion facilitates mathematical modeling of the biochemical and mechanical aspects of their motility [21, 22, 23, 24]. Keratocytes comprise the basal layer of the two-layered epidermis of fish and amphibians [25, 26] and are highly specialized for rapid closure of epidermal wounds *in vivo* [27, 28]. Keratocyte movement is rapid (up to 1 μm/s) and yet remarkably persistent, with individual cells typically moving in a nearly straight line at a nearly constant rate over tens or hundreds of cell diameters. This stands in sharp contrast to the irregular actin-based movement characteristic of many other cell types, including neutrophils and fibroblasts. Remarkably, lamellar fragments of keratocytes, which lack nuclei, microtubules and most organelles, move with similar speed and persistence to whole cells (see Figure 2.1; References [29, 30]), further illustrating the power of this model system as a simplified and perhaps minimal system for studying cell motility.

In a primary culture of epidermis from fish scales, individual keratocytes break away from the margin of the adherent epithelial sheet and assume a characteristic canoe-shaped polarized morphology, with a large lamellipodium in the front and a small, round cell body trailing behind (see Figure 2.2; Reference [18, 29]). This polarized morphology can arise spontaneously [31] and does not require any external cue. In fact keratocytes are not known to respond to any chemotactic signal, though they have been shown to galvanotax toward the cathode in a DC electric field [32, 33] via a mechanism whose nature and biological significance are yet unclear. Actin filaments in the keratocyte lamellipodium assemble primarily at the leading edge [34] and are incorporated into a dendritically branched meshwork [20, 35]. Myosin-dependent inward contraction of the lamellipodial meshwork at the rear gathers the actin filaments meshwork toward the cell body, where filaments tend to depoly-

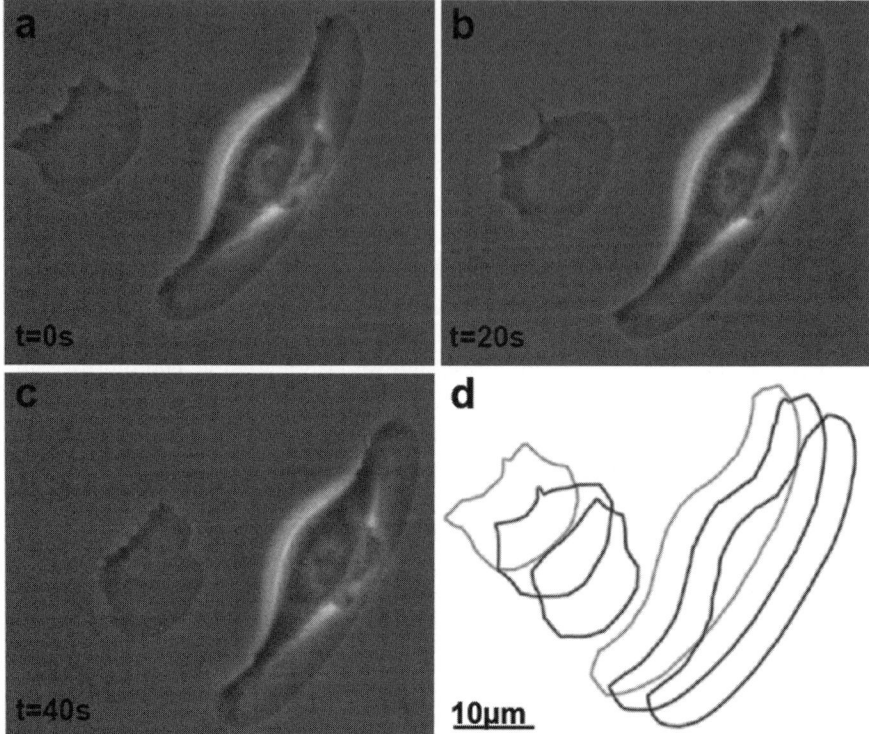

Figure 2.1. Keratocyte cells and lamellipodial fragments exhibit similar persistent motility. (a-c) Time-lapse phase contrast image sequence of a keratocyte cell and a lamellipodial fragment. (d) The cell outlines from the images were overlaid on top of each other. Image sequence courtesy of Greg Allen.

merize [30, 36, 37, 38]. The precise balance between protrusion at the leading edge and inward contraction at the rear enables smooth continuous movement of the cells such that they appear to glide forward without changing shape [39, 40, 41].

Although individual cells in a keratocyte population may move over a broad range of net speeds [43] and take on variable shapes [24, 43, 44], the processes contributing to motility are precisely balanced within each cell such that fast-moving cells and slow-moving cells are both able to maintain persistent shape and behavior. In this chapter, we describe progress toward understanding the mechanisms of this large-scale coordination, using the geometrically simple, rapidly moving keratocytes as a model system. We emphasize the role of mechanical processes and their interplay with biochemical processes in this coordination. For example, global biophysical parameters such as the membrane tension and traction forces introduce effective coupling between biochemical processes occurring at different regions within the cell. We

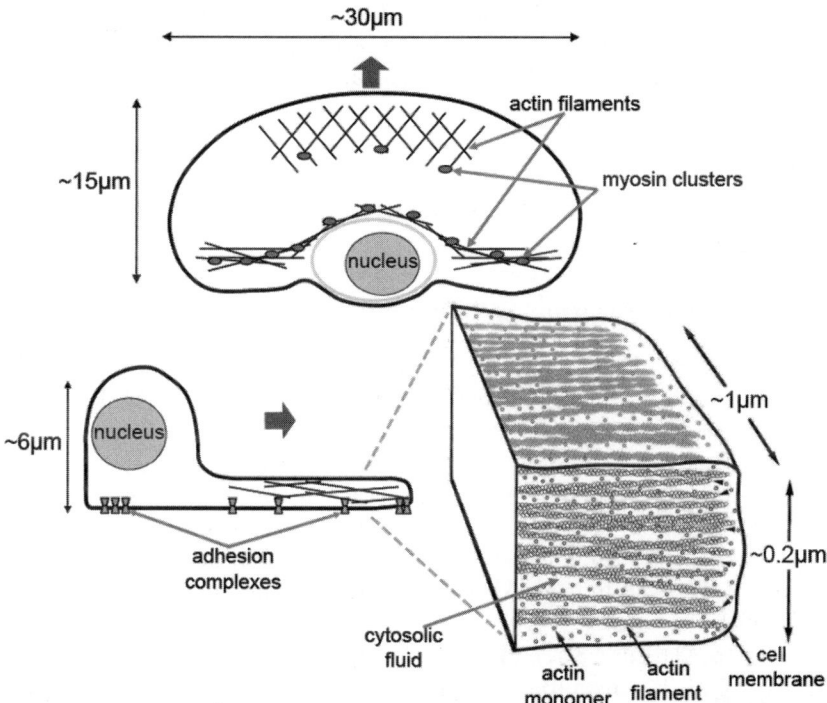

Figure 2.2. Schematics of a keratocyte cell. A top-view (upper scheme) and side-view (lower left) of a keratocyte showing the cell body that contains the nucleus and all the organelles, and the broad, thin, fan-shaped lamellipodium up front. The gray arrow indicates the direction of cell movement. The scheme illustrates the localization of various components, including the dendritic actin network in the lamellipodium, myosin clusters in the lamellipodium and in an actin-myosin bundle at the rear, adhesion complexes, the cell membrane, and the cytosolic fluid. An enlarged scheme of the leading edge (lower right) depicts, roughly to scale, the number of barbed ends of actin filaments at the leading edge, and the black arrowheads indicate monomer assembly onto barbed ends. The enlarged scheme of the leading edge was reproduced and modified from [42] with copyright permission of The Biophysical Society.

discuss the role of the three main mechanical modules involved in keratocyte motility, which include the actin-myosin cytoskeleton, the cell membrane, and the cytoplasmic fluid (see Figure 2.2), while providing only a limited account of the molecular basis of cell motility (the interested reader is referred to other reviews, e.g., [1, 5, 6, 7, 45]). We describe our current understanding of the properties of each module and discuss the coupling between them. At the same time we highlight the gaps in our current knowledge and point out areas where further experimental and theoretical efforts are needed. The chapter is ordered as follows. In Section 2.2 we discuss the actin-myosin cytoskeleton and adhesion to the substrate. Section 2.3 focuses on the cell membrane, and in Section 2.4 we discuss the role of the cytosolic fluid in cell motility. Finally, in Section 2.5 we discuss feedback mechanisms that couple these modules and the large-scale integration.

2.2 The Actin-Myosin Cytoskeleton and Adhesion to the Substrate

2.2.1 Actin Organization and Dynamics

Actin is by far the most abundant protein involved in cell motility with $\sim 10^8$ actin molecules per non-muscle cell [45]. Monomeric actin assembles into intrinsically structurally polarized filaments with one end commonly called the barbed end and the other the pointed end, based on the appearance of filaments bound to myosin fragments observed with an electron microscope [46]. The structural asymmetry of actin filaments is translated into kinetic asymmetry, so that the polymerization kinetics are considerably faster at the barbed ends [47]. Actin is an ATP-binding protein and monomeric actin subunits, which are mostly ATP-bound due to the high concentration of ATP compared to ADP in cells, add on to the tips of actin filaments. The rate of ATP hydrolysis is substantially enhanced within filaments compared to actin monomers [45]. Some of the energy released from ATP hydrolysis is stored in the polymer lattice so that ADP-bound filaments are less stable. At the barbed ends, the growth rate is typically faster than the hydrolysis rate so the filament tip is maintained in an ATP-bound form, whereas at the pointed end hydrolysis can "catch up" with growth so the filament will be in the less stable ADP-bound form. This situation can lead to "treadmilling": steady state elongation at the barbed end, balanced by steady state depolymerization at the pointed end (reviewed in [48]). Within a cell, the system is poised far from equilibrium with a high concentration of unpolymerized actin maintained by a combination of actin monomer binding proteins such as profilin and thymosinβ4, and capping of filament barbed ends. The free energy associated with this excess of monomers provides the driving force for protrusion [49]. The chemical energy from ATP hydrolysis is required for disassembly of actin filaments which is essential for regenerating this large monomeric actin pool, and ultimately provides the energy required to sustain cell motility.

The ability of the actin polymerization motor to generate protrusive forces in the absence of myosin motors has been demonstrated in several different systems. Pure actin was incorporated into lipid vesicles and induced to polymerize by addition of K^+ or Mg^{2+} ions [50, 51, 52]. The polymerization-induced shape changes and distortions in the vesicle demonstrate that actin polymerization can produce sufficient force to deform vesicle membranes. The reconstitution of actin-based motility of bacteria with purified proteins [9] similarly demonstrates that actin network growth and polymerization are sufficient for generating protrusive forces.

The maximal force that can be generated by the polymerization motor can be estimated from the differences in the free energy of an actin subunit in solution and bound to a filament. Taking the intracellular actin monomer concentration to be $\sim 10 - 50\mu$M [45], this amounts to a few pico newtons per filament (reviewed in [48]), comparable to other molecular motors. The elastic Brownian ratchet model [53] provides a mechanism for how this free energy can be translated into mechanical force that can push the cell membrane forward. In this model, thermal fluctuations cause an actin filament to bend away from the membrane, introducing transient gaps that occasionally allow an additional monomer to be incorporated into the filament. The elastic force of the lengthened filament then exerts a pushing force on the membrane.

The intrinsic polar behavior of a large number of individual actin filaments and their ability to generate protrusive forces must be coordinated to enable large-scale protrusion of the cell as a whole. The dendritic nucleation model describes how this is achieved at the leading edge of a moving cell. Actin filaments are oriented with their barbed ends towards the leading edge, so that the addition of new subunits to the growing barbed ends provides the protrusive force that drives the lamellipodium forward [38, 53]. New filaments are nucleated mainly in the vicinity of the leading edge by branching off the sides of existing actin filaments through the action of the Arp2/3 branching protein complex [20, 54]. This results in the formation of a dense dendritic actin meshwork. As shown in Figure 2.3, this meshwork is highly ordered in keratocytes with filaments oriented with their barbed ends at approximately $\pm 35°$ with respect to the leading edge [55]. This organization allows the intrinsically polar behavior of individual actin filaments to be translated into large-scale treadmilling of the cell as a whole, with assembly of the dendritic meshwork at the front and disassembly toward the rear (reviewed in [1, 7]).

The actin in the lamellipodium is highly dynamic, constantly being assembled and disassembled from the meshwork with a half-life of tens of seconds [34, 56, 57]. With such a rapid turnover rate, it is obviously essential to investigate the dynamics of the actin meshwork and not just its static structure. Initially, the pattern of filamentous-actin flow was studied by photoactivation of fluorescent actin that enabled tracking of the movement of a spatially defined subset of actin filaments within a rapidly moving keratocyte. The actin network in the front of the lamellipodium was found to remain nearly stationary with respect to the substrate (i.e., in the laboratory frame of reference), and therefore moved rearward with respect to the cell's leading edge (i.e., in the frame of reference that moves with the cell) at the same rate as net cell translocation [34, 39]. More detailed mapping of the actin meshwork flow has been facilitated by fluorescent speckle microscopy (FSM) [56, 57, 58, 59, 60]. FSM allows simultaneous mapping of actin meshwork flow across the entire lamellipodium with high temporal and spatial resolution. As shown in Figure 2.4 the introduction of low levels (estimated to be $< 0.1\mu$M final concentration

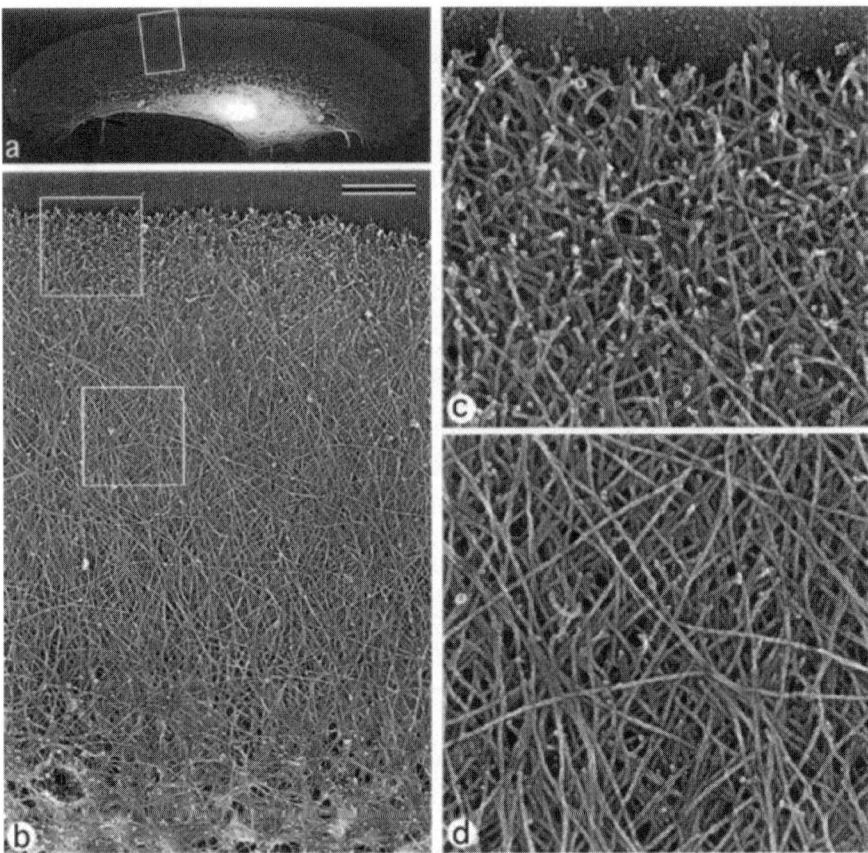

Figure 2.3. Actin meshwork structure in motile keratocytes. Electron microscopy images of the actin meshwork in detergent-extracted cells prepared using the platinum replica technique. (a) Overview of a motile cell; (b) actin network in the lamellipodium from the leading edge (top) to the transitional zone (bottom); (c) brushlike zone at the leading edge with numerous filament ends; (d) smooth actin filament network in the middle part of lamellipodium. Bar, 1μm. Reproduced and modified from [38]. Copyright (1997) permission of The Rockefeller University Press.

inside the cell) of a fluorescent phalloidin[3] results in moving speckles that are readily visible (see Figure 2.4b). After translating the movie into the cell frame of reference [61], a correlation based approach [31, 59] was used to extract the filamentous actin flow patterns from the dynamics of the moving speckles (see Figure 2.4c in the lab frame of reference; see Figure 2.4d in the cell frame of reference). As observed earlier [34], the rate of actin flow in the central part

---

[3] Phalloidin is a toxin made by the poisonous mushroom Amanita phalloides which strongly binds to polymerized actin and at high concentrations prevents depolymerization and kills the cell.

of the lamellipodium in the cell frame of reference was nearly equal to cell speed with only a small retrograde flow of the actin meshwork relative to the substrate [31, 62, 63]. These results show that in keratocytes the actin cytoskeleton is firmly anchored to the substrate so that actin polymerization is directly coupled to protrusion at the leading edge. Note the coherence of the observed flow across the lamellipodium, indicating that the actin meshwork moved as a single, relatively rigid, mechanical unit. The actin network at the rear of the cell displayed a rapid inward flow toward the cell body, with most of the motion in the cell frame of reference perpendicular to the direction of motion (see Figure 2.4c). This inward flow was probably due to forces generated by myosin contraction pulling the actin meshwork (see below), whereas the retrograde flow in the central lamellipodium in keratocytes was largely caused by actin protrusion at the leading edge, although some contribution from myosin mediated contraction is possible.

The observed kinetics of actin assembly and disassembly within a cell are dramatically accelerated compared to the rates measured for pure actin *in vitro*, due to the action of a large number of actin-associated proteins, many of which have been shown to be required for the protrusion process. These include actin capping proteins, profilin, and ADF/cofilin. Capping proteins bind to barbed ends and terminates their growth. This activity regulates the length and the number of active filaments and contributes to the formation of the brushlike zone in the actin network at the leading edge (see Figure 2.3c). Profilin binds actin monomers with a one-to-one stoichiometry and acts as a monomer sequestering protein, to allow higher actin monomer concentrations. At the same time, profilin-actin complexes can readily be incorporated at the barbed ends of growing filaments [64, 65]. ADF/cofilin is required for rapid filament disassembly [66]. Together, these proteins allow the cell to maintain a high concentration of actin monomers that can rapidly add on to the barbed ends of actin filaments. These components, together with actin and Arp2/3, were originally shown to be necessary and sufficient for reconstituting the formation of actin "comet tails" and motility of *Listeria monocytogenes* [9], and have since been found to be required for the formation and protrusion of lamellipodia in tissue culture cells [67]. Protrusion at the leading edge is further regulated by a number of other auxiliary proteins, such as the Ena/VASP protein family [68, 69, 70]. Further details regarding the biochemistry of actin protrusion are described elsewhere [5, 6, 7, 45].

The spatially regulated dynamics of the actin network are responsible for protrusion at the leading edge. The local rate of protrusion will depend on the load force acting on a growing filament, as well as other force-independent factors. The load force is determined by the pressure imposed by the membrane tension and the differences in osmotic/hydrostatic pressures across the cell membrane. The load force on an individual filament is inversely proportional to the local filament density, because the load is distributed among the filaments. Force-independent factors affecting protrusion include the local concentration of actin monomers (as well as the local concentrations of acces-

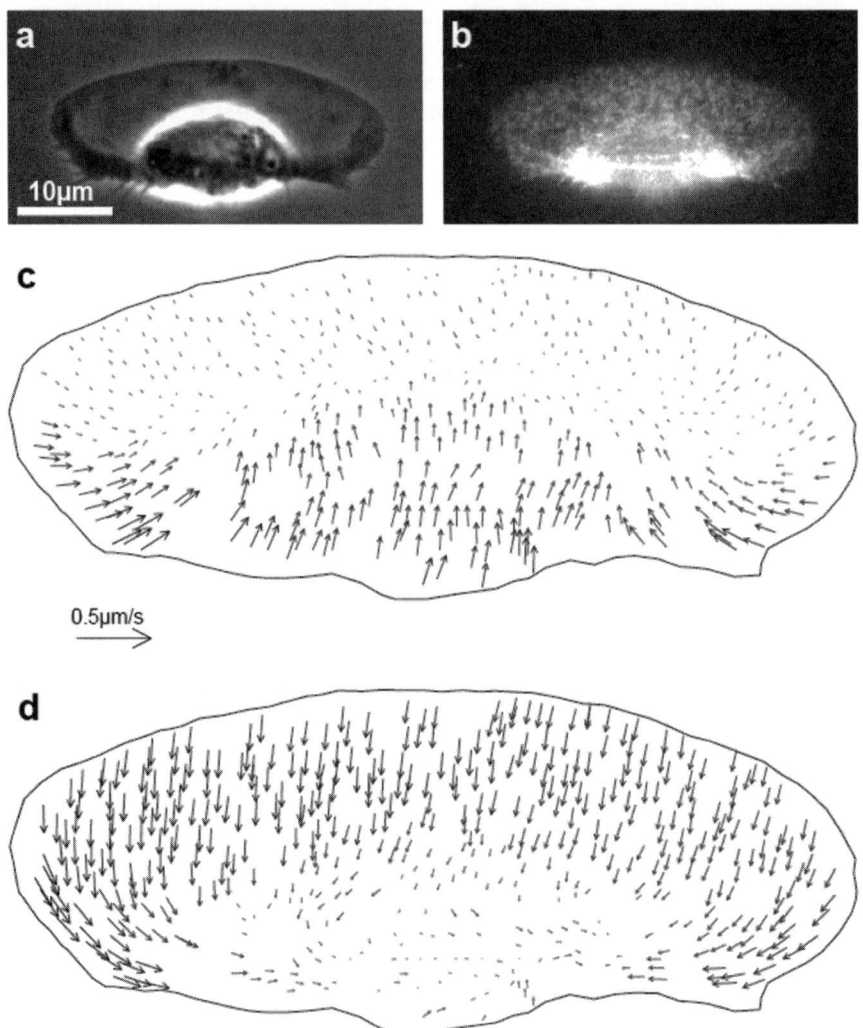

Figure 2.4. Actin network flow in motile keratocytes. Phase contrast (a) and fluorescence (b) images of a cell with Alexa Fluor-546 phalloidin speckles. The actin network movement was calculated from the displacement of phalloidin speckles between a set of consecutive frames. The calculated actin flow field is shown in the lab (c) and cell (d) frames of reference. In the lab frame of reference the actin network had a small retrograde flow in the lamellipodium and large inward movements at the rear sides. In the cell frame of reference, the actin network moved uniformly from the leading edge toward the cell body, with inward movements at the rear sides. Courtesy of Cyrus Wilson.

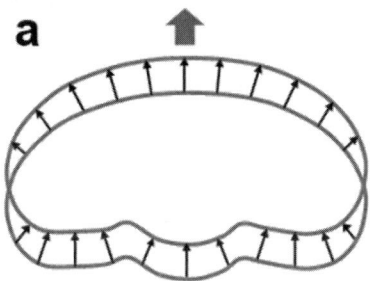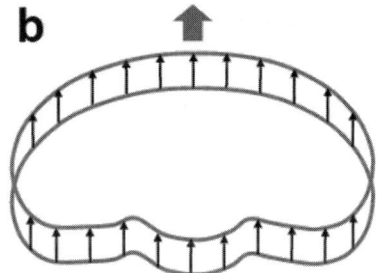

Figure 2.5. The graded radial extension model. (a) Schematic illustration of a cell moving forward while maintaining constant shape by graded radial extension. Protrusion at each point along the leading edge proceeds in a direction locally perpendicular to the edge, at a protrusion rate which decreases away from the center. (b) For comparison, a schematic illustration of a cell moving forward by constant extension which is everywhere equal in magnitude and direction to the net cell speed. The gray arrows depict the direction of cell motion.

sory proteins such as Arp2/3 and VASP), and the inherent time required for actin polymerization. At present, it is not clear which factor is rate limiting in the protrusion process at the leading edge, and the answer may well be different depending on the cell type and/or on the specific conditions within a particular cell.

The local protrusive force generated by the leading edge of moving keratocytes has recently been measured by placing a cantilever in the path of a protruding lamellipodium [71]. The measured force velocity relationship displayed a surprising sharp velocity drop upon contact at very small loads (< 100pN/μm), the nature of which is unknown. At higher loads, the velocity was insensitive to load over a broad range of loads until eventually there was a sharp decrease in velocity and protrusion stalled. The measured stall force was ~ 1nN, with some variation among different cells, which amounts to ~ 4pN force per filament. The force velocity relationship of an *in vitro* dendritic actin meshwork was also found to be insensitive to load force over a wide range of loads [72]. Interestingly, the force velocity relationship measured *in vitro* was dependent on the load history, so more than one possible velocity was observed for a given load. Both results [71, 72] are suggestive of feedback mechanisms between the mechanical load imposed on a protruding actin network and the molecular processes responsible for network growth and organization. Such force-dependent network remodeling may play an important role in the ability of moving cells to respond to mechanical variations in their *in vivo* environment.

Precise control and coordination of local protrusion rates is required to maintain the coherent leading edge of motile keratocytes (see Figure 2.1). A phenomenological model which attempts to address this issue is the graded radial extension (GRE) model (see Figure 2.5a; Reference [39]). According to the GRE model, a cell maintains constant shape by having graded extension along the leading edge with protrusion perpendicular to the edge at each location (see Figure 2.5a), rather than extending at a constant rate everywhere in the direction of cell movement like a rigid object (see Figure 2.5b). This model is supported by experimental observations of the predicted curved trajectories of both actin and surface lectins as the cell moves forward [39]. The GRE model is a phenomenological model and, as such, does not attempt to account for the underlying mechanism leading to the observed graded radial extension. In particular, while it is clear how local information can be used to identify the direction of protrusion normal to the edge at that position, it is an open question how the local protrusion rate is correlated with the overall position of a region within the cell. More recently, there have been attempts to describe the coupling between the underlying biochemical processes at the molecular level and the shape of the leading edge at the cellular level [22, 24]. In particular, both the actin barbed end distribution along the leading edge and the profile of the leading edge were shown to depend on the relative rates of filament elongation, branching, and capping. These studies represent initial steps toward understanding how the local behavior of molecules is coordinated at the cellular level to yield coherent cell movement.

2.2.2 Adhesion

For the polymerization of actin at the leading edge to generate protrusion, a cell must be anchored to the substrate. Many of the important molecular players in the formation of cell-substrate contacts have been identified and the signaling mechanisms that couple cytoskeletal dynamics with adhesion are beginning to be elucidated (reviewed in [3, 8, 73]). Of the many different receptors involved in cell adhesion, the integrin protein family appears to play a central role. These receptors form a link between intracellular adaptor proteins (such as vinculin) to the external environment of the cell, which can be either extracellular matrix proteins, other cells or the substrate in *in vitro* experiments. The adaptor proteins in turn bind to the cytoskeleton, so that together with the adhesion proteins they form complex hierarchal adhesion complexes [3], which mediate coupling between the intracellular actin-myosin cytoskeleton and the extracellular environment. It is important to note that while adhesion is necessary to allow cells to exert force on the substrate, deadhesion is also essential to allow cells to detach and move forward. Thus, cell adhesion must be regulated dynamically in space and time to allow the newly protruding lamellipodium to attach to the substrate at the front of the cell, and at the same time allow the rear of the cell to detach and translocate forward. This is particularly true for fast-moving keratocytes that maintain a constant shape as they advance, because attachment and detachment must occur simultaneously in a fast, yet highly coordinated, fashion.

The pattern of contacts a moving cell makes with the substrate can be visualized by interference reflection microscopy in which image contrast is determined by the distance between the ventral surface of the cell and the substrate [74]. The pattern of adhesion in keratocytes typically includes a rim of very close contact at the leading edge, which appears simultaneously with lamellar extension (see Figure 2.6a; Reference [19]). The rest of the lamellipodium also maintains relatively close contact to the substrate, although the pattern and extent of regional variation differ among cells. Investigation of the molecular composition of adhesion complexes in keratocytes showed that $\beta 1$ integrin adhesion molecules were localized within a narrow rim near the leading edge (see Figure 2.6b; Reference [19]), and hence are probably involved in the establishment of close contact at the rim (see Figure 2.6a-c). Note that in order to maintain such localization, the interaction between $\beta 1$ integrins and the substrate must be highly dynamic since the leading edge in keratocytes advances forward rapidly. $\beta 1$ integrin proteins were also found in small foci throughout the lamellipodium. Vinculin, an adaptor protein that associates actin filaments with integrin complexes at the cell membrane [3] was not found near the leading edge, but rather in small foci throughout the lamellipodium (see Figure 2.6e; References [19, 75]). These foci were shown to be stationary with respect to the substrate [75]. Vinculin was also found in larger foci at the cell rear (see Figure 2.6e), where it appeared to slide inward from the sides toward the cell body. These larger foci were typically confined to the retracting cell rear. The observed distribution of adhesion molecules is consistent with a model in which formation of focal contacts is initiated at the leading edge. These contacts remain stationary with respect to the substrate, and therefore move away from the leading edge. As they move away, the focal contacts mature with additional components such as vinculin being added to them. Finally, upon reaching the cell rear, the focal contacts disassemble. At the same time, new focal contacts are continuously formed at the leading edge to maintain close contact there.

The formation of adhesions at the rim of the leading edge was recently shown to be crucial for forward protrusion. The application of very weak forces $\sim 10\text{pN}/\mu\text{m}^2$ by fluid flow at the leading edge locally stalled forward protrusion [76]. As expected, these forces were much too weak to stall filament growth, and indeed actin polymerization continued nearly unaffected. However, the weak forces appeared to disrupt the formation of nascent adhesions at the leading edge, which resulted in local network growth upward rather than forward. These results suggest a direct coupling between forward protrusion at the leading edge and contact formation.

Mechanical force regulates adhesion size and stability [4]. The amount of tension generated across adhesion complexes depends both on the intracellular force generation primarily by myosin contraction, and on the mechanical properties of the substrate. Experiments in other cell types, such as fibroblasts, showed that force enhanced focal adhesion assembly; the size of focal adhesions correlated with the local traction force [77], and furthermore appli-

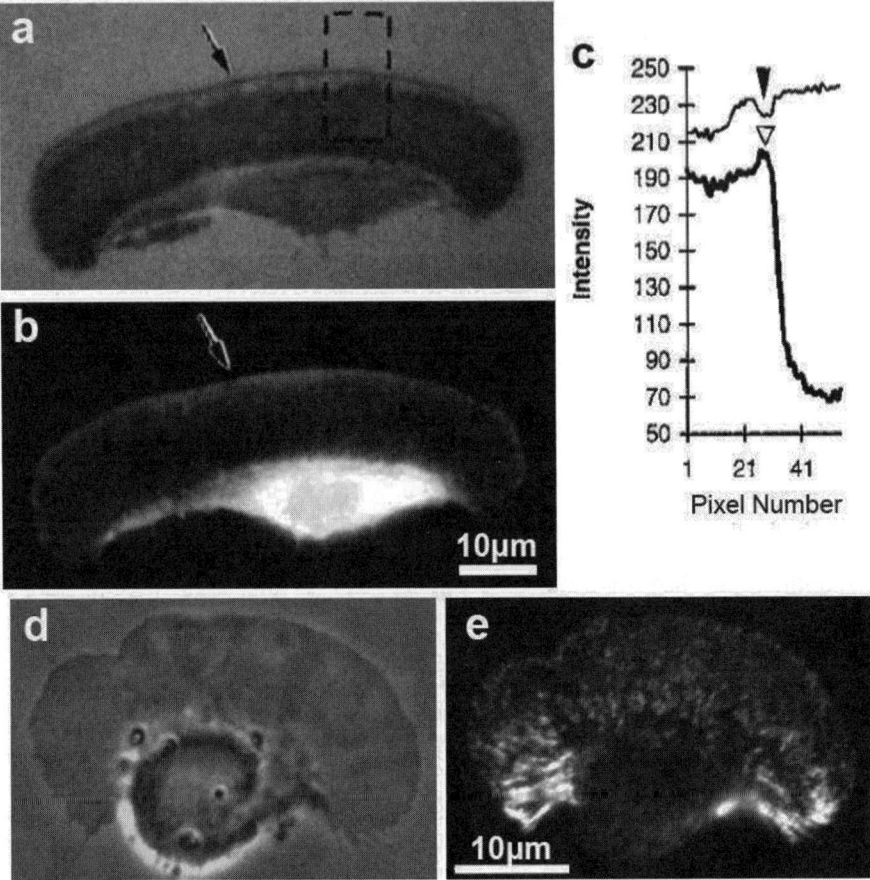

Figure 2.6. Pattern of adhesion in motile keratocytes. Interference reflection microscopy (IRM) image (a) and immunofluorescence image (b) of a moving keratocyte fixed and stained with anti-$\beta 1$ integrin. A rim of increased fluorescence corresponds to a narrow region of very close contact along the leading edge as seen by IRM. This is shown more clearly by the averaged line intensity scans (c) for $\beta 1$ integrin taken along the length of the rectangle shown in (a). A pronounced minimum (black arrowhead) in the graph of light intensity of the IRM image (thin line) marks the position of the rim of very close contact. This coincides with the peak (open triangle) in the graph of fluorescence intensity (bold line). Phase contrast (d) and immunofluorescence (e) images of a moving keratocyte fixed and stained with anti-vinculin. Small vinculin foci are apparent throughout the lamellipodium, while larger foci are restricted to the cell rear. (a-c) Reproduced and modified from [19] with permission from The Company of Biologists. (d-e) Courtesy of Erin Barnhart.

cation of external forces on cells induced focal adhesion assembly and growth [78]. At the same time, it is thought that high-tension forces promote adhesion disassembly and rupture, which are essential for retraction of the cell rear. The regulation and coordination of these processes in fast and smooth moving cells such as keratocytes is still not clear.

Several experiments using deformable substrates or micromachined devices to measure the traction forces generated by moving keratocytes have shown that cell pulling on the substrate is negligible under the front part of the lamellipodium but that keratocytes pull inward very strongly at the sides (see Figure 2.7; References [36, 41]). This tension is in the same direction as the inward actin network flow in that part of the cell (see Figure 2.4c,d). Keratocytes moving on soft substrates such as gelatin demonstrate large inward traction force and thereby generate wrinkles under the cell body, moving forward in a cycle of fast and slow movements where each phase of rapid movement correlates with a sudden release of adhesions to the substrate and a loss of surface wrinkling as the elastic substrate snaps back to its initial shape [79, 80]. The strong perpendicular traction forces are attributed to myosin contraction (see Section 2.2.3). However since these forces are mostly directed perpendicular to the direction of keratocyte motion, their importance to steady state motility is unclear.

2.2.3 The Role of Myosin

As discussed earlier, substantial experimental evidence from a variety of systems established that actin polymerization alone is capable of generating protrusive forces. However, this does not rule out contributions from other force-generating molecular motors, such as myosins, in cell motility. Myosins are a superfamily of molecular motors that bind filamentous actin, and use ATP hydrolysis to generate force by stepping along the filament. Of the different classes of myosins, non-muscle myosin II has been most strongly implicated to play a role in actin-based cell motility. Individual myosin II molecules can come together and form supramolecular assemblies called thick filaments which are $\sim 0.4\mu m$ long [38] and have multiple actin-binding heads that function cooperatively to locally contract the actin meshwork. The distribution of myosin II in keratocytes is enhanced at the cell rear and is most pronounced along the actin bundle in the back of the cell and in the transition zone between the lamellipodium and the cell body (see Figure 2.8; Reference [38]). Discrete myosin clusters are apparent throughout the lamellipodium with their size and density increasing toward the cell body. Live cell microscopy revealed that these clusters assemble in the lamellipodium, are stationary with respect to the substrate, and progressively grow as they move toward the cell body. A similar distribution of myosin II was found in lamellipodial fragments [30].

Several models suggest that myosin activity is required for steady state motility of keratocytes, both for forward translocation of the cell body and for maintenance of cell polarity [37, 38, 39]. The distribution of myosin II

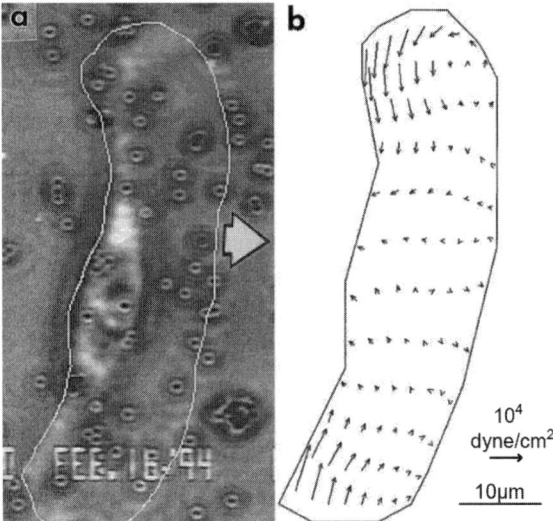

Figure 2.7. Traction force measurements in keratocytes. (a) Phase contrast image of a keratocyte moving (in the direction indicated by the large arrow) on a silicone rubber substratum with embedded beads. The cell is outlined in white for clarity. (b) Traction map for the same cell. Each small arrow represents the magnitude and direction of the traction stress at the location corresponding to the tail of the arrow. Reproduced and modified from [41]. Copyright (1999) permission of The Rockefeller University Press.

and the fact that myosin clusters remain stationary with respect to the actin cytoskeleton, indicate that direct transport by myosin moving along actin tracks as a mechanism for pulling the cell body is not a plausible model. Another model [37] suggests that the cell body moves forward by rolling, which results from tension generated by myosin contraction along the actin bundles at the rear of the cell. While rolling of the cell body is observed in moving cells [37], its magnitude varies between different cells and, in most cases, is not sufficient to account for all the forward translocation of the cell body [38]. Yet another model [38] suggests that contraction of the actin-myosin cytoskeleton at the transition zone between the lamellipodium and the cell body induces forward translocation of the cell body.

While it is obvious that myosin contraction in keratocytes leads to reorientation of the actin network in the transition zone between the cell body and the lamellipodium [38] and is responsible for the observed inward flow of actin at the rear of the cell, the importance of these processes for steady state motility is not clear. Experiments in which myosin II activity was eliminated either by specific drugs or by deletion indicate that myosin II is not essential for steady state motility. For example, keratocytes treated with blebbistatin, a specific inhibitor of myosin II, continue to move although with reduced per-

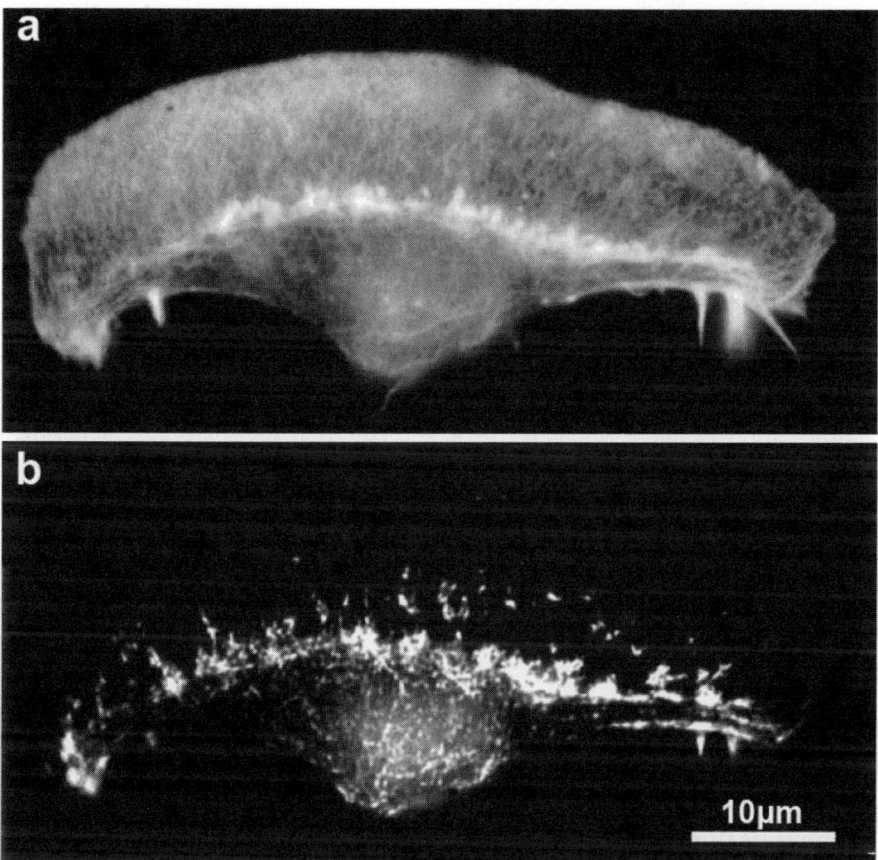

Figure 2.8. Myosin II distribution in motile keratocytes. Fluorescence images of a moving keratocyte fixed and stained with phalloidin and anti-myosin showing filamentous actin distribution (a) and myosin II distribution (b). Discrete myosin spots among the continuous actin network in the lamellipodium as well as accumulation of both actin and myosin at the lamellipodium/cell body boundary are apparent. Reproduced and modified from [38]. Copyright (1997) permission of The Rockefeller University Press.

sistence [81]. Similarly, *Dictyostelium* cells in which the myosin II gene was disrupted are still motile but show several defects [82, 83]. Thus, further work is required to elucidate the role of myosin in cell motility and establish whether the polarized dynamics of actin network assembly and disassembly alone are sufficient for maintaining steady state motility.

While the role of myosin in steady state motility is unclear, myosin II activity does appear to be necessary for detachment when keratocytes are temporarily stuck to the substrate [84]. The mechanical stretching of the membrane when one region of the cell is stuck and the rest of the cell continues to protrude forward activates stress-activated calcium channels, resulting in transient increases in intracellular calcium concentration. These calcium transients are postulated to further activate actin-myosin contractility and allow the cell to detach. While calcium transients can be observed in smoothly moving cells, their importance for steady state motility is unclear because inhibition of calcium transients with intracellular calcium chelators does not have a significant effect on moving cells (G. Allen, unpublished observations).

It is interesting in this context to compare actin-based motility to the mechanistically similar, but biochemically distinct, crawling of the nematode sperm cell [85, 86]. There, motility is driven by assembly and disassembly of non-polar filaments of a protein called major sperm protein (MSP) rather than actin. As no analogs to the myosin motors are known in this system (moreover since the MSP filaments are apolar, the existence of analogous directional motors appears highly unlikely), it is thought that contraction is generated solely through disassembly of MSP filaments [87]. Thus, in this system, both protrusion and contraction forces are generated by assembly and disassembly of filaments without the action of any myosin-like molecular motors. Even though the basis for the spatial segregation of MSP filament assembly and disassembly in the nematode sperm is not entirely clear, this system shows that, at least in principle, myosin-based contraction need not be essential for cell motility.

2.3 The Cell Membrane

The cell membrane has several important roles in cell motility. First, because actin is constantly pushing from within, the cell membrane is stretched leading to membrane tension that exerts an inward force perpendicular to the cell surface. In particular, this force will act on the barbed ends of actin filaments at the leading edge, retarding their growth. Experiments using pharmacological agents to perturb physical properties of the cell membrane in fibroblasts and endothelial cells have implicated membrane tension or membrane microviscosity as possible factors limiting the speed of cell migration [88, 89]. Second, the cell membrane plays a crucial role in regulating the intracellular environment and cell volume through the action of various channels in the cell membrane, including ion channels and aquaporins (reviewed in [90]). Finally, the cell membrane serves to localize various proteins and lipids that have important roles in adhesion, regulation of actin dynamics, and coordination of signaling information (reviewed in [8]).

The lamellipodium of moving cells in general, and keratocytes in particular, is very thin ~200nm (see Figure 2.2; References [42, 91]) compared to the cell body ($5 - 10\mu$m). This thin structure is maintained by restricting protrusion to the apex of the leading edge. The mechanisms responsible for generating and maintaining this unique dynamic interface between the membrane and the cytoskeleton at the leading edge are entirely unclear, but most likely involve localization of factors regulating actin nucleation at the leading edge. More specifically, it is thought that activators of the Arp2/3 complex (which nucleates actin filaments branching off existing filaments), such as the WASP family of proteins, are localized there although the detailed identity of the molecular complexes involved is unknown. A possible mechanism for localization of proteins to highly-curved regions of the membrane such as the leading edge may be related, at least partly, to the spontaneous membrane curvature of the proteins [92, 93]. Interestingly, it was found that the leading edge in keratocytes acts as a lipid diffusion barrier, blocking lateral mobility of lipids in the outer membrane leaflet [94]. While the physical basis for this observation is not clear, it was postulated that this diffusion barrier might be a result of the high concentration of proteins complexes at the leading edge and their extensive connections to the actin cytoskeleton. In addition, it has been shown that channel proteins, including NHE1 sodium channels and aquaporins, are localized to the leading edge in various types of moving cells (reviewed in [90]). It has been suggested that these channels may play an active role in the protrusion process by generating hydrostatic/osmotic pressures that would assist the protrusion process (see Section 2.4 below; [90, 95]).

Various models have been suggested for the behavior of the cell membrane as the cell moves forward. These include the "tank-tread" model in which the membrane flows forward on the dorsal surface and rearward on the ventral surface, as well as rearward flow on both dorsal and ventral surfaces made possible by addition of lipids at the leading edge and their removal closer to the cell body. However, experimental measurements by single particle-tracking of particles on the dorsal surface of the lamellipodium in keratocytes [96, 97] and by photobleaching of labeled lectins [39] indicate that the cell membrane moves forward with the cell passively on both the dorsal and ventral surfaces. Thus, in the cell frame of reference, there is no net lipid flow in the membrane. Note that this is not the case in other systems such as the neuronal growth cone, where membrane flow from the growth cones toward the cell body is observed [98, 99]. The lack of lipid flow, together with the presence of a diffusion barrier at the leading edge, imply that physical trapping may be sufficient for maintaining the localization of various essential membrane-bound components there. The lack of lipid flow also indicates that there cannot be steady gradients in the membrane tension across the cell as these would lead to membrane flow. Thus, membrane tension acts as a global mechanical parameter that can introduce coupling between biochemical events occurring at different regions in the cell. Specifically, growing actin filaments at different regions along the leading edge will experience the same membrane load.

2.4 The Cytosolic Fluid

Cytosolic fluid dynamics are important for understanding cell motility, both because of the hydrodynamic forces they induce and because of their influence on transport of components of the actin machinery to the leading edge. The intracellular fluid flow inside the cell will be affected by pressure differences within the cell [100], as well as the boundary conditions imposed by the cell membrane, in particular, the membrane permeability to water molecules and ions and the osmotic/hydrostatic pressure across the membrane. The simplest scenario for fluid behavior in a moving cell is that fluid moves passively with the cell (see Figure 2.9a), similar to the observed behavior of the cell membrane. In this case, there would be zero net fluid flow in the cell frame of reference, while in the lab frame of reference the fluid would be moving forward with cell speed. Because the actin meshwork is nearly stationary in the lab frame of reference (see Figure 2.4c), this implies that there will be a relative velocity between the cytosolic fluid and the actin meshwork in the lamellipodium, comparable to the velocity of the cell. This situation is referred to as *d'Arcy flow* and will be accompanied by a pressure gradient proportional to the relative velocity between the fluid and the meshwork, with higher pressure found at the cell body [23].

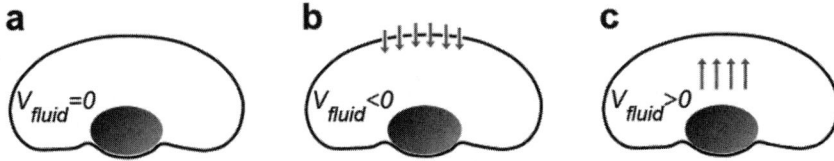

Figure 2.9. Possible schemes for fluid flow in moving cells shown in the cell frame of reference. (a) The fluid moves forward passively with the cell. (b) The fluid flows away from the leading edge driven by water influx at the leading edge. The flow assists protrusion by reducing membrane load on growing actin filaments. (c) The fluid flows toward the leading edge, assisting protrusion at the leading edge by promoting recycling of components of the actin machinery back to the leading edge.

Another possible flow scheme suggested many years ago involves persistent fluid inflow at the leading edge of moving cells driven by an osmotic pressure gradient (see Figure 2.9b; References [95, 101]). Such water influx at the leading edge would partially relieve the membrane load opposing polymerization, and lead to an increase in the rate of actin polymerization at a protruding edge. Thus, in this case, hydrodynamic forces would work together with the polymerization motor to power protrusion. Localized water influx preferentially at the leading edge is consistent with the localization of aquaporins at the leading edge [95], which are known to increase membrane water permeability by more than an order of magnitude [102]. Moreover, it was shown that aquaporin presence enhanced endothelial cell motility [95]. In addition, inward water influx at the leading edge of neutrophils was inferred from intensity measurements of a self-quenching dye that showed enhanced signal near the leading edge attributed to dilution by incoming water there [103]. Note that such water influx at the leading edge would also reduce the concentration of proteins such as actin monomers required for assembly of the actin meshwork.

Alternatively, it has been implied that intracellular fluid flow toward the leading edge (see Figure 2.9c) might contribute to motility by expediting transport of actin and other soluble proteins to the leading edge. Forward transport of actin into protruding regions at a speed of more than $5\mu m/s$ was measured by fluorescence localization after photobleaching of GFP-tagged β-actin in fibroblasts [104]. While the results excluded simple diffusion as the sole transport mechanism, and pointed to the existence of some form of active transport, its nature was not determined. The authors suggested that directed transport could be driven by forward hydrodynamic flow with a magnitude of several $\mu m/s$ generated by myosin contraction at the cell rear that they postulated "squeezed" fluid forward. Our own observations in fish keratocytes that move more than an order of magnitude faster than fibroblasts show that such rapid forward flows do not exist in keratocytes (K. Keren et al., unpublished observations). Intracellular fluid flow of a smaller magnitude that assists recycling of monomeric actin to the leading edge was predicted in a multiscale 2D computational model of the lamellipodium [23]. However, the relative importance of this fluid flow for recycling was small and the model showed that diffusion should be sufficient for recycling monomeric actin back to the leading edge even, in rapidly moving cells such as keratocytes.

We have described several schemes for the pattern of fluid flow in moving cells (summarized in Figure 2.9). All possible scenarios imply a relative velocity between the cytosolic fluid and the actin network. However, the various models assign very different roles for fluid flow. Fluid flow is projected to play an active role in promoting motility by either relieving membrane tension albeit at the price of reducing protein concentration at the leading edge (see Figure 2.9b), or by increasing protein concentration at the leading edge (see Figure 2.9c). Alternatively in the passive model (see Figure 2.9a) fluid moves along inactively with the cell. These mutually excluding scenarios are each supported by several indirect results. However, direct measurements of fluid flow in moving cells, which would resolve the controversy, are lacking due to the difficulty of measuring relatively small flows of fluid that interpenetrates a dense meshwork characterized by a typical pore size of $\sim$30-50nm (see Figure 2.3). Thus, the important question of how the fluid is behaving in a moving cell, and what role fluid dynamics plays in cell motility, remains for now unanswered.

2.5 Feedback Mechanisms and Large-Scale Integration

Although for the sake of simplicity the sub-processes involved in cell motility (namely, protrusion, contraction, retraction and adhesion) are often treated independently, they are, in fact, highly interrelated [105]. An increasing amount of data suggests that these processes are coupled through complicated feedback mechanisms, which in many cases are at least partially mechanical [4, 77, 106]. For example, it has been shown that intrinsic or extrinsic forces affect focal adhesion strength and turnover [4, 77]. Furthermore, there is evidence for feedback between actin polymerization, myosin activity, and adhesion, where a perturbation in one of these leads to the reorganization and modulation of the behavior of the others [106]. As mentioned earlier, measurements of the protrusive forces in growing dendritic actin networks [71, 72] also suggest that actin network growth and organization are dependent on mechanical parameters. These results indicate that the organization and mechanical characteristics of the individual elements involved in cell motility are not fixed, but rather depend on the dynamics of the entire system. These complex interrelationships between different modules are characteristic of biological systems, and emphasize the need for an integrative system-level approach for understanding cell motility.

The integration of numerous molecular components and multiple modules to achieve coherent cell movement requires a high degree of coordination. Multiple signaling pathways have been shown to play important roles in such coordination, and in establishing and maintaining cell polarity (reviewed in [8]). The Rho signaling pathway is one of the important regulators of protrusion; members of the Rho family such as Rac and Cdc42 are thought to stimulate Arp2/3 activation and protrusion at the leading edge via the WASP proteins, while other members such as Rho are important for defining the rear of the cell. The phosphoinositides, $PtdIns(3,4,5)P_3$ (PIP_3) and $PtdIns(3,4)P_2$ (PIP_2), have been shown to be important for the polarized response of cells to chemoattractant gradients [107]. For example, upon receptor activation during chemotaxis, phospho-inositide 3-kinases (PI3Ks) which catalyze the phosphorylation of PIP_2 to PIP_3, are recruited to the membrane and generate an accumulation of PIP_3 at the leading edge of cells. While these signaling pathways have been shown to be essential in other cell types such as neutrophils and *Dictyostelium*, their role in keratocytes, which are not responsive to external chemical cues, is not clear. For example, Rho-kinase dependent myosin activity is required for establishing polarity and motility initiation in keratocytes [31], but its inhibition had essentially no effect on steady state motility, indicating that this pathway is not essential for maintaining polarity in keratocytes. Similarly, PIP_3 does not localize to the leading edge of moving keratocytes, and inhibition of PI3Ks does not have a significant effect on cell speed (P.T. Yam and N.A. Dye, unpublished observations). As illustrated by several examples throughout this chapter (e.g., the mechanosensitivity of adhesions and the load dependence of protrusion) mechanical processes also play an important role in large-scale coordination. Further work is required to determine the relative role of mechanical coupling and the inherent dynamics of the system and signaling processes in large-scale coordination in keratocyte motility.

2.6 Conclusions

The underlying dynamics of numerous molecular components determine behavior at the cellular level and dictate properties such as cell shape and speed. Understanding the mechanisms responsible for this remarkable self-organization is a central challenge in cell motility research. Keratocytes are one of the fastest moving animal cells, and as such are characterized by rapid dynamics at the molecular level. At the same time, their behavior at the cellular level is remarkably robust, exhibiting nearly constant shape and speed. This combination makes keratocytes an excellent model system for studying the nature of large-scale coordination in cell motility. Further work exploring the natural variation in cell behavior among a population of cells and its correlation with variations in the molecular composition [24], as well as the effect of various mechanical and molecular perturbations on cellular behavior, will help elucidate the self-organizational principles underlying cell motility.

The interplay between biochemical processes and mechanical processes plays a central role in cell motility. The coupling between biochemical and biophysical processes is evident at all levels of organization. Molecular processes such as actin polymerization or adhesion formation are dependent on the mechanical forces acting on the molecules. At the same time, biochemical processes and regulation determine the mechanical properties of cellular subsystems such as the cell membrane and the cytoskeleton. In addition, mechanical coupling provides the means for rapid large-scale coordination. For example, membrane tension effectively couples biochemical processes at the front and at the rear of the cell. Mechanical coordination is more apparent in keratocytes, in which motility is rapid and not dependent on external cues. The relative role of such mechanical coordination and biochemical signaling for large-scale coordination in keratocyte motility remains to be determined.

Understanding how motility at the cellular level arises from the action of numerous interacting molecular components is inherently difficult and non-intuitive. This emphasizes the role of mathematical modeling as a means for integrating the large amounts of experimental data into a systematic understanding of this process. Indeed, cell motility has been a subject of mathematical modeling for a long time. These studies have contributed substantially to our understanding of the process of cell motility, but until recently most of them focused on individual reactions or modules involved in cell motility [21, 55, 93, 108, 109]. As noted above, quantitative understanding of cell motility must involve an integrative view of the cell as a whole. Initial attempts to model motility at the cellular level were mostly phenomenological [39]. However, in the last few years, several models have been put forth that attempt to link the physical and biochemical understanding at the molecular level with a cellular level description [22, 23, 24]. While a complete model that quantitatively accounts for the spatio-temporal dynamics of a moving cell is still lacking, progress toward this goal proceeds both by more rigorous and quantitative experimental measurements of the biochemical and biophysical characteristics of a moving cell and by increasingly refined models. Such models should account for the dynamics of an individual cell, as well as explain the nature of the variation observed between individual cells and among different cell types based on their underlying biochemical and biophysical properties. In light of recent advances, it seems likely that in the not-too-distant future, cell motility will become one of the first cellular level systems to be understood in a systematic quantitative manner.

Acknowledgements

We thank Greg Allen, Natalie Dye, Patricia Yam, and Cyrus Wilson for sharing unpublished data. We thank Cyrus Wilson for preparing Figure 2.4, and Greg Allen and Erin Barnhart for images included in Figures 2.2 and 2.5 respectively. We thank Alex Mogilner, Patricia Yam, Zach Pincus, Erin Barnhart, Greg Allen and Cyrus Wilson for critical reading of the manuscript. K.K. is a Damon Runyon Fellow supported by the Damon Runyon Cancer Research Foundation (DRG-# 1854-05). J.A.T. is supported by the National Institutes of Health and the American Heart Association.

References

1. Lauffenburger, D & Horwitz, A. (1996) *Cell* **84**, 359–69.
2. Alberts, B & et al. (2002) *Molecular Biology of the Cell, 4th ed.* (Garland, New York).
3. Geiger, B, Bershadsky, A, Pankov, R, & Yamada, K. (2001) *Nat. Rev. Mol. Cell. Biol.* **2**, 793–805.
4. Bershadsky, A, Balaban, N, & Geiger, B. (2003) *Annu. Rev. Cell. Dev. Biol.* **19**, 677–95.
5. Pantaloni, D, Le Clainche, C, & Carlier, M. (2001) *Science* **292**, 1502–6.
6. Pollard, T & Borisy, G. (2003) *Cell* **112**, 453–65.
7. Rafelski, S & Theriot, J. (2004) *Ann. Rev. Biochem.* **73**.
8. Ridley, A, Schwartz, M, Burridge, K, Firtel, R, Ginsberg, M, Borisy, G, Parsons, J, & Horwitz, A. (2003) *Science* **302**, 1704–9.
9. Loisel, T, Boujemaa, R, Pantaloni, D, & Carlier, M. (1999) *Nature* **401**, 613–6.
10. van Oudenaarden, A & Theriot, J. (1999) *Nat. Cell. Biol.* **1**, 493–9.
11. Gerbal, F, Chaikin, P, Rabin, Y, & Prost, J. (2000) *Biophys. J.* **79**, 2259–75.
12. Bernheim-Groswasser, A, Wiesner, S, Golsteyn, R, Carlier, M, & Sykes, C. (2002) *Nature* **417**, 308–11.
13. Giardini, P, Fletcher, D, & Theriot, J. (2003) *Proc. Natl. Acad. Sci. USA* **100**, 6493–8.
14. Upadhyaya, A, Chabot, J, Andreeva, A, Samadani, A, & van Oudenaarden, A. (2003) *Proc. Natl. Acad. Sci. USA* **100**, 4521–6.
15. Wiesner, S, Helfer, E, Didry, D, Ducouret, G, Lafuma, F, Carlier, M, & Pantaloni, D. (2003) *J. Cell. Biol.* **160**, 387–98.
16. Marcy, Y, Prost, J, Carlier, M.-F, & Sykes, C. (2004) *Proc. Natl. Acad. Sci. USA* **101**, 5992–5997.
17. Alberts, J & Odell, G. (2004) *PloS Biol.* **2**, e412.
18. Goodrich, H. (1924) *Biolog. Bull.* **46**, 252–262.
19. Lee, J & Jacobson, K. (1997) *J. Cell. Sci.* **110**, 2833–44.
20. Svitkina, T & Borisy, G. (1999) *J. Cell. Biol.* **145**, 1009–26.
21. Mogilner, A & Edelstein-Keshet, L. (2002) *Biophys. J.* **83**, 1237–58.
22. Grimm, H, Verkhovsky, A, Mogilner, A, & Meister, J. (2003) *Europ. Biophys. J.* **32**, 563–77.
23. Rubinstein, B, Jacobson, K, & Mogilner, A. (2005) *Multisc. Model. Simul.* **3**, 413–439.
24. Lacayo, C. I, Pincus, Z, van Duijn, M, Wilson, C, Fletcher, D, Gertler, F, Mogilner, A, & Theriot, J. (2007) *PLoS biology, in press.*
25. Kunzenbacher, I, Bereiter-Hahn, J, Osborn, M, & Weber, K. (1982) *Cell Tiss. Res.* **222**, 445–57.
26. Strohmeier, R & Bereiter-Hahn, J. (1991) *Cell Tissue Res.* **266**, 615–21.
27. Radice, G. (1980) *J. Cell. Sci.* **44**, 201–23.
28. Radice, G. (1980) *Develop. Biol.* **76**, 26–46.
29. Euteneuer, U & Schliwa, M. (1984) *Nature* **310**, 58–61.
30. Verkhovsky, A, Svitkina, T, & Borisy, G. (1999) *Curr. Biol.* **9**, 11–20.
31. Yam, P, Wilson, C. A, Ji, L, Hebert, B, Barnhart, E, Wiseman, P, Danuser, G, & Theriot, J. (2007) *Submitted.*
32. Cooper, M & Schliwa, M. (1985) *J.Neurosci. Res.* **13**, 223–44.
33. Cooper, M & Schliwa, M. (1986) *J. Cell. Biol.* **102**, 1384–99.

34. Theriot, J & Mitchison, T. (1991) *Nature* **352**, 126–31.
35. Small, J, Herzog, M, & Anderson, K. (1995) *J. Cell. Biol.* **129**, 1275–86.
36. Lee, J, Leonard, M, Oliver, T, Ishihara, A, & Jacobson, K. (1994) *J. Cell. Biol.* **127**, 1957–64.
37. Anderson, K, Wang, Y, & Small, J. (1996) *J. Cell Biol.* **134**, 1209–18.
38. Svitkina, T, Verkhovsky, A, McQuade, K, & Borisy, G. (1997) *J. Cell. Biol.* **139**, 397–415.
39. Lee, J, Ishihara, A, Theriot, J, & Jacobson, K. (1993) *Nature* **362**, 167–71.
40. Burton, K, Park, J, & Taylor, D. (1999) *Mol. Bio. Cell.* **10**, 3745–69.
41. Oliver, T, Dembo, M, & K., J. (1999) *J. Cell. Biol.* **145**, 589–604.
42. Abraham, V, Krishnamurthi, V, Taylor, D, & Lanni, F. (1999) *Biophys. J.* **77**, 1721–32.
43. Ream, R, Theriot, J, & Somero, G. (2003) *J. Exp. Biol.* **206**, 4539–51.
44. Pincus, Z & Theriot, J. (2007) *J. Microsc.* **227**, 140–156.
45. Pollard, T, Blanchoin, L, & Mullins, R. (2000) *Annu. Rev. Biophys. Biomol. Struct. A* **29**, 545–76.
46. Huxley, H. (1963) *J. Mol. Biol.* **7**, 281–308.
47. Pollard, T & Mooseker, M. (1981) *J. Cell Biol.* **88**, 654–659.
48. Theriot, J. (2000) *Traffic* **1**, 19–28.
49. Hill, T & Kirschner, M. (1982) *Internat. Rev. Cyt.* **78**, 1–125.
50. Cortese, J, Schwab, B, Frieden, C, & Elson, E. (1989) *Proc. Natl. Acad. Sci. USA* **86**, 5773–7.
51. Miyata, H, Nishiyama, S, Akashi, K, & Kinosita, K. J. (1999) *Proc. Natl. Acad. Sci. USA* **96**, 2048–53.
52. Limozin, L & Sackmann, E. (2002) *Phys. Rev. Lett.* **89**, 168103.
53. Mogilner, A & Oster, G. (1996) *Biophys. J.* **71**, 3030–3045.
54. Mullins, R, Heuser, J, & Pollard, T. (1998) *Proc. Natl. Acad. Sci. USA* **95**, 6181–6.
55. Maly, I & Borisy, G. (2001) *Proc. Natl. Acad. Sci. USA* **98**, 11324–9.
56. Salmon, W, Adams, M, & Waterman-Storer, C. (2002) *J. Cell. Biol.* **158**, 31–7.
57. Watanabe, N & Mitchison, T. (2002) *Science* **295**, 1083–6.
58. Waterman-Storer, C, Desai, A, Bulinski, J, & E.D., S. (1998) *Curr. Biol.* **8**, 1227–3026.
59. Ji, L & Danuser, G. (2005) *J. Microsc.* **220**, 150–67.
60. Danuser, G & Waterman-Storer, C. (2006) *Annu. Rev. Biophys. Biomol. Struct.* **35**, 361–87.
61. Wilson, C & Theriot, J. (2006) *IEEE Trans. Image. Process* **15**, 1939–51.
62. Jurado, C, Haserick, J, & Lee, J. (2005) *Mol. Biol. Cell.* **16**, 507–18.
63. Vallotton, P, Danuser, G, Bohnet, S, Meister, J.-J, & Verkhovsky, A. (2005) *Mol. Biol. Cell* **16**, 1223–1231.
64. Tilney, L, Bonder, E, Coluccio, L, & Mooseker, M. (1983) *J. Cell. Biol.* **97**, 112–24.
65. Pollard, T & Cooper, J. (1984) *Biochem.* **23**, 6631–6641.
66. Carlier, M, Laurent, V, Santolini, J, Melki, R, Didry, D, Xia, G, Hong, Y, Chua, N, & Pantaloni, D. (1997) *J. Cell. Biol.* **136**, 1307–22.
67. Rogers, S, Wiedemann, U, Stuurman, N, & Vale, R. (2003) *J. Cell. Biol.* **162**, 1079–88.
68. Bear, J, Loureiro, J, Libova, I, Fassler, R, Wehland, J, & Gertler, F. (2000) *Cell* **101**, 717–28.

69. Bear, J, Svitkina, T, Krause, M, Schafer, D, Loureiro, J, Strasser, G, Maly, I, Chaga, O, Cooper, J, Borisy, G, & Gertler, F. (2002) *Cell* **109**, 509–21.

70. Han, Y, Chung, C, Wessels, D, Stephens, S, Titus, M, Soll, D, & Firtel, R. (2002) *J. Biol. Chem.* **277**, 49877–87.

71. Prass, M, Jacobson, K, Mogilner, A, & Radmacher, M. (2006) *J. Cell. Biol.* **174**, 767–72.

72. Parekh, S, Chaudhuri, O, Theriot, J, & Fletcher, D. (2005) *Nat. Cell. Biol.* **7**, 1219–23.

73. Sastry, S & Burridge, K. (2000) *Exp. Cell Res.* **261**, 25–36.

74. Izzard, C & Lochner, L. (1976) *J. Cell. Sci.* **21**, 129–59.

75. Anderson, K & Cross, R. (2000) *Curr. Biol.* **10**, 253–60.

76. Bohnet, S, Ananthakrishnan, R, Mogilner, A, Meister, J, & Verkhovsky, A. (2006) *Biophys. J.* **90**, 1810–20.

77. Balaban, N, Schwarz, U, Riveline, D, Goichberg, P, Tzur, G, Sabanay, I, Mahalu, D, Safran, S, Bershadsky, A, Addadi, L, & Geiger, B. (2001) *Nat. Cell Biol.* **3**, 466–72.

78. Riveline, D, Zamir, E, Balaban, N, Schwarz, U, Ishizaki, T, Narumiya, S, Kam, Z, Geiger, B, & Bershadsky, A. (2001) *J. Cell. Biol.* **153**, 1175–86.

79. Doyle, A & Lee, J. (2002) *Biotechn.* **33**, 358–64.

80. Doyle, A & Lee, J. (2005) *J. Cell. Sci.* **118**, 369–379.

81. Straight, A, Cheung, A, Limouze, J, Chen, I, Westwood, N, Sellers, J, & Mitchison, T. (2003) *Science* **299**, 1743–7.

82. De Lozanne, A & Spudich, J. (1987) *Science* **236**, 1086–91.

83. Knecht, D & Loomis, W. (1987) *Science* **236**, 1081–6.

84. Lee, J, Ishihara, A, Oxford, G, Johnson, B, & Jacobson, K. (1999) *Nature* **400**, 382–6.

85. Wolgemuth, C, Miao, L, Vanderlinde, O, Roberts, T, & Oster, G. (2005) *Biophys. J.* **88**, 2462–2471.

86. Bottino, D, Mogilner, A, Roberts, T, Stewart, M, & Oster, G. (2002) *J. Cell. Sci.* **115**, 367–384.

87. Miao, L, Vanderlinde, O, Stewart, M, & Roberts, T. (2003) *Science* **302**, 1405–7.

88. Raucher, D & Sheetz, M. (2000) *J. Cell. Biol.* **148**, 127–36.

89. Ghosh, P, Vasanji, A, Murugesan, G, Eppell, S, Graham, L, & Fox, P. (2002) *Nat. Cell. Bio.* **4**, 894–900.

90. Schwab, A, Nechyporuk-Zloy, V, Fabian, A, & Stock, C. (2006) *Pflugers Arch.*

91. Bereiter-Hahn, J, Strohmeier, R, Kunzenbacher, I, Beck, K, & Voth, M. (1981) *J. Cell Sci.* **52**, 289–311.

92. Odell, E & Oster, G. (1994) in *Some Mathematical Problems in Biology, vol. 24.*, eds. Goldstein, B & Wofy, C. (American Mathematical Society., Providence, RI.), pp. 23–36.

93. Gov, N & Gopinathan, A. (2006) *Biophys. J.* **90**, 454–69.

94. Weisswange, I, Bretschneider, T, & Anderson, K. (2005) *J. Cell. Sci.* **118**, 4375–80.

95. Saadoun, S, Papadopoulos, M, Hara-Chikuma, M, & Verkman, A. (2005) *Nature* **434**, 786–92.

96. Kucik, D, Elson, E, & Sheetz, M. (1989) *Nature* **340**, 315–7.

97. Kucik, D, Elson, E, & Sheetz, M. (1990) *J. Cell. Biol.* **111**, 1617–22.

98. Popov, S, Brown, A, & Poo, M. (1993) *Science* **259**, 244–6.

 99. Dai, J & Sheetz, M. (1995) *Cell* **83**, 693–701.
100. Charras, G, Yarrow, J, Horton, M, Mahadevan, L, & Mitchison, T. (2005) *Nature* **435**, 365–9.
101. Oster, G & Perelson, A. (1987) *J. Cell. Sci. Suppl.* **8**, 35–54.
102. Maurel, C. (1997) *Annu. Rev. Plant. Physiol. Plant. Mol. Biol.* **48**, 399–429.
103. Loitto, V, Forslund, T, Sundqvist, T, Magnusson, K, & Gustafsson, M. (2002) *J. Leuk. Biol.* **71**, 212–22.
104. Zicha, D, Dobbie, I, Holt, M, Monypenny, J, Soong, D, Gray, C, & Dunn, G. (2003) *Science* **300**, 142–5.
105. Schwartz, M & Horwitz, A. (2006) *Cell* **125**, 1223–5.
106. Gupton, S & Waterman-Storer, C. (2006) *Cell* **125**, 1361–74.
107. Wang, F, Herzmark, P, Weiner, O, Srinivasan, S, Servant, G, & Bourne, H. (2002) *Nat. Cell. Biol.* **4**, 513–8.
108. Janmey, P, Hvidt, S, Kas, J, Lerche, D, Maggs, A, Sackmann, E, Schliwa, M, & Stossel, T. (1994) *J. Biol. Chem.* **269**, 32503–13.
109. Carlsson, A. (2001) *Biophys. J.* **81**, 1907–23.

3

Directed Motility and *Dictyostelium* Aggregation

Herbert Levine and Wouter-Jan Rappel

Center for Theoretical Biological Physics, University of California San Diego, 9500 Gilman Drive, La Jolla, CA 92093

3.1 Introduction

In recent years, physicists have begun to tackle the dynamical mechanisms underlying the behavior of living systems. This is a major undertaking, as the complexity of these biological processes is apparently well beyond that familiar from the study of naturally occurring abiotic analogs. In fact, the major questions seem to concern how multiple levels of feedback and control interact with the basic physics and chemistry of soft condensed matter to produce robust yet flexible behavior. Our task is a bit like trying to infer both Bernoulli's law and the design of digital logic circuits by careful observations of a modern jet aircraft.

Given the above situation, it is critical to pick the most appropriate biological system for investigating a specific capability of living matter. This volume is concerned with cell motility and we believe that the *Dictyostelium discoideum* amoeba is by far the best laboratory for studying directed motion. There are several reasons for this. First, *Dictyostelium* uses directed motion as an indispensable part of its survival strategy, and hence the circuitry implementing this capability is presumably highly efficient. *Dictyostelium* cells are easily manipulated, both biophysically and genetically. There is a critical mass of both biological data and continuing biological study of this microorganism, and thus there is a constant feedback between modeling studies and experimental findings. Finally, at the level of interest to physicists, the results for *Dictyostelium* will almost definitely apply more broadly to many other motile eukaryotic cells, including mammalian examples such as neutrophils and fibroblasts of clear medical importance.

The outline of this chapter will be as follows. First, we will briefly introduce basic biology and the lifecycle of *Dictyostelium* and explain why motility is so critical. Next, we will turn to the signaling system that creates the navigational information that the cells use to find their way to aggregation centers. We will then zoom in on the single cell gradient-sensing capability - how a cell can use spatio-temporal chemical concentration patterns to decide which way to go. This problem has been the subject of intense effort over the last few years, mostly because it seems to be the question most amenable to study via the physicist's toolbox. Afterwards, we will touch on the future challenge of combining the gradient-sensing system with the actual mechanics of force generation, shape transformation, and cell translocation. Our goal is to provide a concise introduction to these issues in lieu of a comprehensive review. Hopefully we will motivate readers to use the cited articles to fill in all the details that we will leave out.

3.2 Why Dicty?

Dictyostelium discoideum, "Dicty" to its fans, is a 10μm-sized soil-dwelling eukaryotic microorganism, which has become an obvious choice for the study of directed motility. Its mode of motion is amoeboid, by which we mean that cell propulsion occurs by the protrusion of a leading "pseudopod" which then grabs onto the substrate and a myosin-based trailing edge that then pulls in the rear. The pseudopod itself is associated with the active generation of a cross-linked actin network, which is, of course, coupled to membrane deformation. The cell can move at a maximum speed of around 10 μm/min. It is a good model system for actin-based basal motion, in the same league as keratocytes or *Listeria*.

What makes Dicty special is the use to which it puts directed motility. By directed or equivalently chemotactic motion, we mean that the normally random motion can be suppressed in favor of an operating mode whereby external chemical information in the form of the concentration of the small organic molecule cyclic AMP (cAMP) directly controls the decision regarding where to place the leading-edge pseudopod. This chemical information is sensed by binding to a surface-resident protein called CAR1, and then processed via a signal transduction network whose output feeds into the aforementioned actin dynamics. This system is turned on in response to starvation. The response of the individual amoebae to starvation is to set up a self-organized pattern of cAMP waves progressing through the population. As we will see, the waves are formed as individual cells detect cAMP and, after a short delay, manufacture more of it for secretion so as to pass the information along downstream. It takes several hours for the wave patterns to become fully developed, and the details depend to some degree on the cell density. Under typical laboratory conditions, the cells are placed at reasonably high monolayer coverage on an agar surface and the concomitant developed wavefield is dominated by spiral waves emanating from signaling centers - this type of pattern is depicted in Figure 3.1.

Once the waves are formed, the cells begin using the cAMP concentration field as roadmap for figuring out how to make their way towards a signaling center, thereby aggregating with their neighbors. As they receive the directional information in the form of the specific cAMP wave pattern, they are

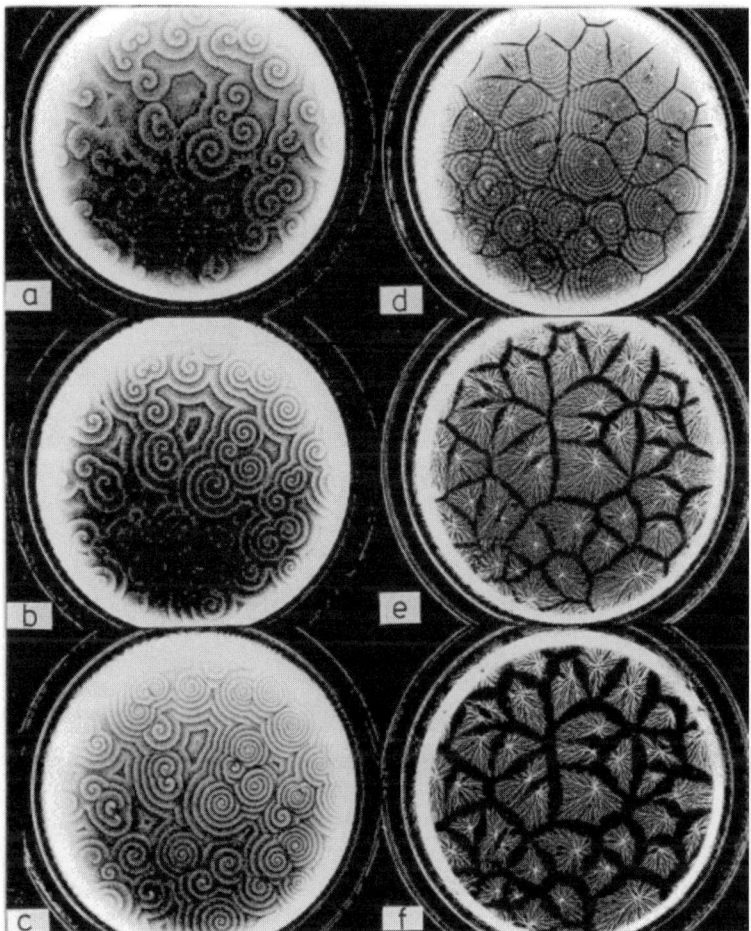

Figure 3.1. Transition from well-developed spiral field (seen in the upper left in a darkfield image) to streaming pattern (seen on the bottom right), as the cells begin to move chemotactically and the density becomes highly nonuniform. Courtesy of P. Newell. For similar pictures, see also [1].

able to detect the angle to the wave source and move towards it. As the motion progresses, the cells become increasingly polarized, with molecular composition and mesoscale structure becoming increasing different in the front as compared to the back of the cell. Thus, the natural progression appears to be that naive cells at around 5-6 hours post development are instructed by the cAMP gradient-sensing system to move towards the nearest center, and to become more polarized so as to optimize that motion. As this occurs, the combined effects of the signaling dynamics and the motility response leads the population to adopt a streaming pattern of the type shown in Figure 3.1 and in the closeup in Figure 3.2, eventually leading to an aggregate with $10^4 - 10^5$ cells. After aggregation, the cells form a rudimentary multicellular organism, cooperating via cell-cell contact, as well as chemical information exchange. This later stage is more complicated and will not be discussed further in this chapter.

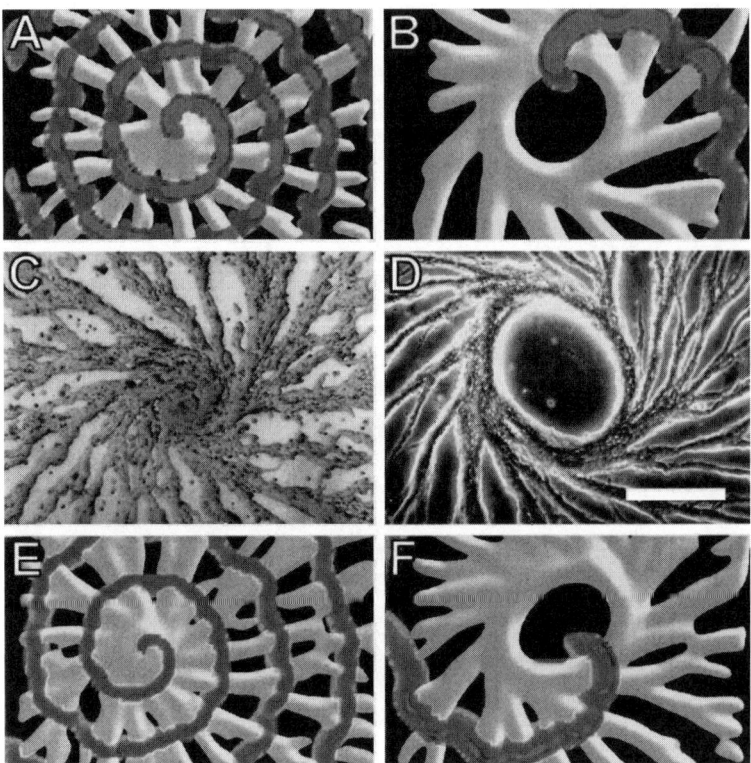

Figure 3.2. Close-up view of the aggregation center in experiments (C, D) and a model (A,B; also E,F) due to Vasiev et al. [2]. The hole has been enhanced in going from A to B and C to D by lowering the excitability of the system, accomplished experimentally by adding caffeine. (courtesy of C. Weijer). (See color insert.)

Given the aforementioned phenomenology, there are many questions that are worth addressing with the help of mathematical models. First, we would like to understand the signaling system responsible for the cAMP relay dynamics underlying the waves. This will then allow us to explain how the wavefield is able to organize itself so as to create the desired spatial information. Next, we would like to understand the mechanism that enables the chemotactic response. This requires modeling how the receptor occupancy data is processed, via a spatially-extended reaction network, to lead to a symmetry-breaking decision. Once the single-cell aspects of the motion are in place, we must still figure out the nature of the collectively-produced streaming pattern that has already been mentioned.

Each of these topics could serve as the focus of a long, comprehensive review covering all the relevant data and all the different models. This is not the purpose here. Instead, we will focus on one representative model for each aspect as a way of showing what types of dynamical mechanisms are necessary to produce the observed behavior. None of these models are perfect, and the shortcomings will be used as a way to motivate alternatives (also not perfect) in the literature. By this strategy, we can illustrate the types of questions that can be answered with the help of models and, perhaps more importantly, how to use models to figure out what key experiments still need to be done.

3.3 Signaling Models

We have just seen why Dicty is such a rich system for modelers. It is thus not surprising that the modeling of the cAMP signaling machinery has attracted considerable attention from the physics and mathematics communities. Here we will mostly focus on one of the earliest models, due to Martiel and Gold-beter [3]. Afterwards, we will talk about shortcomings, other approaches, and needed research.

The basic step underlying cAMP waves is the detection by the cell and the resultant signal relay. This means that an initially small (but not infinitesimal) stimulus will be amplified and give rise to a propagating disturbance; looked at over the millimeter scales, these disturbances become organized in wave patterns. The waves can be visualized by the technique of darkfield microscopy in which changes in cell shape (sometimes called cringing) as the wave passes by are detected via changes in light scattering. Note that the cells do not need to be physically translocating to be imaged in this manner - all that is needed is a shape alteration.

Goldbeter and colleagues realized that the experimentally observed amplification of a cAMP stimulus must originate from a positive feedback loop. They also realized that a mechanism that eventually shuts off the positive feedback was needed. After all, without such a mechanism, the positive feedback loop would continue indefinitely, leading to a singularity. In early attempts, they proposed that such a mechanism could be offered by the depletion of one of the ingredients involved in the machinery [4]. They postulated that this ingredient was ATP, the substrate of the enzyme adenylate cyclase (ACA) that generates cAMP production, and that feedback would terminate once ATP was used up. However, subsequent experimental data ruled out this possibility because it demonstrated that the internal ATP level does not significantly alter during an oscillation cycle [5, 6].

In an updated version of the model, Martiel and Goldbeter (hereafter denoted by MG) removed this depletion step and assumed that the cAMP receptor CAR1 has two distinct states [3]. In this model, schematically represented in Figure 3.3, cAMP can bind both to an activated form of the receptor (R) and a desensitized form (D). Upon stimulation, cAMP can bind to both R (which activates ACA resulting in cAMP production), and to D (which is unable to activate ACA). Both bound and unbound R can be converted into the bound and unbound D, respectively, with the conversion thought off as a phosphorylation step. The produced cAMP is secreted into extracellular space and its degraded by both internal and external phosphodiesterases.

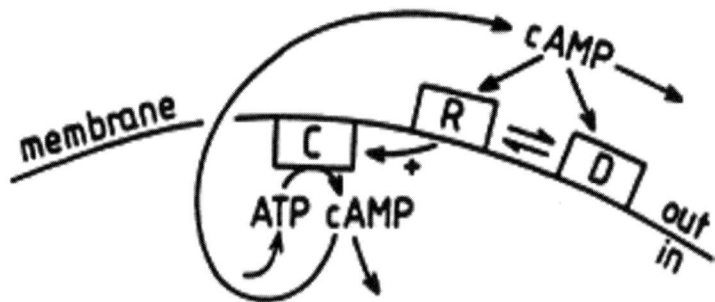

Figure 3.3. A schematic representation of the MG model. cAMP can bind to both the desensitized form (D) and the active form of the receptor (R), which can activate ACA (C). Activated ACA can generate cAMP, using ATP as a substrate, and the resulting internal cAMP gets secreted to the extracellular space.

The MG model includes a nonlinearity via the requirement that two active receptor complexes bind to ACA before it gets activated. This is one of the weaknesses of the model, as will be mentioned later.

After casting all the reaction steps into ordinary differential equations, the MG model consists of nine equations. Fortunately, this number can be reduced significantly by realizing that some processes occur on a much smaller time scale than others. Then, via singular perturbation methods, the MG model can be reduced to only four equations. One of these is describing the ATP concentration, and when examining the dynamical behavior of this concentration one realizes that it is safe to assume that it is constant. The MG model thus reduces to the following three variable model

$$\frac{d\rho_T}{dt} = -f_1(\gamma)\rho_T + f_2(\gamma)(1 - \rho_T)$$

$$\frac{d\beta}{dt} = q\phi(\rho_T, \gamma, \alpha) - (k_i + k_t)\beta$$

$$\frac{d\gamma}{dt} = k_t\beta/h - k_e\gamma \tag{3.1}$$

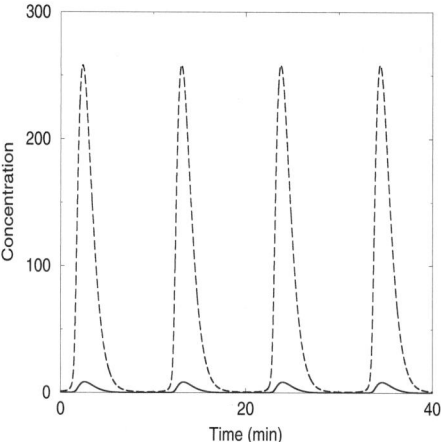

Figure 3.4. Typical oscillations in the MG model for the internal cAMP concentration (dashed line) and the external cAMP concentration (solid line). Parameters are as in [3].

where ρ_T, β, and γ denote the total fraction of receptor in the active state, the normalized concentration of intracellular cAMP, and the normalized concentration of extracellular cAMP, respectively. The difference between extracellular and intracellular volumes are taken into account by the scaling factor h (which therefore depends on cell density), and q is a constant which scales the ATP utilization rate Φ. Furthermore, α is the normalized (constant) ATP concentration, k_i, k_t, and k_e are rate constants and f_1, f_2, and ϕ are specific nonlinear functions. These can be found in the original MG publication, and will also be given explicitly in the next section on cAMP waves.

The MG model displays, for a relatively large part of the parameter space of the equations, periodic oscillations. This is demonstrated in Figure 3.4 where we have plotted the normalized internal cAMP concentration as a function of time. The period of the oscillation depends, of course, on the exact parameter values but can easily be brought within the experimentally observed range (5-10 min). Qualitatively, the oscillation cycle is described as follows: external cAMP binds to the receptor which is mostly in its active form. This leads to significant cAMP production providing a positive feedback. The active form of the receptor, however, desensitizes, and together with the phosphodiesterase degradation, leads to a decrease in the cAMP levels. The desensitized form can then revert back to the active form, and once a sufficient number has become active, the residual cAMP level can start the cycle again.

In addition to oscillations, the MG model exhibits excitability in the sense that a small increase in γ (the external cAMP concentration) can be amplified and can lead to a much larger response in β (the internal cAMP concentration). Through secretion, this in turn leads to a significant increase in γ. Through this amplification mechanism, the MG model permits propagation of pulses in spatially extended systems, which we will discuss in more detail in the next section. We should point out here that excitability occurs in many different systems including cardiac excitation and nerve impulse, and the theory and modeling of these phenomena are extremely active fields.

The MG has become the de facto model, in particular when examining spatially extended systems. This, however, does not mean that the MG model captures the precise workings of the relay mechanism and a number of objections can be raised. First of all, there is the aforementioned assumption of dimeric activation, for which, no experimental evidence exists to date. Second, experiments clearly indicate that there is a 30-60 s time lag between the binding of cAMP to CAR1 and the ACA activation. The origin of this time lag is not well understood but it is clear that a large number of steps exists between the ligand binding and ACA activation. Third, the original MG model is not able to capture the adaptation of Dicty cells to continuing and varying cAMP stimuli. As demonstrated by Devreotes and Steck [7], when cells in suspension were presented with extracellular levels of cAMP that were held constant for 225 s and increased stepwise from zero to 10^{-6} M, the excreted cAMP levels rose rapidly after each increase in external cAMP but reduced to previous levels before the next increase. Thus, the cAMP relay machinery adapts within roughly four minutes, after which it can respond again to a higher level of extracellular cAMP. Furthermore, recent data on mutants that contain non-phosphorable receptors cast serious doubt on the role phosphorylation plays in adaptation [8]. In these mutants, adaptation of ACA activity was found to be identical to the one in wild-type cells, thus indicating that the primary mechanism for adaptation does not depend on the phosphorylation of cAMP receptors.

Because of these shortcomings, several other groups have produced alternative models, and Martiel and Goldbeter themselves have attempted to extend their work to include adaptation. Of particular note is the Tang-Othmer approach which dispenses with the suspect dimerization step, but nonetheless appeals to unknown biology in the form of a receptor that binds cAMP and generates an inhibitory signal [9]. This extra pathway is needed to get the adaptation. Also, a strong nonlinearity is postulated for the cAMP secretion, with no obvious biological motivation. A third approach is that of Laub and Loomis, who used only known biological components to produce a cAMP oscillation [10]. There is still no adaptation and still no delay in cAMP production. More importantly, this model does not have an excitable phase and hence cannot transmit waves if the cells would not be spontaneously oscillating by themselves. It is also important to realize that even this more detailed molecular approach ignores many known facts regarding the internal spatial locations of various parts of the chemical pathway (for example, the fact that activated ACA is localized at the rear of the cell); how important these will prove to be is uncertain.

Thus, the situation with regard to models of the signaling system is somewhat mixed. On the one hand, there are several good phenomenological models available for use in constructing theories of wave pattern dynamics and streaming response; these will be presented in subsequent chapters. On the other hand, there is no one model that starts from known biochemistry and recovers all the desired features. This remains a challenge for future work.

3.4 The cAMP Wavefield

The cAMP signaling system studied in the last Section forms the basis upon which we can understand the cAMP waves that are formed during the early hours of the aggregation phase and thereafter guide the cells to the nascent mounds. To get from the signaling models to models for the spatially-extended system, we need to add transport of cAMP from cell-to-cell. We will do this within the context of one of the simpler forms of the MG model, a four variable reduction that keeps the ATP concentration α as a dynamical variable in addition to the three given above in Equation 3.1,

$$\frac{\mathrm{d}\rho_T}{\mathrm{d}t} = -f_1(\gamma)\rho_T + f_2(\gamma)(G - \rho_T)$$

$$\frac{\mathrm{d}\alpha}{\mathrm{d}t} = \nu - k'\alpha - \Phi(G, \rho_T, \gamma, \alpha)$$

$$\frac{\mathrm{d}\beta}{\mathrm{d}t} = q\Phi(G, \rho_T, \gamma, \alpha) - (k_i G^3 + k_t)\beta$$

$$\frac{\mathrm{d}\gamma}{\mathrm{d}t} = \frac{k_t}{h}\beta - k_e G^2 \gamma + D\nabla^2\gamma \qquad (3.2)$$

where

$$f_1(\gamma) = \frac{k_1 + k_2\gamma}{1 + \gamma}$$

$$f_2(\gamma) = \frac{k_1 L_1 + k_2 L_2 c\gamma}{1 + c\gamma}$$

$$\Phi(G, \rho_T, \gamma, \alpha) = \frac{\sigma\alpha G^2(\lambda\theta + \epsilon G^2 Y^2)}{1 + \alpha\theta + \epsilon G^2 Y^2(1 + \alpha)}. \qquad (3.3)$$

and $Y = \frac{\rho_T \gamma}{1 + \gamma}$. For ATP, ν and k' are the production and decay rates, respectively. All the various constants appearing in Equation 3.3 are given in Table 3.1. At the moment, we will imagine G to be fixed; the dynamical system is excitable (i.e., can propagate waves) for $1.0 < G < 1.11$, self-oscillatory for $1.11 < G < 1.65$, and again excitable for $1.65 < G < 2.24$. Later, we will introduce a dynamical equation for G that changes in time due to the feedback between the waves and the genetic development program. Finally, in the MG model, density changes are incorporated by scaling the ratio of extracellular to intracellular volume by the parameter δ, which then affects

Table 3.1. Parameter values for modified MG model.

parameter	value
k_1	0.036 min^{-1}
k_2	0.666 min^{-1}
L_1	10
L_2	0.005
k_i	0.958 min^{-1}
k_e	3.58 min^{-1}
h	5
k_t	0.9 min^{-1}
ϵ	0.108
σ	0.57 min^{-1}
λ	0.01
θ	0.01
q	4000
ν	12
k'	4
c	10
D	0.024 mm^{2}min^{-1}
δ	1

the volume ratio h and the extracellular cAMP decay rate via $h \to h\delta$ and $k_e \to k_e/\delta$; none of the other parameters are affected.

It is important to understand how an excitable system can propagate waves. In the above model, the cAMP secreted by a given cell diffuses to that cell's neighbors and hence can excite them into emitting cAMP. Each cell in turn relaxes back to the quiescent state, but there are always downstream cells that are just then being excited. The net effect is that a single localized emission of cAMP can lead to a pulse propagating throughout the system. Alternatively, a region in which the cAMP system is in the spontaneously oscillating range of parameters can serve as an emitter that sends periodic pulse trains. The dispersion relationship for the waves (i.e., the relationship between the temporal forcing period and the spatial wavelength) can be calculated and compared to direct measurements [11].

Often, the visualized field contains rotating spiral patterns. The general theory of excitable media predicts that spirals are a stable, self-sustaining pattern that often arises via the breaking of wavefronts. Specifically, we can imagine that a propagating pulse is disturbed, perhaps by passing through a regime of reduced excitability. The wavefront will then break, leading to two points, at which the phase of the wave wraps around 2π as we go around them. Each of the phase singularities can potentially form the core of a spiral; hence, a typical such event will lead to a counter-rotating pair. This type of generic occurrence is shown in Figure 3.5. A spiral formed from our basic MG model equations is shown in Figure 3.6.

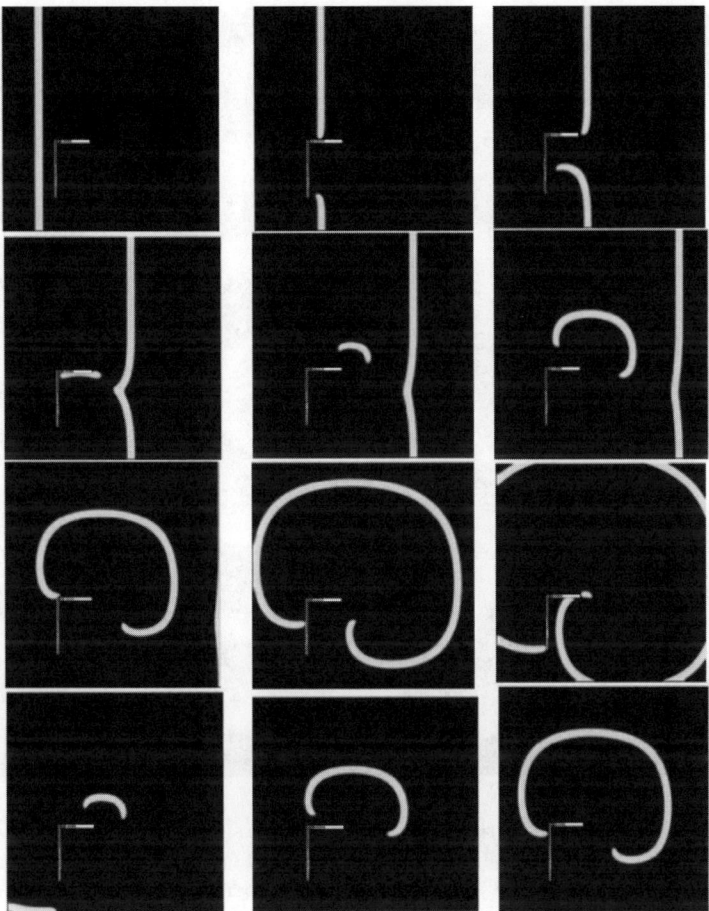

Figure 3.5. Formation of a spiral by obstacle-induced wave breaking and subsequent wave detachment from the obstacle boundary. This was generated using the Fitzhugh-Nagumo model; see [12] for more details. Figure is reprinted from [12]. (Copyright (2000), with permission from Taylor & Francis Ltd).

Figure 3.6. The final state of the formation of a single spiral domain from a counter-rotating spiral pair, due to an infinitesimal initial gradient in G (see [13] for details). Figure is reprinted from [13]. (Copyright (1998), with permission from The American Physical Society).

Understanding the typical structures seen in a fully developed wavefield is more straightforward than understanding the dynamical process whereby the wavefield is formed from an initially quiescent set of cells. Observations reveal that immediately after starvation sets in, there is no wave activity among the *Dictyostelium* cells. After a few hours, a few bursts can be detected; these waves propagate short distances and then die away. Subsequently, this activity becomes more pronounced and, depending on the cell density, one of two patterns emerge. At low density, some of the sites responsible for the initial bursts go into a spontaneously oscillating mode and become periodic pacemakers, sending out circular waves and leading to target patterns that govern the eventual chemotactic motion. If the cell density is higher, the initial bursts lead to rotating spiral waves that entrain the local pacemakers and thereafter dominate the wavefield. A typical set of wavefield snapshots, taken from the experimental studies of Lee, Goldstein and Cox [14], are shown in Figure 3.7.

That targets dominate at low density [14, 15] is easy to understand. Low cell density corresponds to low excitability, and here spirals have low frequencies. These should therefore be suppressed relative to the aforementioned pacemakers. These pacemakers are presumably cells (or clumps of cells) that happen to have their signaling parameters fall within the oscillatory regime of the cAMP response dynamics. Their frequencies remain fixed as we lower the overall density of the system; eventually, the spirals will not occur at all even though the system can still support waves [16]. Somewhere in this progression, targets will dominate. This pattern selection method is also apparent in other cases where the excitability is low. For example, one can delete the gene for phosphodiesterase inhibitor (PDI) which then leads to over-secretion of phosphodiesterase, the enzyme that degrades cAMP. This, too, leads to targets rather than spirals [17]. This type of reasoning can be extended to make sense of the wave patterns of a variety of mutants, as in the recent work of the Cox lab [18].

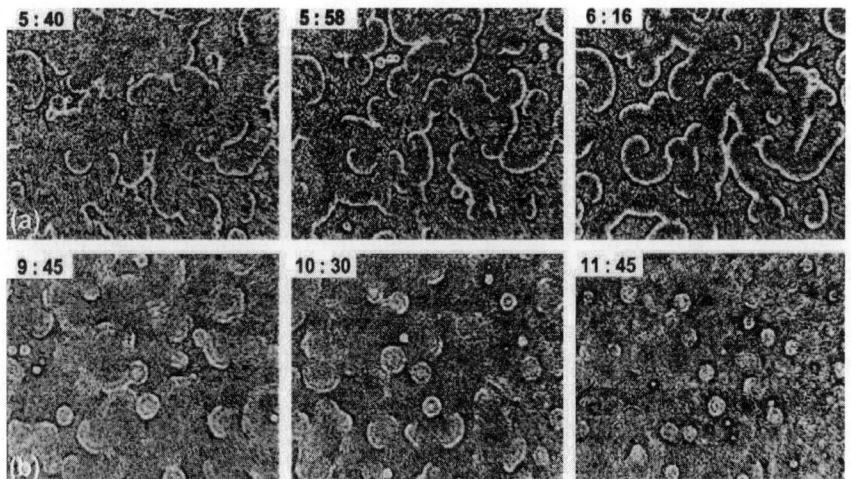

Figure 3.7. Darkfield visualization of cAMP waves during early aggregation in *Dictyostelium*. The top row corresponds to aggregation at high density, the lower row at low density where targets dominate the wave pattern. Taken from [14]. (Copyright (1996), with permission from The American Physical Society).

So, the real issue is the mechanism that causes spirals to form. Spiral formation requires wave breaking, which in turn requires some type of sufficiently strong heterogeneity in the medium. We suggested [19] that this process required changing the view that Dicty should be thought of as a *static* excitable medium. Instead, spiral formation relied on the fact that the system itself is changing due to the increasing expression of the genes required for the cAMP signaling system. In particular, the system clearly goes through a temporal evolution from being nonexcitable (right after starvation) to very weakly excitable (short wave bursts) to fully wave-competent. During the transition, small inhomogeneities can indeed break barely stable wavefronts. There are two classes of models that make use of this developmental dynamics. In the work of Laurezal et al. [20], the developmental dynamics proceeds autonomously and each cell progresses from nonexcitable to excitable to oscillatory. One can easily obtain spiral-dominated fields if one inputs small initial inhomogeneities in the developmental age of the cells. But, it is clear from a variety of experiments [21, 22] that, in fact, the genetic development is coupled to the cAMP system; cells need to receive pulses to up-regulate the genes for the signaling components.

To take this genetic feedback [19] into account, we return to our MG model and now let the parameter G (for gene expression) be governed by the new equation

$$\frac{dG}{dt} = \frac{1}{80}\left(\frac{Y^4}{0.25^4 + Y^4} - \frac{Y^4}{2^4 + Y^4}\right) \tag{3.4}$$

where Y, given explicitly above, reflects a scaled version of the cAMP signal. The details as to why this particular form was picked, as well as the reasoning behind the way that G enters into the MG equations is given in [13].

We start our simulations with cells with G values normally distributed with upper and lower cutoffs (see [13] for more details), the remaining variables set at the stationary state. Centers appear due to some regions being self-oscillatory, and generate outgoing rings that are typically broken into many small pieces (see Figure 3.8a). This should be compared to what happens at high density where spirals invariably develop from these broken pieces (see Figure 3.8b). The rotation frequency of these spirals is higher than the oscillation rate of the centers, and hence the spirals suppress the centers, thereby controlling the overall pattern. Furthermore, many initial spiral tips (formed at the ends of broken pieces of waves) are suppressed by other spirals; this can happen in this class of models because the spatial inhomogeneity in G can cause different spirals to have differing rotation frequencies. This competition mechanism creates large aggregation territories and is one of the most striking predictions of the genetic feedback model. On the other hand, aggregation at low cell density does not lead to spirals. The agreement with the experimental data is quite satisfactory.

This extended MG model naturally explains one other experimental result, one that directly implicates the time dependence of the excitability as the cause of spirals. Specifically, [23] describes experiments in which the entire wave pattern is temporarily extinguished by spraying cAMP on the system; the excess cAMP is degraded by the phosphodiesterase, typically in a few minutes, and the reformation of the wavefield is monitored as a function of the time of cAMP addition. Early in the aggregation process, spirals reappear

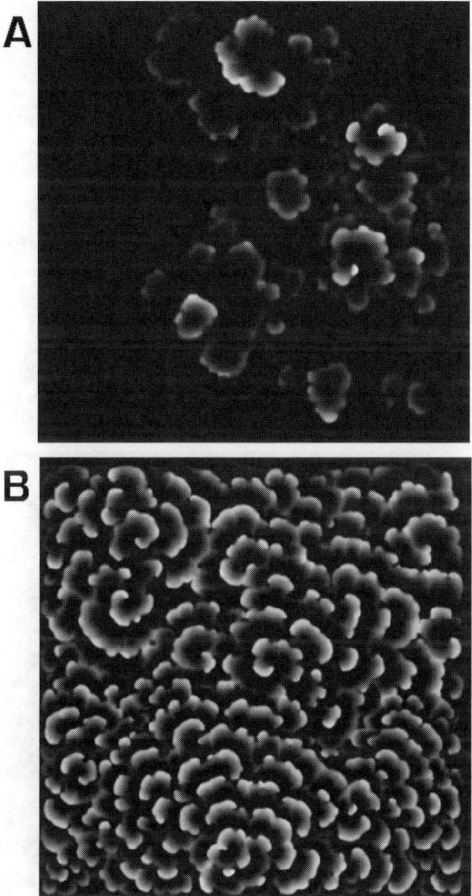

Figure 3.8. Simulations of the wavefield evolution at low density after 600 min (1a) and at a threefold higher density after 376 min. Figure is from [13]. (Copyright (1998), with permission from The American Physical Society).

via the same mechanism (i.e., curling up of the ends of wave pieces) as before. However, a few hours later, when spirals are well-formed and encompass large fields, spraying with cAMP led to patterns without spirals even at high cell density. Figure 3.9 shows a simulation based on this protocol, with the results in striking agreement with the experimental data.

What remains to be done for the understanding of the wavefield? We are convinced that our ideas have established the basic mechanism at work, namely the slowly increasing excitability of the system due to the feedback from the cAMP dynamics to the expression levels of the signaling genes. However, the MG model does not really correctly capture the detailed biology of the signaling circuitry, so it is impossible at this stage to directly compare

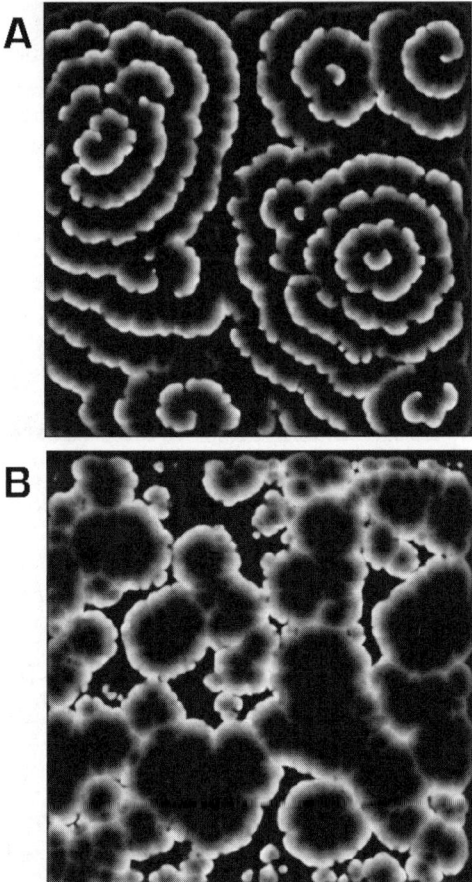

Figure 3.9. Simulation of the wave-resetting experiments. Shown is the concentration of extracellular cAMP for reset times of 120 min. (a) and 240 min. (b), 360 min and 347 min resp. after resetting. The density is the same as for the high density aggregate shown in Figure 3.8(b). Figure is from [13]. (Copyright (1998), with permission from The American Physical Society).

aberrant wave patterns from mutants to those predicted by simulations such as the ones presented here. This is unfortunate, given the nice experimental progress on characterizing these patterns [18]. One would have hoped that the Laub-Loomis could fill this void, but results to date on using this model to study wave patterns have been disappointing, mostly because the model does not really have an excitable phase; perhaps some augmented version that incorporates a more nonlinear response would do better.

3.5 Chemotactic Response

We now turn to the question of how a cell can process gradient information to decide which way to move. As already explained, Dicty uses this capability to move towards the nearest aggregation center, guided there by the emitted cAMP waves. It is thus natural to study chemotaxis during this temporal epoch; here the typical velocity of 10 μm/minute means that it takes about a minute for the amoeba to translate by one body length. This speed is more than one order of magnitude slower than the cAMP wave discussed in the last Section; however, it is faster than the cell speed at other epochs. A typical cell has a mean directional change of roughly 40°-50°/minute when it is not being guided by an external signal.

These measurements are best carried out by the tracking of cells (and their shapes) as they move over various surfaces. Software programs that specialize in this task have been developed by Soll and coworkers [24]. Extensions of these movements to three dimensions (by optical sectioning, followed by computer reconstruction) have revealed finer details of this basic motility process [25]. Usually, cell motion exhibits a roughly one minute periodicity. During each cycle, a primary pseudopod moves along the surface (thereby translocating the cells center-of-mass) and then lifts upward. A new pseudopod is then formed at a nearby angle, the previous one shrinks and the new one becomes primary. Given the high quality of this data, a reasonable long-term goal would be to predict the sequence of shape changes and concomitant motility properties.

Once the cell detects a chemoattractant, its behavior changes dramatically. The simplest experimental approach to studying this change is to let cAMP leak out of a pipette located close to a single cell. Because there are no other cells nearby, the cAMP signal can be safely assumed to arise solely from the pipette, at least until the cell turns on its own ACA production of cAMP roughly 30-40 seconds after initial stimulation. The basic finding is that an amoeba is capable of changing direction within a few tenths of a second to reorient itself to move towards the pipette.

cAMP released from a pipette passes over the cells in a fairly complex "diffusion wave". It thus mixes a spatial signal (a concentration gradient across the cell at any given instant of time) with a temporal signal (the concentration of each point goes up in time, eventually reaching a constant for the case of allowing the pipette to continue leaking) and furthermore is difficult to control very quantitatively. Because of this, various researchers have endeavored to devise alternate protocols. The simplest stimulation is, of course, uniform across the cell but varying in time. This has been called a "temporal wave", especially when the time course is chosen to mimic the average concentration that a cell might feel during a "natural wave" of the type governing chemotaxis *in vivo*, as discussed in the preceding chapter. Doing this reveals a natural cycle of events, with a periodicity imposed by the temporal wave, of seven minutes. The cycle consists of choosing a primary pseudopod (in a random direction just as for basal motility, because there is no directional information available in the signal), suppressing lateral pseudopods and moving forward (and thereby eliminating the one minute basal motility cycle), rounding up at the cAMP peak and then dithering about with no net motion during the downward part.

In a spatially extended setting, cells experience natural cAMP waves, generated via the already described relay mechanism, that have rising and falling parts of roughly equal duration. These temporal results suggest that cells only experience a net motion during the rising part of the wave and thereby we can explain why cells do not move backwards during the falling part of natural waves, thereby undoing their net positive motion towards the aggregation center. The shape near the peak, where the cell has rounded up by withdrawing its leading pseudopod, can be tentatively identified as a "cringed" state which we discussed earlier in connection with dark-field imaging of the aggregation stage wave dynamics.

To complement the purely temporal stimulation, many experiments have tried to impose a static spatial gradient across a single cell. In fact, there is a long history of chemotaxis chambers of various designs that have aimed to measure the steady-state response of the cell [26, 27, 28]. The results are usually quantified in the form of a chemotactic index, which is just the linear length traversed divided by the total path length. One popular device is the Zigmond Chamber [26], in which a bridge that supports the cells lies between troughs filled with buffer (the sink) and chemoattractant (the source), respectively. Of course, spatial gradients in this chamber will form rapidly and then flatten; thus it is not really fair to think of this as a purely spatial signal. More complex chambers were designed by Tani and Naitoh [27] and Korohoda et al. [28]. Most recently, there has been a great deal of excitement about the possibility that microfluidic devices could enable designer concentration patterns to be applied to cells. Some preliminary results using these devices are just now becoming available [29]. Results from one such study showing a threshold for gradient detection are shown in Figure 3.10.

Finally, there have been many studies of *in vivo* chemotaxis during the natural aggregation process. The preponderance of the data suggests that the sequence of events is roughly similar to that of the temporal wave protocol, with the obvious exception that now the direction is chosen by the wave propagation direction, not randomly. As we will see in the next Section, putting this chemotactic response together with the wave dynamics causes the formation of high-density streams which efficiently transport the cells to the nascent aggregate and allows for the later stages of the developmental process.

So far, our discussion has been at the behavioral level. It is obviously of great interest to trace the signal processing strategy that is used by the cell to produce this response. The current picture can be summarized as follows: external cAMP binds to CAR1 receptors, which leads to the dissociation of G proteins into α and $\beta\gamma$ subunits. This event leads to a large set of downstream biochemical changes, eventually leading to the polymerization of actin and the concomitant creation of a protruding pseudopod. Some of these downstream

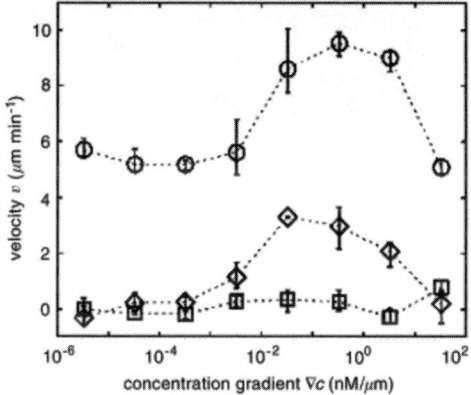

Figure 3.10. Chemotactic velocity versus gradient strength as measured in a microfluidics chamber. Plotted are average velocity components v_x (squares) and chemotactic velocity v_y (diamonds) as well as motility (circles). Figure is reprinted from [30]. (Copyright (2006), with permission from Elsevier).

events include the activation/membrane recruitment of kinases such as PI3K that can phosphorylate various membrane bound phosphoinositides, for example changing PIP_2 into PIP_3. PI3K is cytosolic in unstimulated cells and is recruited to the front of the cell [31, 32]. The kinase action is opposed by a phosphatase, PTEN, whose membrane-bound concentration becomes enriched towards the rear of the cell in the sensing process [31, 32]. The change in lipid composition gives rise to binding sites for Pleckstrin Homology (PH) domain proteins, which are components in the signaling cascade. These translocated PH-domain proteins, which can thus be used as markers for the distribution of PIP_2 and PIP_3, are convenient early indicators of the significant internal asymmetry after the cell is subjected to an asymmetric stimulus. These findings are based on subcellular fluorescence experiments, in which different components are tagged with a fluorescent marker. The temporal and spatial characteristics of these components can then be determined by using confocal microscopy. These types of experiments have revealed that the distribution of both CAR1 receptors and G proteins remain roughly uniform. The distribution of PH-domain proteins, on the other hand, displays a sharp peak at the front, i.e., the side of the cell closest to the source of stimulation [33].

A typical example of an experiment showing the emerging asymmetry is shown in Figure 3.11, where the PH-domain protein PhdA has been labeled with a fluorescent protein. A pipette is held close to the cells and releases high concentration cAMP solution. Before the stimulus is applied, PhdA is localized uniformly in the cytosol. After the stimulus reaches the cell, a more or less uniform fluorescent crest appears but then fades away, leaving behind a crescent only in the forward part of the cell. This is the indication that the cell has correctly decided which direction it should be begin moving towards.

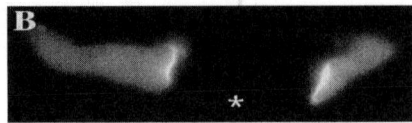

Figure 3.11. Accumulation of PhdA GFP at the leading edge of migrating cells. The asterisk indicates the position of the tip of the micropipette containing the cAMP solution. Figure is from [34]. (Copyright (2001), with permission from The Rockefeller University Press).

So, in the presence of a gradient, we can detect the first signs of cell asymmetry within perhaps 5-10 seconds of stimulation. What happens if we stimulate the cell with a spatially uniform signal? We get the same initial response, but now there is no residual crescent - the cell more or less adapts perfectly. This requires the reversal of the action of the PI3K phosphorylation, presumably via the membrane-bound phosphatase PTEN. But this perfect adaptation may not last; van Haastert's group has reported that after 30 seconds or so, the PH-domain protein CRAC jumps back onto the membrane, this time in relatively small patches [35]. They report that at the locations of each of these patches there is a strong tendency to become a pseudopod, probably using the same machinery that enables the forward patch to become a pseudopod when there is a directional signal to sense. In other words, the system spontaneously breaks the rotational symmetry, picking a set of directions at random and then turning on the same process for each of these as would have turned on in the case of an actual gradient. It is worth noting that the same thing has been observed in neutrophils [36], cells of the mammalian immune system that use chemotaxis to move towards bacterial invaders. There, however, the system seems to be tuned to only pick one direction spontaneously, in other words to make one dominant patch.

3.6 Chemotaxis Modeling Example

Unlike the case for the signal relay circuit and the wave patterning, attempts at creating models of the gradient-sensing circuit are still in their early stages, and it is fair to say that no model explains all the existing data. We will proceed by analyzing one such model (our own [37], naturally) to indicate what these models look like, what they attempt to explain, and what future work needs to be done to test them and to extend their usefulness. The model itself will be referred to as one of balanced inactivation, for reasons that will soon become apparent. The goal of the model is to explain the fluorescent response data, specifically how the external signal leads to a roughly symmetric transient followed by a robust decision. It assumes that the cell does not have perfect adaptation.

The basic idea of our model relies on the general notion of rapid local activation followed by non-local inhibition, originally proposed by Parent and Devreotes [38] and explicitly modeled by Iglesias and Levchenko [39, 40, 41]. The key is the mode of inhibition; we will assume that the inhibitor molecule is produced in exactly the same numbers as the activator, and that the inhibitor acts by binding to the activator and sequestering it. In more explicit detail, we will first ignore the specifics of the binding process of the chemoattractant to the receptors and will assume that the concentration of activated receptors, S, is directly related to the chemoattractant concentration. These activated receptors produce a membrane-bound species A and a cytosolic species B at equal rates k_a; this is absolutely crucial. The cytosolic species diffuses inside the cell and can attach itself to the membrane at a rate k_b, where we will label it B_m. There, it can inactivate A with rate k_i, a process that will be assumed to be irreversible. Thus, A plays the role of activator and B plays the role of inhibitor in our model. Finally, we will allow for the spontaneous degradation of A and B_m at rates k_{-a} and k_{-b}, respectively. These rates will be taken to be small compared to both the activation and the recombination. In mathematical terms, these reactions are written as:

$$
\begin{array}{lll}
\frac{\partial A}{\partial t} = & k_a S - k_{-a} A - k_i A B_m & \text{at the membrane} \\
\frac{\partial B_m}{\partial t} = & k_b B - k_{-b} B_m - k_i A B_m & \text{at the membrane} \\
\frac{\partial B}{\partial t} = & D \nabla^2 B & \text{in the cytosol}
\end{array}
\tag{3.5}
$$

with a boundary condition for the outward pointing normal derivative of the cytosolic component:

$$
D \frac{\partial B}{\partial n} = k_a S - k_b B
\tag{3.6}
$$

Let us first look at the steady state solution resulting from a uniform stimulus, S_0. In this case, the inhibitor concentration B becomes uniform and equals:

$$
B_0 = \frac{k_a S_0}{k_b}
\tag{3.7}
$$

Substituting into the remaining two equations, we find $B_{m,0} = k_b B_0/(k_i A_0 + k_{-b})$ where the concentration of the activator A_0 is given by

$$
A_0 = \frac{-k_{-a} k_{-b} + \sqrt{(k_{-a} k_{-b})^2 + 4 k_a k_i k_{-a} k_{-b} S_0}}{2 k_i k_{-a}}
\tag{3.8}
$$

We always imagine that the dimensionless parameter $K \equiv (k_a k_i \bar{S})/(k_{-a} k_{-b})$ is large, with $\bar{S}$ equaling the average value of S along the cell wall (here being

S_0). This condition assures that the overall "balance" between production of activators and inhibitors is only weakly broken by the decay processes. In this case, Equation (3.8) simplifies to

$$A_0 \sim \sqrt{\frac{k_a k_{-b}}{k_i k_{-a}}} \sqrt{S_0} = \frac{k_{-b}}{k_i} \sqrt{K} \qquad (3.9)$$

Note that because $A_0 \sim \sqrt{S_0}$, the signal response A is not perfectly adapting. However, the transient response can be much larger than the eventual steady state level. This can be seen in Figure 3.12, where we have plotted the response to a ten-fold increase in S at $t = 0\,s$ and again at $t = 50\,s$.

When exposed to a shallow gradient, our model is able to create a large internal asymmetry; furthermore, this internal polarization can be rapidly reversed if the external information changes (see Figure 3.13). The kinetics matches the experimental fluorescence data [33, 42]; upon introduction of the gradient, both the back and the front of the cell respond, followed by a loss of response at the back. The timescale of the response of A at the back of the cell, as well as its dynamics upon a gradient reversal, is determined in large part by the value of the diffusion constant of B. A small value of the diffusion constant will lead to slow dynamics, while a large value leads to fast loss of activation at the back and a fast reversal. Figure 3.13 shows that the timescales in our model can be consistent with experimental findings [42]. Of course, an exact comparison with kinetic experiments requires additional knowledge of the pathways involved.

Contrary to previous theoretical efforts [43, 39, 44, 41], the response of the model, measured in terms of the variable A, is not a simple amplification of the external gradient. Instead, the response should be thought of as a switch: the front (i.e., the side of the membrane closest to the chemoattractant source) has a high level of A, while the back has a very low level. For a wide range of parameters, this leads to an internal asymmetry that is much larger than the external one (Figure 3.14), as reported in the literature [42]. Moreover, it can be shown [37] that the level at the back is independent of the external signal for a large range of gradients. Finally, this model is also able to replicate experiments in which cells are exposed to multiple simultaneous sources; again, see [37] for a discussion.

What is the mechanism underlying this switch-like and rapidly reversible asymmetry? The key elements in our model are the equal production of A and B, together with a sufficiently large diffusion constant of the cytosolic component and sufficiently slow decays. The inclusion of cytosolic diffusion for B leads to an almost constant concentration of B throughout the cell. Thus, the initial value of A will be higher than B at the front but lower at the back. When B jumps onto the membrane as B_m, there is more than enough to eliminate A in the back, but residual A's will be left over at the front. The final result is a nearly complete inactivation at the back but not at the front. And, because this asymmetry is a driven response (i.e., is not created

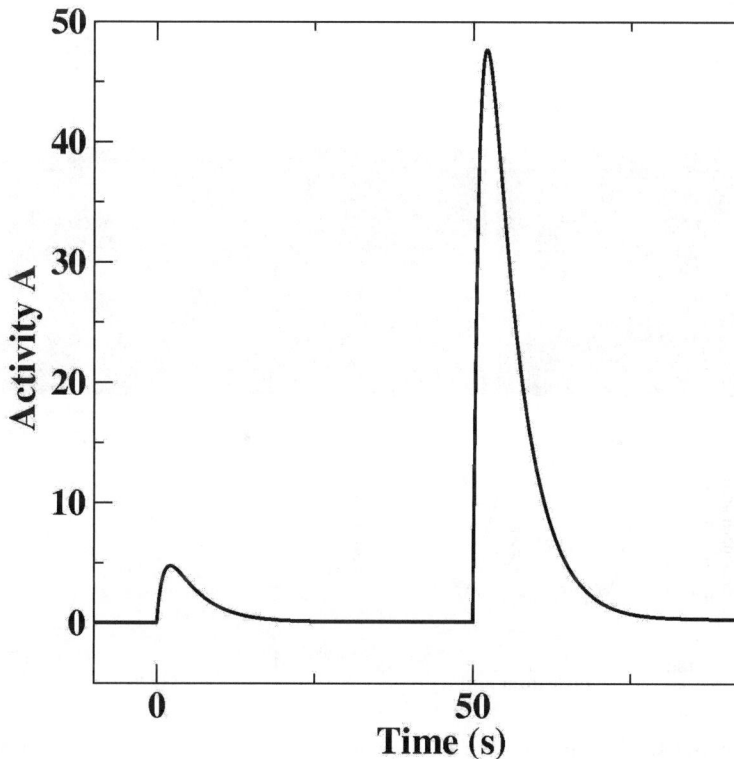

Figure 3.12. The response of our model to step-wise changes in uniform concentration of activated receptors. The initial concentration of activated receptors is $S = 1\,molecule/\mu m$ and is changed ten-fold at $t = 0\,s$ and another ten-fold at $t = 50\,s$.

spontaneously) there is very little hysteresis when the external gradient is changed. This latter behavior stands in contrast to what is observed with instability-based models of gradient sensing [45, 46, 47].

Experimental results to date (see [37] for a full discussion) suggest that we might be able to identify the model component A with G_α and the model component B with $G_{\beta\gamma}$. $G_{\beta\gamma}$ would then be the inhibitor and G_α, in both its GTP-bound and GDP-bound form, would play the role of activator and

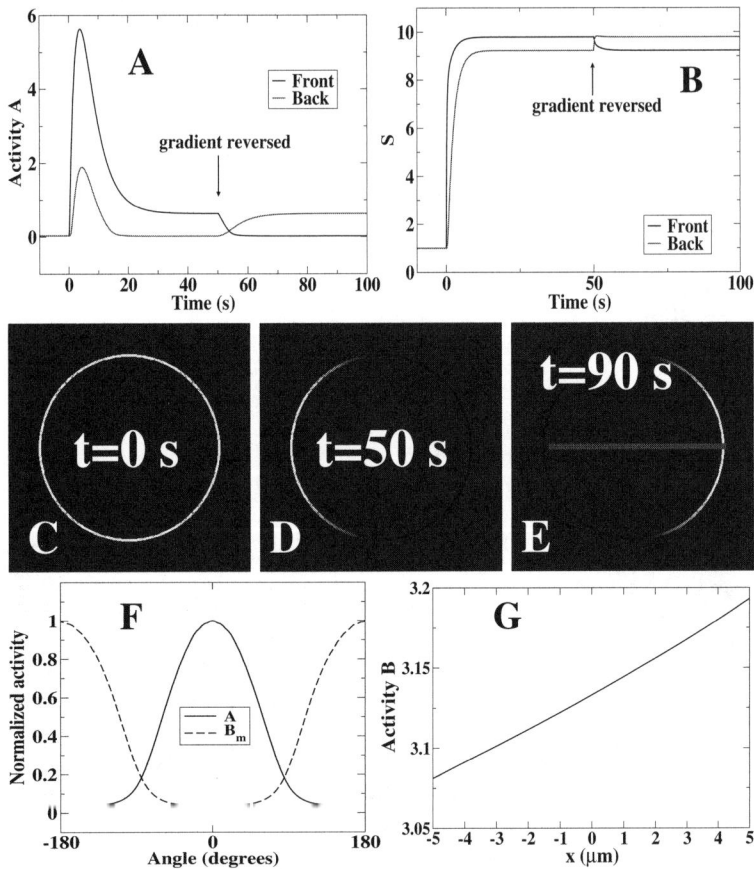

Figure 3.13. A: The value of A as a function of time at the front of the cell (black line) and the back of the cell (red line) as a response to a gradient reversal experiment. B: The external concentration (S) before and after gradient reversal at t=50 s. C-E: Value of A along the perimeter of the cell in a linear gray scale at different times. F: Values of A and B_m, normalized by their maximum value, along the perimeter at t=90 s. G: Value of B as a function of space measured along the symmetry axis of the cell, parallel to the gradient (drawn as a red line in E). (See color insert.)

would be coupled to the downstream modules responsible for Ras and PH-domain protein localization. This specific realization of our model naturally solves the problem of how to ensure the proper balance between activation and inhibition in the system - it occurs naturally because these are both created by G-protein disassociation. We should point out the alternate possibility of an additional feedback mechanism that ensures balanced inactivation without equal A and B production rates.

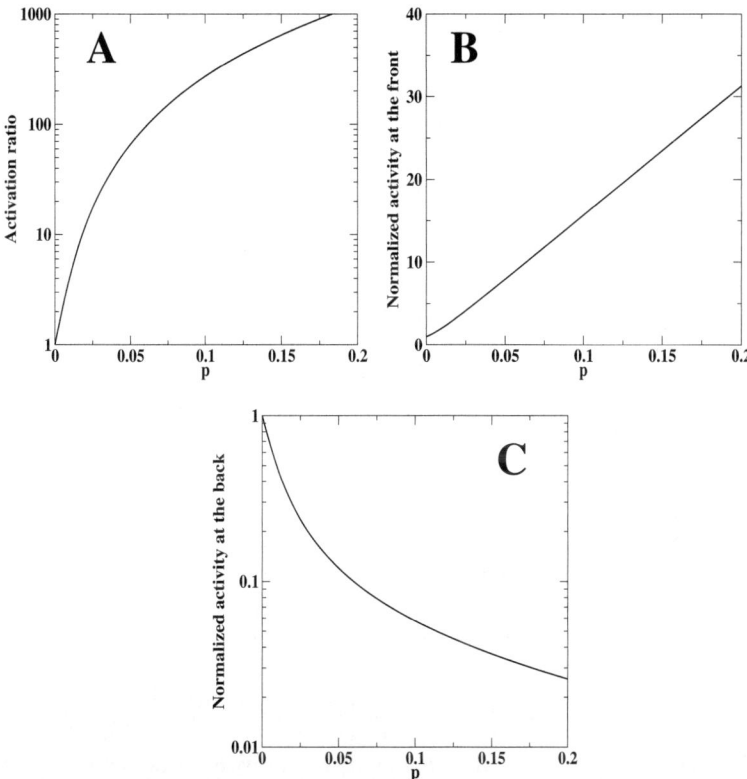

Figure 3.14. A: The ratio of A at the front and the back (A), the value of A at the front (B) and the value of A at the back (C) as a function of the steepness of the gradient parameterized by p. The curves are normalized by the activity of A for a uniform field.

As already mentioned, it is not possible at this stage to definitively decide whether or not this modeling approach is correct. This effort was motivated by focusing on some subset of the existing data - lack of perfect adaption, limited hysteresis, switch-like fluorescence patterns, transient dynamics in which the initial excitation is roughly uniform - and by ignoring other aspects - detailed molecular interaction, the spontaneous emergence of patches, the difference between cells with and without Latrunculin treatments (which interfere with actin polymerization), and any coupling between actual cell motion and the signaling pathway. Other models adopt a different set of assumptions and thereby come to different conclusions about how the processing could be occurring. Some models stick to the level of abstraction that we have utilized, whereas others try to be much more explicit about the components. Most importantly, all models make specific semi-quantitative predictions that can be tested for their accuracy. Progress will emerge via combining new experimental protocols with a better understanding of the mapping between models and predicted cell phenomenology.

Finally, we need to mention one future direction that is necessary for all modeling efforts. In the discussion so far, we have taken the signal S to be deterministic, exactly following a prescribed pattern in space and time laid down by the experimentalist. Of course, the cells can only determine S imperfectly, via the stochastic process of receptor-ligand binding. This introduces noise into the processing and is, we believe, ultimately responsible for the existence of a gradient detection threshold. In terms of equations of the model presented here, we need to generalize to allow the signal to be stochastic, with statistics governed by the known properties of the CAR1 receptor, and see how the noise is propagated through to the final activator concentration. There also in principle could be noise in the processing chain itself (for example due to finite numbers of A and B molecules), but we believe that this is probably less important in practice.

3.7 Motility Mechanics

After our discussion of what we have referred to as the *informatics of cell motility*, we will be brief in discussing the actual physical aspects of shape change and force generation. The reasons for this are that little work has been done on the former and the latter gets into the entire complicated field of actin-based propulsion, a full review of which would take us far afield.

There has, of course, been tremendous progress in understanding how cytoskeletal proteins can be manipulated to cause motion. The most exciting results come from the study of *Listeria*, an intracellular bacterial pathogen that hijacks the cytoskeletal machinery to propel itself across a eukaryotic cell (for a review see [48]). The basic lesson is that actin polymerization (i.e., the assembly of actin monomers into actin filaments) can give rise to forces on neighboring objects such as particles or membranes. The polymerization dynamics is controlled by a set of actin binding-proteins that govern capping, cutting, branching, cross-linking and nucleation of the filaments. Dicty has analogs of almost all of these proteins and there is every reason to believe that pseudopod extension is governed by similar principles.

Aside from the leading-edge machinery, there is the issue of coupling to the cytoskeleton from the chemotactic signal. This probably takes place via activation of ARP2/3, a protein complex that nucleates actin branches and thereby helps organize actin networks; activation requires a Dicty analog of WASP, which in turn is probably bound and activated by RAC (a small GTPase from the rho family). The presence of PIP_3 apparently helps to localize the activity of these small GTPases. The action of these on the cytoskeleton helps to stabilize PIP_3; this is probably an important part of the positive feedback loop that we have already seen postulated in some of the models. The details of how all this works remain murky and there may be some differences between Dicty and other chemotaxing cells with regard to the exact role of differing connector elements - for example, cdc42 is important in neutrophils and fibroblasts to sharply define the pseudopod [49], but Dicty has no obvious cdc42 homologue. Anyone who wishes to make a model of this process would be well advised to keep up with the extremely rapidly evolving experimental data, as relevant information arrives every month.

As mentioned earlier, efficient chemotaxis requires not only the formation of a pseudopod in the correct direction but the active suppression of later pseudopods that would otherwise form. This is accomplished at least partially by myosin reinforcement of the actin cortex, preventing cell extension. Myosin attraction to the cortex is governed by phosphorylation, which appears to rely on internal cAMP-activating protein kinase A (PKA). A different isoform of myosin acts not as a stiffener but instead as an active motor to pull the rear of the cell forward. Mutants with various changes in the differing myosins have, as expected, a variety of motility defects.

How all these pieces work together to finally determine the sequence of cell shapes is far from being solved. There have been some attempts to write down equations for actin-based motility of other types of cells [50, 51], but these are very phenomenological and need extensive development and testing. One requirement for progress is an understanding of the forces arising from the lipid membrane itself. Historically, membrane dynamics has been studied in liposomes, i.e., protein-free vesicles surrounded by lipid double layers. These systems can undergo large shape deformations (such as vesicle budding) and there is an ongoing effort to create computational tools that can handle what turns out to be a complicated free surface problem coupling fluid flow to membrane elasticity. A very promising direction is afforded by adapting the phase-field approach, which now dominates computational studies of free surface problems during thermodynamic phase changes [52, 53, 54, 55], to this problem.

The idea behind the phase-field method is the replacement of the sharp interface (in this case, the cell membrane) by a diffuse self-consistently determined interface in an auxiliary "phase" field ϕ. The creation of such a boundary is accomplished by a double well potential

$$V(\phi) = V_0 \left(\phi^2 - 1\right)^2$$

The energy associated with the interface is encoded by adding terms to this free energy that depend on the gradient of ϕ. In its original application, the only such term was surface energy. For membranes, on the other hand, the most important term is the bending energy contribution $\frac{1}{2}B \int d^2s \, \kappa^2$, where κ is the local mean curvature and B is a bending modulus. Misbah has shown how to modify the phase-field model to include bending energies, and has applied his ideas to the deformations of vesicles due to imposed flow [56].

To our knowledge, no one has yet tried using this natural approach to model and simulate cell shape dynamics. One difficulty, of course, is that the membrane dynamics is invariably coupled to cytoplasmic flow inside the cell, which in turn depends on the dynamic state of the cytoskeleton. This is in addition to the already mentioned direct coupling between the cytoskeleton and forces on the membrane. Thus, making even a semi-quantitative model will be a significant undertaking. On a positive note, it has been shown, however, that this formalism can naturally accommodate the type of reaction-diffusion model that we have reviewed for directional sensing [55]. In particular, modifying the diffusion equation (say, for an intracellular species C) to

$$\frac{\partial}{\partial t}\left[\left(\frac{1-\phi}{2}\right)C\right] = D\boldsymbol{\nabla}\cdot\left(\frac{1-\phi}{2}\boldsymbol{\nabla}C\right)$$

will naturally guarantee that C can never leave the region where $\phi = +1$ (inside the cell) to the region with $\phi = -1$ (outside), but reduces to normal diffusion in the cell's interior. In fact, several of the simulations shown so far have used this method, employing a pre-imposed static field ϕ. In the future, one should be able to solve similar models inside cells that are dynamically changing their shapes, and therefore offer a more realistic approach to the actual biological system.

3.8 Cell Streaming

As we stressed throughout, the entire purpose of the cAMP wavefield is to provide guidance for the cells as they move towards aggregation centers. Observationally, the cells start to move chemotactically at around the time that the wavefield becomes fully developed. This motion quickly destroys the nice darkfield patterns, and instead produces an inhomogeneous density field that can be directly imaged. In this section, we discuss the modeling of cell aggregation, focusing on the dynamics responsible for stream formation.

In Figure 3.1, we saw the basic phenomenon that has been termed cell streaming [1]. Instead of collapsing in an axisymmetric matter, the density becomes concentrated into thin (a few cells wide) streams that flow into the nascent aggregate as something like a river pattern. There are a few possible reasons why such a dramatic change could occur. One alternative is that the behavior is directly caused by a specific set of genes turning on at this temporal stage; operationally, one asks whether mutant with specific gene knockouts could exhibit signaling and chemotaxis, but have no streaming. The other possibility is that the behavior follows directly from the collective physics of the cells with their pre-existing responses- simply put, streaming is inevitable given signaling and chemotaxis. Reality is often somewhere in between. For streaming, the cells do at this stage become more adhesive [57] via expressing some specific genes, and this stickiness clearly influences the integrity of the streams [58]. Nevertheless, it has been established [59, 2, 60, 61, 62, 63] that the basic mechanism responsible is traceable to a physics process, that of an instability in the coupled density-signaling dynamical system.

To understand this new instability, we can again use the MG framework, but this time augmented by a cell-density field acting as a new dynamical variable. This was first done by Levine and Reynolds (LR) [59], who analyzed an augmented MG kinetics model with cell density acting as a new dynamic field. The density obeyed the advection equation

$$\frac{\partial \sigma}{dt} + \boldsymbol{\nabla} \cdot (\sigma \, \boldsymbol{v}) = 0 \qquad (3.10)$$

with a velocity determined by

$$\frac{d\boldsymbol{v}}{dt} = -\Gamma \, \boldsymbol{v} + k(\gamma) \, \boldsymbol{\nabla} \, \psi \qquad (3.11)$$

with Γ equal to a relaxation time. The rate $k(\gamma)$ is chosen to be a (rapidly) decreasing function of the cAMP concentration γ. Because of the nature of the cAMP pulse, this motion rule corresponds to giving the cells a positive kick every time a wavefront (but not a waveback) passes. LR showed that this model exhibits a linear instability to density fluctuations, a sort of excitable medium version of the Keller-Segel mechanism that leads to the density collapse of bacterial colonies due to chemotactic signaling [12]. That is, regions of high cell density are more excitable and the waves move more quickly through them. This deforms the wavefront so that cells tend to detect waves coming primarily from high-density locations and move accordingly. The motion leads to even higher density, closing the feedback loop. This instability is strong enough that within a few wave cycles, noticeable inhomogeneities can be seen in the density.

The basic mechanism relies on the dependence of wave speed on cell density. This dependence has been directly measured by Van Oss et al [62] and the results were in agreement with predictions based on the MG framework and utilized in LR; namely, speed increases with density. Of course, we have already discussed the relationship between excitability and density in the context of the wavefield pattern selection process, and the results there are consistent as well with what we have used here to get streams.

Past onset, one can carry out numerical simulations of continuum models to study the nonlinear streaming state - an example of one such computation is shown in Figure 3.15. A variety of groups have done this with fairly consistent findings. To take one specific example, Hofer et. al. [60, 61, 64] utilized the equation

$$\frac{\partial \sigma}{\partial t} = \boldsymbol{\nabla} \left(\mu(\sigma) \boldsymbol{\nabla} \sigma \right) - \boldsymbol{\nabla} \left(\chi(r) \sigma \boldsymbol{\nabla} \gamma \right) \qquad (3.12)$$

Here, $\mu(\sigma)$ is a non-linear diffusivity to take into account random cell motion and χ, the chemotaxis coefficient, depends explicitly on the state of the receptor variable r. The fact that χ decreases at large r will accomplish the required rectification of the traveling wave. These authors showed in a series of papers that this model has the streaming instability and that simulations can produce realistic streaming structures. For similar work by a different group, see [65]. There are also discrete models based on the same mechanism [66, 63], that have similar instability-driven streaming and are often easier to simulate. All told, there seems to be broad agreement between results obtained with differing modeling approaches (but the same assumptions), and these results seem to match up reasonably well with the experimental findings.

More recently, a new experiment by Kriebel et al. [67] discovered the *a priori* surprising fact that adenylyl cyclase (ACA), which of course is the enzyme that makes cAMP, is localized to the cell membrane in the *back* of Dicty cells. The proposed explanation for this was that the secretion of cAMP from the rear (as opposed to uniformly) would be necessary to account for stream formation because cells would chemotax specifically to the rear of other cells. While we have no problems with the experimental data per se, we believe that this interpretation has severe problems. First, we have already seen that such a localization is not needed to produce streams. And, it is far from obvious that localization even abets streaming. Using the known diffusion constant of cAMP ($250 \ \mu m^2/sec$), one can easily determine that any concentration difference due to a precise emission spot on a 10-micron size cell would be severely reduced as the chemoattractant spreads to a distance of several cell diameters towards neighboring cells. It is very unlikely that a cell could resolve this small difference in the face of many fluctuations and head precisely for the back of the signaling cell. Conversely, emission from all the cells in a stream would naturally create a gradient that would point towards the stream. If the cell happens to be located close to the end of the stream, this could easily result in a movement towards the back of the stream, even without ACA localization. Thus, cells can and would move towards both existing and nascent streams (and even preferentially to the back of the stream) without being sensitive to the precise source location; hence, ACA localization may affect some of the details of stream formation but not stream existence.

3.9 Reprise

As even this brief review has indicated, Dicty offers numerous opportunities for physicists interested in interfacing their expertise with the real world complexity of biology. We have emphasized here that there needs to be a balance between relatively simple models that can be used to elucidate the basic mechanisms underlying the signaling dynamics and the motility response thereto, and very detailed models that attempt to come to grips with the overall complexity of actual biological systems. A fully detailed model can indeed make very detailed predictions, but is almost impossible to analyze; simplified models can lead to great insight, but one must always strive to understand which of the predictions are likely to remain valid for the real system.

Models of different aspects of the aggregation process have attained different degrees of maturity. cAMP signaling is the most advanced, even though

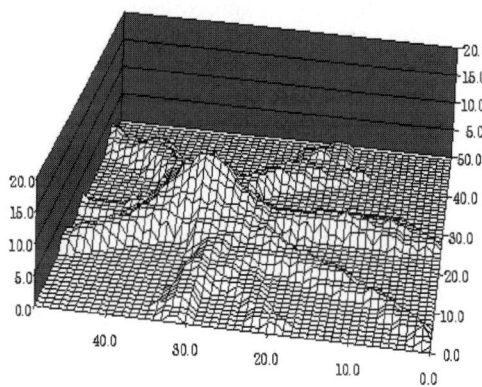

Figure 3.15. Three-dimensional depiction of cell density field during streaming in a coupled signaling-chemotaxis model. Cells are modeled as points that move in response to both chemical signals (arising from cAMP relay waves) and adhesive forces of other cells. Each unit on the axes corresponds to a computational point and thus a cell. Assuming a typical cell size of 10 μm along the horizontal direction, and 2 micron along the vertical direction, the computational domain represents 500x500x40 μm. Many researchers have obtained similar results using a variety of models; one must therefore conclude that streaming is a generic occurrence in this coupled system, independent of fine details such as the precise emission geometry.

there is still uncertainties regarding molecular details. Wavefield evolution models and related efforts to predict streaming patterns still need to be compared more quantitatively with experiments, but we expect no real surprises as we refine our thinking. On the other hand, there are still fundamentally unresolved questions regarding single-cell chemotactic decision-making. Finally, there are only a few initial attempts at modeling the actual mechanics of amoeboid motion (and none applied specifically to Dicty), not to mention the coupling between this complex process and the signaling system.

As should be evident from our discussion, ongoing progress requires active collaboration between biologists and physicists, both experimental and theoretical. It is very exciting to see how the tools of complex-systems physics can help unravel living behavior, and it is our contention that Dicty remains one of the best examples of progress in this newly emerging interdisciplinary field.

Acknowledgments

This work has been supported in part by the NSF-sponsored Center for Theoretical Biological Physics (grant numbers PHY-0216576 and PHY-0225630) and by NIH P01 GM078586.

References

1. Newell, P. (1986) in *Biology of The Chemotactic Response*, ed. Lackie, J. M. (Cambridge Univ. Press).
2. Vasiev, B, Siegert, F, & Weijer, C. J. (1997) *J. Theor. Biol.* **184**, 441.
3. Martiel, J & Goldbeter, A. (1987) *Biophys. J.* **52**, 807–828.
4. Goldbeter, A & Segel, L. A. (1977) *Proc. Natl. Acad. Sci. USA* **74**, 1543–1547.
5. Roos, W, Scheidegger, C, & Gerish, G. (1977) *Nature* **266**, 259–261.
6. Theibert, A & Devreotes, P. N. (1983) *J. Cell Biol.* **97**, 173–177.
7. Devreotes, P. N & Steck, T. L. (1979) *J. Cell Biol.* **80**, 300–309.
8. Kim, J. Y, Soede, R. D, Schaap, P, Valkema, R, Borleis, J. A, Haastert, P. J. V, Devreotes, P. N, & Hereld, D. (1997) *J. Biol. Chem.* **272**, 27313–27318.
9. Tang, Y & Othmer, H. G. (1995) *Philos. Trans. R. Soc. Lond. B Biol. Sci.* **349**, 179–195.
10. Laub, M. T & Loomis, W. F. (1998) *Mol. Biol. Cell* **9**, 3521–3532.
11. Tyson, J. J & Murray, J. D. (1989) *Development* **106**, 412–6.
12. Ben-Jacob, E, Cohen, I, & Levine, H. (2000) *Adv. Phys.* **49**, 395–554.
13. Falcke, M & Levine, H. (1998) *Phys. Rev Lett.* **80**, 3875–78.
14. Lee, K. J, Cox, E. C, & Goldstein, R. E. (1996) *Phys. Rev. Lett.* **76**, 1174–7.
15. Lee, K. J. (1997) *Phys. Rev. Lett.* **79**, 2907–10.
16. Winfree, A. T. (1991) *Chaos* **1**, 303–34.
17. Palsson, E, Lee, K. J, Goldstein, R. E, Franke, J, Kessin, R. H, & Cox, E. C. (1997) *Proc. Nat. Acad. Sci. USA* **94**, 13719–23.
18. Sawai, S, Thompson, P. A, & Cox, E. C. (2005) *Nature* **433**, 323.
19. Levine, H, Aranson, I, Tsimring, L, & Truong, T. V. (1996) *Proc. Nat. Acad. Sci. USA* **93**, 6382–6.
20. Lauzeral, J, Halloy, J, & Goldbeter, A. (1997) *Proc. Nat. Acad. Sci. USA* **94**, 9153–8.
21. Loomis, W. F. (1979) *Dev. Biol.* **70**, 1.
22. Wang, M, van Haastert, P. J. M, Devreotes, P. N, & Schaap, P. (1988) *Dev. Biol.* **128**, 72.
23. Petrich, D. M & Goldstein, R. E. (1994) *Phys. Rev. Lett.* **72**, 1120–3.
24. Soll, D. R, Wessels, D, Voss, E, & Johnson, O. (2001) *Methods Mol. Biol.* **161**, 45–58.
25. Soll, D. R, Wessels, D, Heid, P. J, & Voss, E. (2003) *Scient. World J.* **3**, 827–841.
26. Zigmond, S. H. (1977) *J. Cell. Biol.* **75**, 606–616.
27. Tani, T & Naitoh, Y. (1999) *J. Exp. Biol.* **202**, 1–12.
28. Korohoda, W, Madeja, Z, & Sroka, J. (2002) *Cell. Motil. Cytosk.* **53**, 1–25.
29. Dertinger, S. K. W, Chiu, D. T, Jeon, N. L, & Whitesides, G. M. (2001) *Anal. Chem.* **73**, 1240.
30. Song, L, Nadkarni, S, Bodeker, H, Beta, C, Bae, A, Franck, C, Rappel, W, Loomis, W, & Bodenschatz, E. (2006) *Eur. J. Cell Biol.* **85**, 981–9.
31. Funamoto, S, Meili, R, Lee, S, Parry, L, & Firtel, R. A. (2002) *Cell* **109**, 611–623.
32. Iijima, M & Devreotes, P. (2002) *Cell* **109**, 599–610.
33. Parent, C. A, Blacklock, B. J, Foehlich, W. M, Murphy, D. B, & Devreotes, P. N. (1998) *Cell* **95**, 81–91.
34. Funamoto, S, Milan, K, Meili, R, & Firtel, R. A. (2001) *J. Cell. Biol.* **153**, 795–810.

35. Postma, M, Roelofs, J, Goedhart, J, Gadella, T. W, Visser, A. J, & Haastert, P. J. V. (2003) *Mol. Biol. Cell* **14**, 5019–5027.
36. Servant, G, Weiner, O. D, Herzmark, P, Balla, T, Sedat, J. W, & Bourne, H. R. (2000) *Science* **287**, 1037–1040.
37. Levine, H, Kessler, D, & Rappel, W. (2006) *Proc. Natl. Acad. Sci. USA* **103**, 9761–6.
38. Parent, C. A & Devreotes, P. N. (1999) *Science* **284**, 765–770.
39. Levchenko, A & Iglesias, P. A. (2002) *Biophys. J.* **82**, 50–63.
40. Krishnan, J & Iglesias, P. A. (2003) *Bull. Math. Biol.* **65**, 95–128.
41. Ma, L, Janetopoulos, C, Yang, L, Devreotes, P. N, & Iglesias, P. A. (2004) *Biophys. J.* **87**, 3764–3774.
42. Janetopoulos, C, Ma, L, Devreotes, P. N, & Iglesias, P. A. (2004) *Proc. Natl. Acad. Sci. USA* **101**, 8951–8956.
43. Meinhardt, H. (1999) *J. Cell. Sci.* **112**, 28672874.
44. Rappel, W. J, Thomas, P. J, Levine, H, & Loomis, W. F. (2002) *Biophys. J.* **83**, 1361–1367.
45. Meinhardt, H. (1999) *J. Cell. Sci.* **112**, 28672874.
46. Narang, A, Subramanian, K. K, & Lauffenburger, D. A. (2001) *Ann. Biomed. Eng.* **29**, 677–691.
47. Gamba, A, de Candia, A, , Di Talia, S, Coniglio, A, Bussolino, F, & Serini, G. (2005) *Proc. Natl. Acad. Sci. USA* **102**, 16927–32.
48. Cameron, L. A, Giardini, P. A, Soo, F. S, & Theriot, J. A. (2000) *Nat. Rev. Mol. Cell. Biol.* **1**, 110–119.
49. Weiner, O. D. (2002) *Curr. Opin. Cell Biol.* **14**, 196–202.
50. Mogilner, A & Rubinstein, B. (2005) *Biophys. J.* **89**, 782.
51. Gracheva, M. E & Othmer, H. G. (2004) *Bull. Math. Biol.* **66**, 167–93.
52. Karma, A & Rappel, W.-J. (1998) *Phys. Rev. E* **57**.
53. Folch, R, Casademunt, J, Hernandez-Machado, A, & Ramirez-Piscina, L. (1999) *Phys. Rev. E* **60**, 1724–33.
54. Karma, A, Kessler, D. A, & Levine, H. (2001) *Phys. Rev. Lett.* **87**, 045501.
55. Kockelkoren, J, Levine, H, & Rappel, W. J. (2003) *Phys. Rev. E* **68**, 037702.
56. Biben, T & Misbah, C. (2003) *Phys. Rev. E* **67**, 031908.
57. Noegel, A. A & Luna, J. E. (1995) *Experientia* **51**, 1135–1143.
58. Xu, X. S, Kuspa, A, Fuller, D, Loomis, W. F, & Knecht, D. A. (1996) *Develop. Biol.* **175**, 218–26.
59. Levine, H & Reynolds, W. N. (1991) *Phys. Rev. Lett.* **66**, 2400–3.
60. Hofer, T, Sherratt, J. A, & Maini, P. K. (1995) *Proc. Roy. Soc. B* **259**, 249–57.
61. Hofer, T, Sherratt, J. A, & Maini, P. K. (1995) *Physica D* **85**, 425.
62. Oss, C. V, Panfilov, A, Hogeweg, P, Siegert, F, & Weijer, C. (1996) *J. Theor. Biol.* **181**, 203–213.
63. Dallon, J. C & Othmer, H. G. (1997) *Phil. Trans. Roy. Soc. B* **352**, 391–417.
64. Hofer, T & Maini, P. K. (1987) *Phys. Rev. E* **56**, 2074–80.
65. Vasiev, B, Hogeweg, P, & Panfilov, A. (1994) *Phys. Rev. Lett.* **73**, 3173–6.
66. Kessler, D. A & Levine, H. (1993) *Phys. Rev. E* **48**, 4801–4804.
67. Kriebel, P, Barr, V, & C., P. (2003) *Cell* **112**, 549–60.

4

Microtubule Forces and Organization

Marileen Dogterom, Julien Husson, Liedewij Laan, Laura Munteanu, and
Christian Tischer

FOM Institute for Atomic and Molecular Physics (AMOLF), Kruislaan 407, 1098
SJ Amsterdam, The Netherlands. Tel: +31 20 6081234, dogterom@amolf.nl

4.1 Introduction

Eukaryotic cells contain a network of semi-flexible protein polymers that provide the cell with a mechanical framework for the control of cell shape, cell
locomotion as well as intracellular transport, and the spatial organization of
intracellular organelles [1]. Microtubules are the stiffest component of this
so-called cytoskeleton. They consist of hollow tubes that are 25 nm wide
and typically built from 13 protofilaments of tubulin protein dimers [2]. The
way microtubules are spatially organized in the cell depends strongly on the
cell type and its cell cycle state (see Figure 4.1). For example, in a typical
animal cell that is in interphase (i.e., not dividing) the microtubules are nucleated by a single centrosome, which acts as microtubule organizing center
(MTOC) close to the cell nucleus. From this organizing center, stiff dynamic
microtubules radiate with their fast growing plus end towards the periphery
of the cell. This particular microtubule organization serves several purposes:
motor proteins such as kinesin and dynein follow the orientation of the microtubules and transport cargo either towards the plus end of the microtubules
(the cell periphery) or the minus end of the microtubules (the cell interior).
During mitosis, microtubules are drastically reorganized to form the mitotic
spindle. In this configuration dynamic microtubules are responsible for faith-
fully dividing the duplicated chromosomes between the two newly forming
daughter cells, first by positioning duplicated chromosome pairs in the middle of the cell and then by pulling exactly half of the chromosomes to each
cell pole. Dynamic microtubule ends also interact with the cell cortex, where
they can, for example, respond to external cues and help polarize a motile
cell into a particular direction. For many of the cellular functions of microtubules, it is essential that their dynamic properties are precisely controlled.
Microtubules constantly switch between growing and shrinking states, through
events termed *catastrophes* and *rescues* (this process is called dynamic instability). Control of these switching events allows the cell to locally control the

microtubule length distribution and rearrange their organization depending on the activity of the cell.

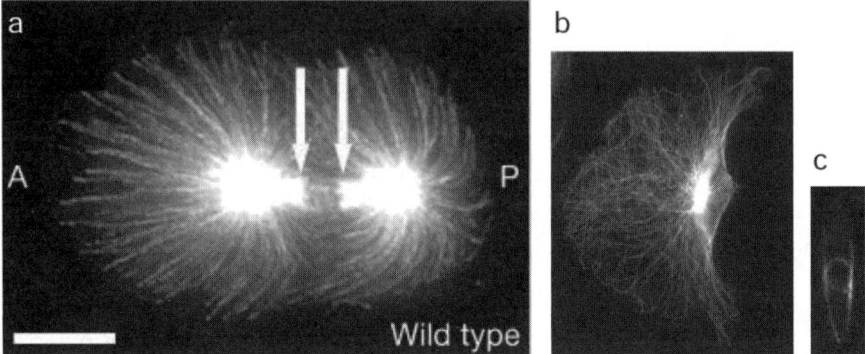

Figure 4.1. Functional microtubule organization in cells. Microtubules are organized differently depending on the cell type or cell cycle state. (a) Mitotic spindle in a first cell stage of a *C elegans* embryo. Microtubules are nucleated by centrosomes. They keep the chromosomes aligned in the middle of the cell (arrows; not visible) and interact with the cell cortex to keep the spindle asymmetrically positioned with respect to the posterior (P) and anterior (A) axis of the embryo. Scale bar is 10 micron (approximately also the scale for b and c). Image adapted from [3]. Copyright (2001), with permission from Macmillan Publishers Ltd; (b) Microtubule organization in an interphase motile mouse fibroblast. Microtubules are nucleated by a single centrosome located near the nucleus on the right and interact with the cortex at the leading edge of the cell (cell moves to the left) to help polarize the cell. Image courtesy of Dr. Anna Akhmanova. (c) Microtubule array during interphase in a fission yeast cell. Microtubules are nucleated by nucleation sites attached to the nuclear membrane and interact with the cell ends to keep the nucleus positioned in the middle of the cell. Image adapted from [4]. Copyright (2001), with permission from The Rockefeller University Press.

Dynamic microtubules are furthermore capable of generating forces that can be used to drive intracellular transport processes. A well-known example is the microtubules that are attached to the kinetochores of chromosomes during anaphase in dividing cells [5]. In this situation, shrinking microtubules are involved in generating the pulling forces that move chromosomes to the new cell poles. The growth of microtubules, in analogy to the growth of actin filaments, can also generate forces that help position organelles and microtubule-organizing centers within the cell. The biophysical aspects of how microtubules (and actin filaments) generate these forces have been studied in experimental and theoretical work by several groups over the last ten years. In this chapter, we summarize the current knowledge on microtubule structure and dynamics, and review our experimental work that has focused on the force-generating properties of growing microtubules. We also discuss some of the theoretical framework that is used to interpret these experiments and provide examples of how force-generating microtubules may be involved in specific cellular processes (see also [6]).

4.2 Microtubule Structure and Dynamics

Microtubules self-assemble from purified $\alpha\beta$-tubulin dimers when supplied with a pool of GTP. At high enough temperatures and tubulin concentrations, nucleation of new microtubules is a spontaneous process [7]. Spontaneously nucleated microtubules contain an average of 14 protofilaments, but this number can range between 11 and 16. At lower concentrations, nucleation needs to be seeded by a nucleation site such as a purified centrosome or axoneme (see Figure 4.2). In that case, the protofilament number follows the structure of the nucleating structure (usually 13), although growing microtubules *in vitro* occasionally switch between different protofilament numbers. Electron micrograph studies reveal that growing and shrinking microtubule tips have structures that are different from what would be the result of a cut through the core of a microtubule (a straight blunt end) [8]. Growing microtubules often terminate in sheet-like structures of laterally connected protofilaments that appear to bend slightly outwards (see Figure 4.2). At the end of shrinking microtubules, individual protofilaments that curl outward more strongly can be observed. This conformational change of tubulin protofilaments at the end of dynamic microtubules is believed to be connected to the hydrolysis of β-tubulin-bound GTP that accompanies the assembly of microtubules. Structural studies of tubulin in different nucleotide states suggest that GTP tubulin dimers build protofilaments with an intrinsic angle between subsequent dimers of about 5 degrees, whereas protofilaments built from GDP tubulin dimers have a preferred angle of 12 degrees between subsequent dimers [9, 10]. The most likely scenario for microtubule assembly and dynamics thus includes a growing phase where GTP-bound tubulin dimers assemble in protofilaments that laterally connect into a sheet that maintains a slight outward curved shape until lateral closure into a hollow tube forces the protofilaments to be straight. When microtubules switch to a shrinking state, individual GDP-tubulin protofilaments that have the tendency to curve outward more strongly lose their lateral connections. Exactly what sequence of molecular events triggers the switches between growing and shrinking states (catastrophes and rescues) remains poorly understood. Recent models and computer simulations based on increasing knowledge about the energetics of tubulin-tubulin bonds and the elastic properties of tubulin protofilaments are able to reproduce the growing and shrinking phases of microtubules [11]. They also provide insight into the intrinsic metastability of growing microtubules [12], but so far no modeling or simulation results have been reported that quantitatively reproduce all aspects of the catastrophe behavior that is observed in experiments.

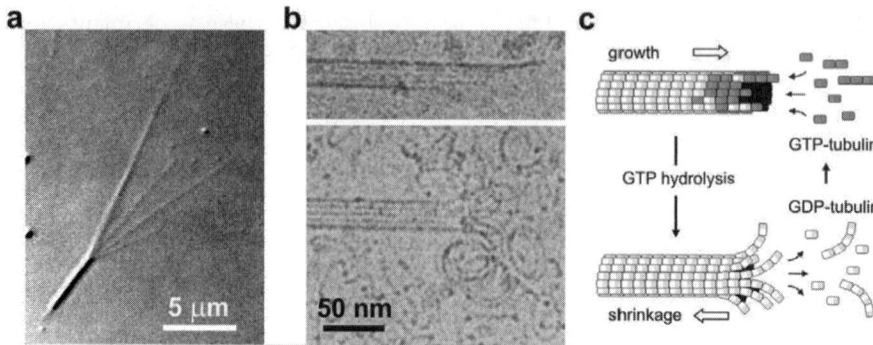

Figure 4.2. Microtubule structure and dynamics. Intrinsic properties of micro-tubules can be obtained from *in vitro* experiments with pure tubulin. (a) Dynamic microtubules as observed by video-enhanced differential interference contrast microscopy (VE-DIC). Multiple microtubules are nucleated by an axoneme that is attached to a microscope slide. The fast growing plus ends point to the upper right. Short microtubules growing from the other (minus) ends can also be seen. (b) Electron microscopy images of what are believed to be growing (top) and shrinking (bottom) microtubules. In addition to blunt ends, slightly outward curved sheet-like structures are often observed at the ends of growing microtubules. Image adapted from [13]. Copyright (1991), with permission from The Rockefeller University Press. Shrinking microtubules are recognized from individual tightly-curved protofilaments pealing of the end. (c) Schematic drawing of how microtubule assembly is believed to be accompanied by GTP-hydrolysis. GTP-tubulin assembles at the end of a microtubule forming a stabilizing structure that prevents microtubules from switching to a shrinking state. A catastrophe occurs when this stabilizing structure is lost (the molecular details of this process are not yet well understood).

Although purified microtubules are able to spontaneously alternate between growing and shrinking phases, the rate at which they do this is vastly different in the presence of microtubule-associated proteins in living cells (see Figure 4.3). Recent years have produced a wealth of information on microtubule-associated proteins (including motor proteins) that interact specifically with microtubule ends (see, for example, [14] and [15]). These proteins regulate microtubule dynamics, for example, by promoting or suppressing switching events and mediate interactions with microtubule interaction sites such as chromosomes and the cell cortex (see, for example, [16]). Controlling the (local) activity of these proteins provides one of many means available to the cell for organizing the microtubule cytoskeleton in a functional way.

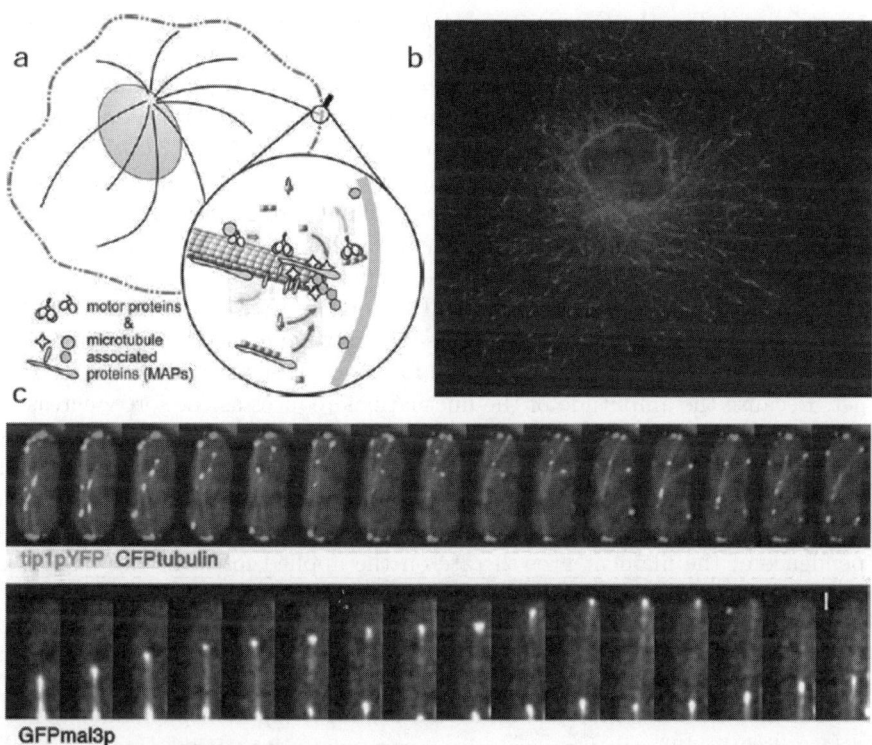

Figure 4.3. Microtubule end-binding proteins. Specialized proteins interact with the dynamic ends of microtubules *in vivo* to control microtubule dynamics, deliver cargo, or mediate interactions with the cell cortex. (a) Artist impression of the protein complexity at microtubule ends in cells. (b) Fluorescent antibody staining of EB1, a catastrophe suppressing end-binding protein in mammalian cells. Image of COS-7 cell courtesy of Dr. Anna Akhmanova. (c) Life imaging using fluorescent protein (FP) labeling of two different end binding proteins, tip1p and mal3p, in interphase fission yeast cells. Images taken every 10.4 sec (top) and 4 sec (bottom); courtesy of Dr. Damian Brunner. When microtubules reach the cell ends, Mal3p disappears before catastrophes occur. (See color insert.)

4.3 Microtubule Assembly Forces in Theory

The notion that the assembly (and disassembly) of cytoskeletal filaments can provide a useful source of energy for mechanical force generation was first put forward in the 1980s by Terrell Hill [17]. Fueled overall by the consumption of GTP, both the assembly and disassembly of microtubules are energetically favorable processes that are available in principle as a source of energy to perform work [2]. Depending on the exact conditions, the gain in free energy (ΔG) connected with the addition of a single GTP-tubulin dimer to a growing microtubule is on the order of $5 - 10k_BT$, where k_B is Boltzmann's constant and T is the temperature (1 $k_B T$ corresponds to 4.1pNnm at room temperature). This means that, in principle, around 50 pN of force can be produced if a microtubule advances $\delta = 8$nm for every $N = 13$ dimers added ($F_{max} = N\Delta G/\delta$). GDP-associated dimers that are rapidly released from the microtubule after a catastrophe release again between 5 and 10 $k_B T$ of free energy.

The simplest mechanistic way to think about force generation by growing microtubules is provided by the Brownian ratchet model, originally proposed in the context of force generation by actin filaments [18] (see Figure 4.4). In this picture, the thermal fluctuations (Brownian motion) of a target allow for the occasional addition of new subunits to a growing filament. This causes the filament to grow even when an external force is resisting the motion of the target. Because the amplitude of the fluctuations reduces as the force increases, the rate at which new subunits can insert slows down. At the stall force, this rate drops to the constant rate at which subunits occasionally detach and no net growth occurs anymore. In its simplest form (for a single filament in a reaction-limited regime), the Brownian ratchet predicts an exponential dependence of the filament growth rate on the applied force. In this case, the force velocity relation is given by:

$$V(F) = \delta \left(k_{on} e^{-F\delta/k_B T} - k_{off} \right). \tag{4.1}$$

For microtubules, more elaborate models involving multiple filaments predict a slightly different force velocity curve, whose functional details depend on the exact assumptions that are made for the structure of the growing microtubule end [19, 20, 21]. These models can be simulated and the obtained force velocity curves can be compared with available force velocity data for single microtubules *in vitro* (see next Section).

4.4 Measuring Polymerization Forces *in vitro*

Experimental evidence that microtubules are able to generate forces *in vitro* was first reported in the early 1990s (reviewed in [5]), followed by quantitative measurements of forces generated by single growing microtubules in the late 1990s [22, 23]. When trying to measure forces generated by pushing filaments, an experimental difficulty that needs to be overcome is the finite rigidity of the filaments. When a growing filament pushes against a rigid object, a compressive force is generated that may cause the filament to bend or buckle. Elasticity theory tells us that a filament of finite length L and stiffness κ will only remain straight as long as the force stays below a critical buckling force given by [24]:

$$F_c = A\kappa/L^2, \tag{4.2}$$

where the prefactor A depends on how the filament is attached at its extremities. For a filament that is clamped at one end and free to pivot at the

a) Single filament Brownian ratchet

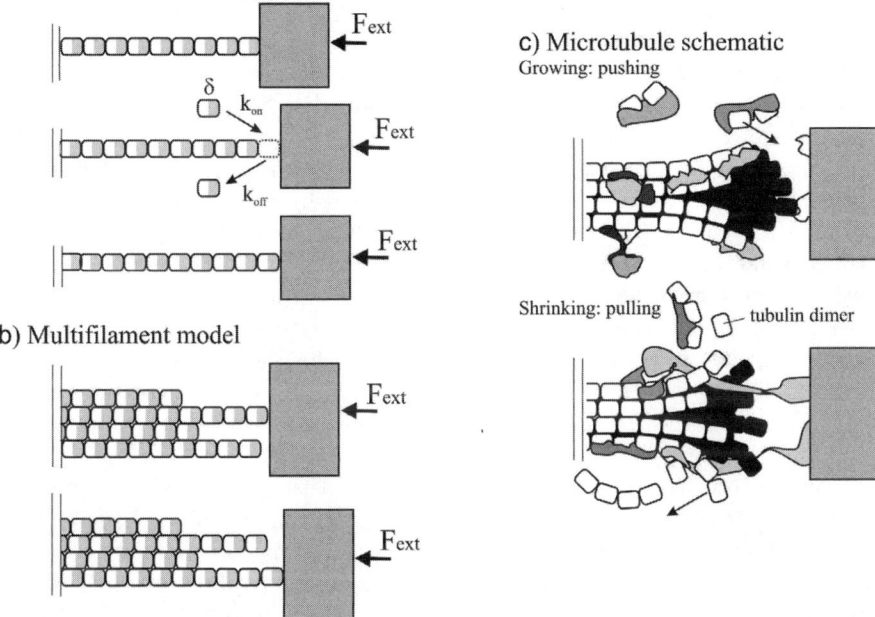

c) Microtubule schematic
Growing: pushing

b) Multifilament model

Shrinking: pulling — tubulin dimer

Figure 4.4. Brownian ratchet models. Models for microtubule force generation and dynamics. Adapted from [6]. Copyright (2005), with permission from Elsevier. (a) The Brownian ratchet principle for polymerization-based force generation (see Section 4.3). Thermal fluctuations of the target allow for the occasional insertion (with rate k_{on}) of a new subunit (with size δ, even when an external force F_{ext} opposes the motion of the target. Subunits detach with a constant rate k_{off}. (b) Generalization of the Brownian ratchet model for multifilament polymers such as the microtubule. Shown is a (arbitrarily chosen) polymer with four protofilaments that, in this case, are assumed to grow independently. (c) Schematic representation of more realistic microtubule end-structures emphasizing the differences between growing and shrinking states. In the cell, growing and shrinking microtubule ends are decorated with end-binding proteins that regulate microtubule dynamics and force generation, and mediate the connection to cellular interaction sites.

other, this prefactor is approximately 20. This means that any measurement technique that relies on a straight growing filament pushing against a force or position sensor can only be applied to sufficiently short filaments. For microtubules ($\kappa \simeq 25\mathrm{pN}\mu\mathrm{m}^2$) the maximum filament length that remains straight when compressive forces of a few piconewton are generated is a few micrometer, whereas actin filaments ($\kappa \simeq 0.06\mathrm{pN}\mu\mathrm{m}^2$) need to remain shorter than a few hundred nanometer. When filaments buckle or bend under an applied force, the increase in length cannot be detected anymore by the response of a position sensor leading to difficulties in measuring the growth velocity of the filament, and thus the force velocity relation. Alternatively, if the buckling shape itself can be measured with high precision, curve-fitting of the buckled filament can provide a measure of both the applied force and the filament length. We used this so-called "buckling" technique to measure force velocity relations for relatively long microtubules [25]. More recently, we developed an optical tweezers-based technique in which growing filaments shorter than 1 micrometer can be followed directly [26, 27].

The experimental set-up for the buckling experiment is shown schematically in Figure 4.5. A microtubule is nucleated by a short stabilized microtubule (a seed) that is attached to the surface of a microscope coverslip. On this coverslip glass barriers of 2 micron high with slight undercuts are built using microfabrication techniques. A growing microtubule that reaches a barrier is caught underneath the undercut and forced to grow against the glass wall. Fluctuations of the microtubule end allow for the continued insertion of tubulin dimers, and the microtubule buckles while its length increases. Fitting of the shape of the growing microtubule to the theoretical shape of an elastic rod under compressive force allows us to determine both the length of the microtubule and the elastic restoring force that is acting on the microtubule end as a function of time (see Figure 4.6a). Using the component of the force that is parallel to the microtubule growth direction, average force velocity relations can be obtained for microtubules growing at different tubulin concentrations (see Figure 4.6b). Although this buckling technique does not allow us to obtain a convincing measurement of the stall force (the external force required to reduce the growth velocity to zero), it is clear from these data that (fairly slow-growing) microtubules *in vitro* can resist forces of at least 5 piconewton, which is comparable to the stall force of molecular motors such as kinesin.

In Figure 4.6b, we compare the normalized force velocity data with predictions of two idealized versions of the multi-filament Brownian ratchet model. The force velocity relation (but not the stall force itself [20]) depends on the assumed details of microtubule growth: optimal force generation (i.e., a smallest decrease in velocity for a given external force) is accomplished when growth occurs in a regular fashion, and each added subunit pushes the target a small, equal distance forward. When growth is more irregular, and not all subunit additions can perform work (see Figure 4.4b), the growth velocity decreases faster with force. Comparing our force velocity data to these two possibilities leads to the suggestion that microtubules, growing *in vitro* without the assistance of cellular microtubule associated proteins, do not in fact efficiently convert all their assembly free energy into work.

These experiments can also be used to measure the effect of force on the catastrophe rate [28]. When microtubules are growing under force, we observe that catastrophes occur more frequently. When we plot the average time it takes before microtubules experience a catastrophe as a function of their reduced velocity under force, we find a more or less linear relationship (see Figure 4.6c). A similar relationship is found when we observe free-growing microtubules and reduce the growth velocity by reducing the tubulin concentration. The simplest explanation for this is that force reduces the growth

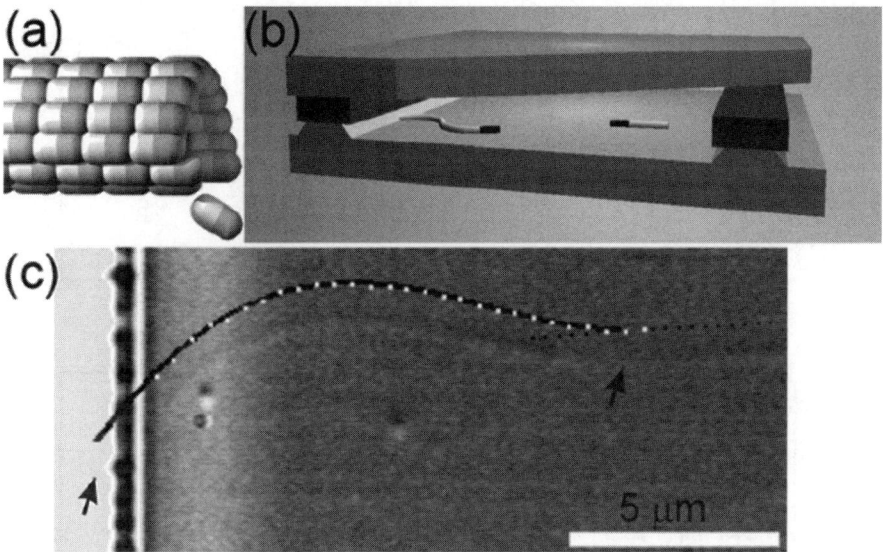

Figure 4.5. Buckling experiment. Growth and subsequent buckling of microtubules against microfabricated barriers can be used to study the force velocity behavior of microtubules *in vitro*. Adapted from [25]. Copyright (2004), with permission from The American Physical Society. (a) Schematic drawing of a blunt microtubule end, showing the lattice structure of tubulin dimers (white/gray) in the microtubule. (b) Schematic drawing of the buckling experiment (side view, not to scale). Short stabilized segments of microtubules (seeds) are attached to the glass surface of a coverslip using biotin-streptavidin bonds. Silicon monoxide (SiO) barriers of 2 micron in height are deposited on the coverslip using photolithography and evaporation techniques. An undercut below the barrier is created using a hydrogen fluoride (HF) etch. When microtubules grow against the barrier, the undercut prevents them from sliding upwards or buckling out of the microscope image plane. (c) VE-DIC image of a buckling microtubule (top view of the experiment) together with (shifted upwards) the positions along the microtubule as detected by a computer algorithm (white dots) and the subsequent fit to the shape of an elastic rod (black line). The left end of the microtubule contacts the barrier underneath the undercut, obscuring the last 1 or 2 micron of the microtubule. The barrier contact point (left arrow) and clamp position of the seed (right arrow) are found by optimizing all shape-fits within one buckling event. The black dotted line indicates the fixed angle of the microtubule seed at the clamped end. The end at the barrier is assumed to pivot freely.

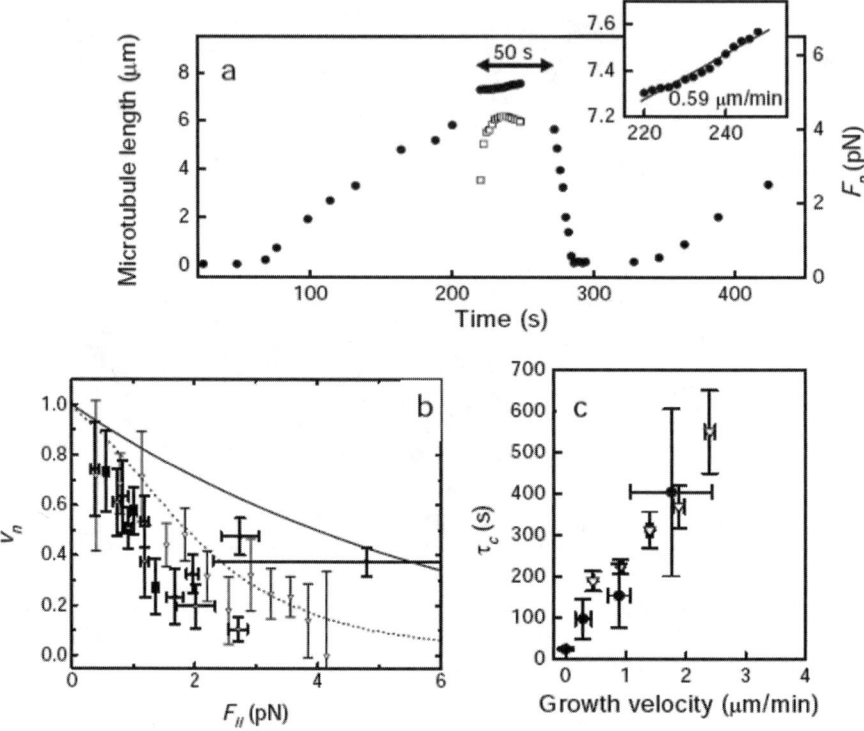

Figuro 1.6. Duckliug data. Microtubule dynamics under force obtained from experiments such as shown in Figure 5. Adapted from [28] (Copyright (2003), with permission from The Rockefeller University Press) and [25] (Copyright (2004), with permission from The American Physical Society). (a) The length (closed round symbols) of an individual growing microtubule before, during and after buckling together with the parallel component of the force F_p (open square symbols) as a function of time. Nucleation from a seed is followed by growth at an average rate of $2.5 \pm 0.1 \mu m/min$. After initiation of barrier-contact, the average growth velocity slows down to $0.59 \pm 0.03 \mu m/min$ (inset). A catastrophe causes rapid shortening at t=270 sec. In this case, the time until catastrophe while in contact with the barrier is 50 sec. (b) Normalized force velocity curves at three different tubulin concentrations. The velocities are normalized with the free-growth velocity before barrier contact. The data are compared with predictions of two idealized versions of the Brownian ratchet model, one for coordinated protofilament growth (solid line) and one for independent protofilament growth (dotted line; Figure 4.4b). (c) Average catastrophe time, τ_c, as a function of measured average growth velocity for microtubules under force (closed round symbols) and freely growing microtubules at different tubulin concentrations (open triangles).

velocity in a way that is similar to the effect of a reduction in tubulin concentration, namely by reducing the rate of tubulin addition. This is consistent with what is predicted by the Brownian ratchet model.

The finding that forces generated by growing microtubules in contact with a barrier increase the catastrophe rate hints at another mechanism (besides the action of microtubule associated proteins) by which cells may locally regulate microtubule dynamics. Depending on the geometry and size of the cell, microtubules should experience compressive forces when they run into the cell boundary that may prevent microtubules from growing longer than they need to be. Indeed, in several cell types it has been observed that microtubule catastrophe rates are specifically enhanced near or at the cell boundary [4, 29, 30]. Whether these effects are due (in part) to force effects remains to be investigated, although we found compelling evidence that in small fission yeast cells, this is indeed the case (see next Section).

4.5 Microtubule Forces in Fission Yeast Cells

In interphase fission yeast cells, microtubule pushing forces have a clear functional role to play. To keep the position of the nucleus near the middle of the cell, microtubules nucleated at the nuclear membrane grow towards the cell ends where they push against the cell membrane for a short time before undergoing a catastrophe and shrinking back to their nucleation site [4]. During this contact time, growth of the microtubules can be correlated with deformation of the nuclear membrane, as well as motion of the complete nucleus away from the cell end. For the correct positioning of the nucleus in this system, it is not only important that forces are generated, but also that microtubules undergo catastrophes some time after they reach the cell end. In the absence of sufficient catastrophes, microtubules that continue to grow eventually curl around the cell ends, which compromises their ability to position the nucleus correctly. One option for the cell, to ensure that catastrophes occur when microtubules reach the cell ends, is to target a protein factor to the cell ends that can locally enhance the rate of catastrophes. Indeed, it is known that in this system proteins that travel at the ends of microtubules accumulate near the cell ends where they could perform such a local role (see Figure 4.3). Alternatively, it is possible that compressive forces generated when microtubules polymerize against the cell end have the same catastrophe-enhancing effect that was observed *in vitro*.

Evidence for such a force effect comes from the quantitative analysis of catastrophe rates as a function of position in the cell. If force has an effect on the catastrophe rate one would expect to find an increase in catastrophes specifically at the cell ends that is correlated with the generation of force. In Figure 4.7, we show an example where near-simultaneous catastrophes occur on microtubules in a bundle that make contact with both cell ends simultaneously. In this situation, a maximum compressive force is generated because the microtubule overlap zone located at the nuclear membrane cannot move in either direction. When we quantify the catastrophe rate of microtubules bundles that are in contact with one or both cell ends, we find a clear increase when contact is made on both sides. This finding strongly suggests that compressive polymerization forces enhance catastrophes *in vivo* as well.

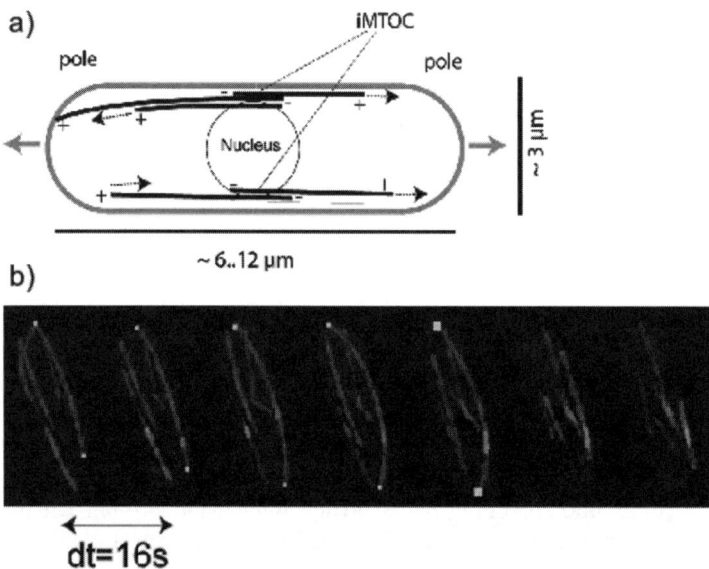

Figure 4.7. Microtubule forces in fission yeast. Nuclear positioning in interphase fission yeast cells relies on microtubule pushing forces. (a) Schematic drawing of microtubule organization in an interphase fission yeast cell. Microtubules are nucleated by interphase microtubule organizing centers (iMTOCs) that are attached to the nuclear membrane. They grow with their plus ends towards the cell poles where pushing forces are generated to keep the nucleus positioned near the center of the cell. (b) Life imaging of microtubule dynamics reveals that catastrophes occur more readily when contact with the cell poles is made on both ends, consistent with a force-induced effect. Small yellow squares indicate growing microtubule ends, large yellow squares indicates near-simultaneous catastrophe events at both cell poles, and red squares indicate shrinking microtubule ends.

4.6 Assembly of Force-Generating Microtubules at Molecular Resolution

A second method that we developed to study microtubule force generation *in vitro* is based on an optical tweezers technique. One of the drawbacks of the original buckling method is that it is hard to obtain data in the vicinity of the stall force. The buckling method relies on growth and subsequent bending of the filament for a measurement of the force, and because growth is slow (and catastrophes are fast), this is inherently difficult at high forces. In addition, progressive buckling of initially fast-growing microtubules does not lead to an increasing, but to a decreasing force because the critical buckling force is inversely proportional to the (increasing) length of the filament squared. In an optical tweezers set-up, a trapped bead instead of filament buckling is used as a force sensor. In this type of experiment, it is important to keep the filament straight so that any length increase of the growing microtubule leads to a proportional displacement of the bead in the trap. In a standard trap set-up without force feedback, this leads to a linearly increasing force on the growing filament. We developed the optical tweezers method for two purposes: to be able to measure dynamics up to the stall force for elongating microtubules, and to be able to follow the assembly details under force at nanometer (i.e., molecular) resolution. Using normal light microscopy, the displacements of a bead can be detected with much higher precision than the length increases of a microtubule. Detecting the position of a bead in this set-up, instead of the end of a growing microtubule directly, therefore gives us a much higher resolution on the growth process than was possible before, potentially allowing us to unravel some of the still unknown molecular details of how microtubules grow and shrink in the presence of force.

In Figure 4.8, we schematically show the optical tweezers set-up. In this set-up, microtubules are no longer nucleated by single microtubule seeds, but by a naturally occurring microtubule bundle, an axoneme. Purified axonemes can nucleate up to 9 parallel growing microtubules depending on the tubulin concentration and temperature used. We tune these parameters such that most of the time zero or one microtubules are growing from the axoneme. A polystyrene bead is attached to one end of the axoneme and this "construct" is positioned in the close vicinity of a 7-micron high microfabricated barrier by what we call a "keyhole" optical trap. This trap consists of a normal trap that holds the bead and is used as a force sensor and a time-shared row of tightly-spaced optical traps of lower power that together form a line trap. This line trap is used to orient the axoneme in the direction of the barrier, thereby forcing the growing microtubule to encounter the barrier. After positioning of the construct in front of the barrier, tubulin dimers are flown into the sample, triggering the growth of microtubules from the axoneme. In (statistically) half of the cases, the minus end of the axoneme is pointing towards the barrier. In these cases, we usually observe no or slow microtubule growth. In the other cases, we observe that after some time the beads start to move away from the barrier at a speed that is comparable to the growth velocity of microtubules under these conditions (see Figure 4.8c). The microtubules remain short and straight so that their length increase can be directly inferred from the displacement of the bead (through a linear relationship that is determined by the trap stiffness and the finite stiffness of the bead-axoneme construct). In these experiments, growth, stalling, catastrophes, and subsequent shrinking of microtubules are readily observed. In principle, collecting sufficient amounts of such data should allow us to again construct force velocity curves that we expect to give us more detailed information of the functional behavior of such curves near the stall force.

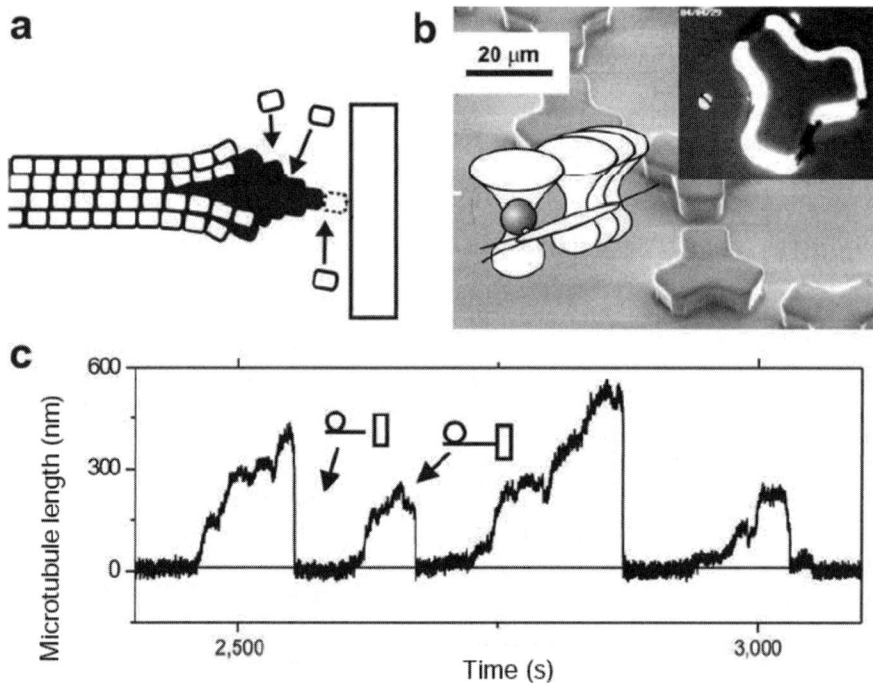

Figure 4.8. Optical tweezers experiment. Optical tweezers can be used to study microtubule length changes under force with nanometer resolution. Adapted from [27]. Copyright (2006), with permission from Macmillan Publishers Ltd. (a) Schematic picture of microtubule assembly in front of a barrier. (b) Set-up of tweezers experiment (schematic). A bead is attached to an axoneme, a naturally occurring bundle of microtubules, which is held in front of a microfabricated barrier using a "keyhole" optical trap. Inset shows a VE-DIC image of the bead-axoneme construct (top view). The nucleating end of the axoneme is held within 1 micron from the barrier. Individual microtubules cannot be seen using these imaging settings. (c) Microtubule growth and shrinkage events as detected from the motion of the bead in the trap. Due to the spring-like nature of the trap, the force on the microtubule end increases linearly with bead displacement.

Until now, however, we have mostly used this technique to obtain new information on the microtubule growth process itself [27]. When zooming in on the growth phases in our trap data, details can be observed that were previously not possible to detect due to the limited resolution with which microtubule assembly can be followed using conventional light microscopy techniques. This has allowed us to observe, for example, that microtubule growth does not always occur through a smooth process but that sometimes fast length excursions are observed that are larger than single tubulin subunits (see Figure 4.9). Because the characteristic length scale of these excursions does not seem to depend on how fast the microtubule grows on average, we propose that growth may sometimes occur through the addition of small tubulin oligomers. In addition, this technique has allowed us to show how, on a molecular scale, tubulin assembly is altered by the presence of the microtubule associated protein XMAP215. This protein was originally isolated from *Xenopus* egg extracts but has homologues in many other systems. It has been shown to be an elongated flexible protein of about 60 nm long that significantly enhances the growth rate of microtubules *in vitro*. When we add XMAP215 to our trap experiment, we frequently observe fast length excursions whose characteristic size is remarkably similar to the size of the XMAP215 protein itself. Again, we propose that XMAP215 may recruit tubulin subunits in solution and thus promote the addition of long tubulin oligomers to the growing microtubule end. An alternative explanation for the fast length excursions that we observe, both in the absence and presence of XMAP215, is that the microtubule occasionally goes into a mode where it rapidly adds subsequent subunits up to a characteristic length. Given the limited spatial and time resolution of our set-up, this would be indistinguishable from the addition of an oligomer.

4.7 Dynamics and Forces of Microtubule Bundles

Up to this point, we have focused on the force-generating capabilities of single growing microtubules. While this is relevant for single microtubules pushing against the cell ends of fission yeast cells, there are other situations where microtubules more likely operate in parallel growing bundles. For example, it has been long known from electron microscopy studies that the ends of multiple dynamic microtubules interact "head-on" with the kinetochores of mitotic chromosomes (see for example [31]). In fact, the first speculations about microtubule (dis)assembly-based force generation were put forward in this context [5]. One natural question is whether arranging multiple growing microtubules in a parallel array allows for pushing forces to be created that are larger than can be achieved with a single microtubule. Another question is whether parallel growing bundles of microtubules can coordinate their catastrophe behavior when in contact with a resisting barrier. This would be relevant to understand how multiple microtubules can push and pull on chromosomes in a coordinated fashion which is what they appear to do when moving chromosomes away and towards the poles of the mitotic spindle.

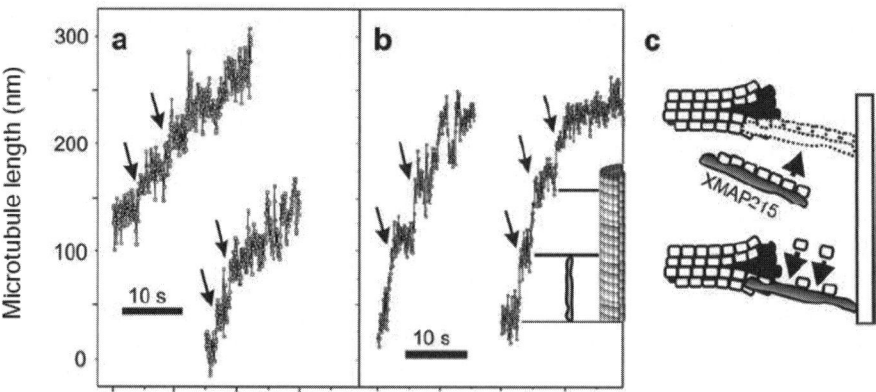

Figure 4.9. Microtubule growth details. With nanometer resolution the mechanisms by which microtubule-associated proteins alter microtubule dynamics can be studied directly. Adapted from [27]. Copyright (2006), with permission from Macmillan Publishers Ltd. (a) Microtubule growth details for pure tubulin growth. Occasional fast length increases are seen with maximum lengths around 30 nanometer (arrows). (b) In the presence of XMAP215, larger length increases are observed with maximum sizes that correspond to the size of the elongated XMAP215 itself (around 60 nanometer; see schematic drawing of microtubule and XMAP215). (c) Possible explanations for the large length increases: XMAP215 recruits tubulin dimers in solution and adds them to the growing microtubule end as a long oligomer (top); XMAP215 binds to a growing microtubule end and stimulates the subsequent addition of tubulin dimers (bottom).

The optical trap set-up described in Section 4.6 can be used to study the dynamics and force-generating capabilities of up to 9 growing microtubules. The axonemes that we use have 9 nucleation sites for microtubules and the actual number of simultaneously growing microtubules can be tuned by changing the tubulin concentration and temperature. While the experiments shown in Figure 4.8 are performed under conditions where (at most) 2 microtubules are growing simultaneously, the experiment we present in Figure 4.10c is performed under conditions where up to 9 microtubules are present at the same time. We find that indeed forces are generated that are much higher (up to 20 pN) than we ever observed for single microtubules (both in the buckling and trap experiments). This implies that microtubules are indeed able to divide a resisting force between them, allowing them to collectively push more strongly than alone.

Interestingly, even though microtubules are not normally known to undergo simultaneous catastrophes in each others vicinity, we sometimes observe fast shrinking of an entire microtubule bundle, indicating simultaneous shrinking of all microtubules in the bundle. In an effort to understand this behavior, we performed simple simulations of multiple parallel microtubules growing against a barrier. On this barrier, a force was applied that linearly

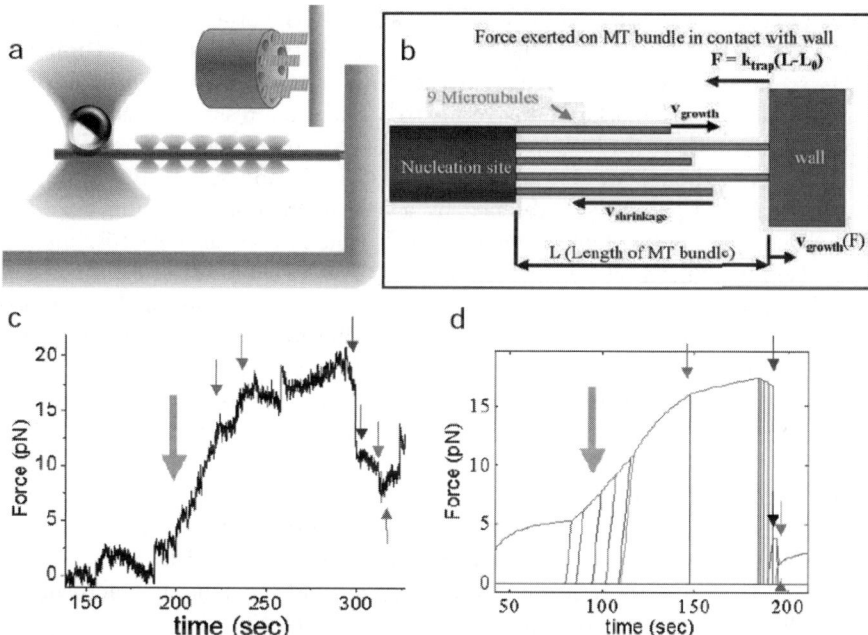

Figure 4.10. Microtubules can coordinate their dynamics under force. (a) Bead-axoneme construct in a keyhole optical trap. Axonemes can nucleate up to 9 microtubules simultaneously (see enlargement) when the tubulin concentration and temperature are high enough. (b) Parameters used to simulate growth of a microtubule bundle in the optical tweezers experiment. (c) Experimental result for force generated by a growing microtubule bundle. Arrows indicate events that likely correspond to events observed in the simulations. (d) Simulation result. Yellow arrows indicate arrivals of new microtubules, red arrows indicate catastrophes of individual microtubules without shrinkage of the bundle, blue and green arrows indicate catastrophes of the entire bundle and black and pink arrows indicate rescues of the bundle due to the arrival of new microtubules. (See color insert.)

increased with the length of the bundle (simulating the situation in our trap set-up). In these simulations we assumed that microtubules grow and shrink at constant deterministic rates, and that catastrophes occur randomly with an average rate that depends on the growth speed of the microtubules as we measured using our buckling experiment (see Figure 4.6c). We further assumed that when multiple microtubules are in contact with the barrier, they evenly share the force, and that for each individual microtubule, the force reduces the growth velocity again following the force velocity relation we measured in Figure 4.6b. Using these simple assumptions, we found that we can qualitatively reproduce the dynamic behavior of the bundle that we observe in experiments (Figure 4.10d). The simulations showed how sequentially nucleated individual microtubules catch up with each other to generate large forces.

When one microtubule experiences a catastrophe, the others suddenly feel an increased force (because now the force is shared between fewer microtubules). This reduces the growth velocity of the remaining microtubules, increasing the probability for another catastrophe to occur. As a result, one catastrophe stimulates another, leading to the occasional disappearance of the complete bundle. Even though we did not observe rescues for individual microtubules, a bundle can apparently be rescued by a new microtubule catching up with the shrinking bundle. This behavior, which we observed in our experiments, was also reproduced by the simulation results (see Figure 4.10d).

To understand the role of microtubule forces in the motion of chromosomes, it is essential to also consider the pulling forces created by shrinking microtubules. In the discussion above we only considered the simplified situation where multiple microtubules push against a resisting barrier. In the kinetochore reality, dynamic microtubules are actually attached by protein complexes to the object they push against. This attachment makes it possible that shrinking microtubules can also exert forces on the chromosomes in a situation where the chromosome resists motion in both directions. Interestingly, in budding yeast cells, a kinetochore protein complex called Dam1 has been identified that appears to form a sliding ring around the dynamic microtubule end [32]. Biophysical experiments have demonstrated that depolymerizing microtubules can indeed exert forces both on beads attached directly to the side lattice of the microtubule [33] and on beads that are coated with the Dam1 complex [34]. The hypothesis is that outward curving protofilaments that are released from shrinking microtubules can provide a mechanism for dragging both beads and Dam1 rings back along the microtubule [35]. It is likely that depolymerizing microtubules that interact with other specialized complexes at, for example, the cell cortex generate forces in a similar way.

4.8 Positioning Strategies: Pushing Versus Pulling

The fission yeast system discussed earlier suggests that pushing forces created in contact with the cell periphery are enough to position microtubule organizing centers with respect to the finite geometry of the cell. Using *in vitro* experiments in microfabricated chambers [36] and theoretical calculations [37], we have confirmed that pushing forces can indeed be used to find the center of a confining geometry. These *in vitro* experiments have also shown that positioning fails once microtubules become long enough to buckle and bend (see Figure 4.11a). One remedy against this is to make sure the microtubules undergo catastrophes frequently enough [38]. This assures that microtubules stay in contact with the cortex long enough to relate pushing forces to the organizing center, while preventing microtubules from becoming long enough to hinder motion of the organizing center due to microtubule-cortex interactions from another direction. It has been shown that even in small fission yeast cells, microtubules buckle and bend around the cell ends in mutant cells that have a reduced cytoplasmic catastrophe rate [39]. Forces may specifically enhance catastrophes when contact is made with the cell ends, but this is apparently not sufficient when the spontaneous catastrophe rate is not high enough in these mutants.

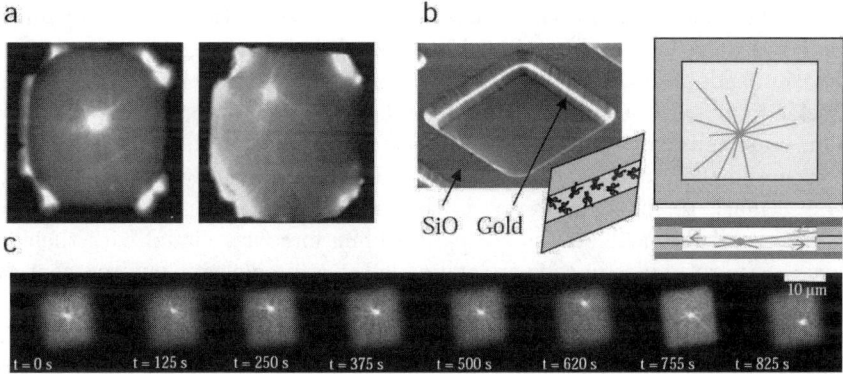

Figure 4.11. Pushing and pulling *in vitro*. Microfabricated chambers can be used to study positioning strategies in a simplified experimental environment. (a) Microtubule pushing combined with catastrophes is sufficient to center microtubule organizing centers in microfabricated chambers (left), but fails when catastrophes occur too infrequently leading to long buckled microtubules that destabilize the positioning process (right) [38]. Square chambers are 20 microns on the side. (b) Modifying the chamber technology to study the competition between pushing and pulling forces. A sandwiched layer of gold can be used to specifically functionalize the side walls of chambers with (in this case) minus-end directed motor proteins. (c) Preliminary data showing how "cortical" pulling forces generated by immobilized dynein molecules appear to repeatedly pull a centrosome off center. Counteracting pushing forces keep redirecting the centrosome to the chamber center.

In other systems, dynamic microtubules are observed to create pulling force in contact with the cell cortex [6]. This pulling can be mediated by minus-end directed motor proteins, such as dynein, or by the cortical attachment of shrinking microtubules as discussed earlier. In fact, pulling forces generated at the cortex seem to be the dominant mechanism for positioning spindles and microtubule organizing centers in larger eukaryotic cells. In single cell-stage *Caenorhabditis elegans* (*C elegans*) embryos, the spindle gets asymmetrically positioned towards the posterior end of the embryo by dynein-related pulling forces generated at the cell cortex. Laser-cutting experiments have not only demonstrated that pulling forces are generated, but also that the unbalance between the two sides is probably caused by a higher number of force-generating entities at the posterior side (as opposed to an increase in the magnitude of the forces generated at each posterior interaction site) [40].

Why may pushing forces be used in certain systems and pulling forces in others? One important aspect is probably the size of the system. It is simply not possible to efficiently generate pushing forces when microtubules become too long. A growing microtubule can move a microtubule-organizing center through the cell, and withstand counteracting forces up to its stall force, as long as the counteracting force is less than the critical buckling force of the microtubule. When a microtubule buckles, the pushing force on the organizing center is decreased and continued growth of the microtubule no longer necessarily leads to forward motion of the organizing center. Long microtubules buckle more easily than short microtubules, which is probably why small fission yeast cells can use microtubule pushing to position the nucleus, whereas spindles in the much larger *C elegans* embryos need to rely on pulling forces. One remark to make, however, is that long microtubules in large cells do not necessarily buckle at the critical buckling force associated with their full length. Microtubules in cells are embedded in a visco-elastic environment due to the presence of (for example) the actin cytoskeleton. Elastic deformation of this network may prevent large amplitude bending of microtubules leading to shorter wavelength buckling and consequent increased resistance to force [41].

Recent modeling efforts have shown that positioning processes based on pulling forces cannot be stable in the absence of restoring (pushing) forces [40, 42]. The intuitive reason is that as soon as an organizing center moves closer to one side of the cell, more microtubule-cortex interactions become possible, leading to larger forces in that same direction (assuming that an equal number of microtubules are nucleated in all directions and that every microtubule-cortex contact, or a fixed percentage thereof, generates only a pulling force). In contrast, in the case of pushing, more interaction automatically leads to larger forces in the opposite direction, and a stable situation is reached when the organizing center sits (on average) in the middle [37]. In *C elegans* embryos, it appears that not all microtubules interact with cortical force generating sites. It is therefore likely that antagonistic pushing forces are generated by the other microtubules and that the balance between them, combined with regulation of the actual number of pulling-force generators on the anterior and posterior sides controls the final (asymmetric) position of the spindle [43].

To study the antagonistic effect of pulling and pushing forces in a simplified experimental setting, we recently modified our microfabricated chamber experiments to be able to introduce localized pulling activities at the chamber boundaries (see Figure 4.11b). Instead of using chambers made from pure silicon monoxide (SiO), as we normally do, we etch these chambers into deposited multilayers of SiO (1.5 micron thick), chromium (thin layer to help adhere the gold), gold (between 100 and 800 nm), chromium, and SiO again. This provides us with chambers with a rim of gold along the edges to which we can specifically react biotin tags after assembling a monolayer of thiol-terminated reactive groups to the gold. Using streptavidin as an intermediate layer, we then attach biotinylated dynein molecules (a gift from Sam Reck-

Peterson and Ron Vale [44]) specifically to the "cortex" of these artificial cell-like chambers. In Figure 4.11c, we show preliminary results on how a microtubule aster that is nucleated by a centrosome (a gift from Michel Bornens) is trying to maintain a position close to the center of the chamber while being subject to constant attempts by dynein molecules to pull the aster towards the chamber edge. To understand what balance of pulling and pushing forces is needed to keep the aster actively centered, we plan to vary the relative of amount of dynein-interacting microtubule ends by reducing or enlarging the thickness of the gold layer (and thereby the amount of surface area available for motor proteins to attach).

4.9 Controlling Microtubule Organization *in vivo*

Forces generated by growing or shrinking microtubules provide only one of a large repertoire of mechanisms by which cells organize the microtubule cytoskeleton. For example, in the organization of the fission yeast interphase array, additional important ingredients are microtubule-associated proteins that control microtubule nucleation at the nuclear membrane and elsewhere, proteins that allow for anti-parallel bundling at mid-cell, motor proteins that slide anti-parallel microtubules towards the cell center, and microtubule end-binding proteins that travel to the cell ends and locally create a different protein environment.

In general, the activity of molecular motors is an essential component of most microtubule-based force generation processes in cells. In the case of chromosomes motions in the mitotic spindle, it is clear that a number of different motor proteins associated with chromosomes, kinetochores, microtubule poles, and anti-parallel microtubule arrays somehow work together with microtubule polymerization and depolymerization processes. Understanding the complex regulation of all these antagonistic forces and the versatility that this creates for a dynamic and responsive control of the microtubule array in different cell types and situations, provides an exciting challenge in cell biology and cell biophysics research. It seems evident, however, that mechanical aspects related to cell shape and size, microtubule elastic properties, and biochemical response to force generation need to be an integral part of understanding functional microtubule organization in cells.

Acknowledgements

We thank former members of our group, Mathilde de Dood, Martijn van Duijn, Cendrine Faivre-Moskalenko, Astrid van der Horst, Marcel Janson, Jacob Kerssemakers, and Guillaume Romet-Lemonne for their contribution to work that is presented in this chapter. We thank Henk Bar, Niels Dijkhuizen, Roland Dries, Johan Herschied, Marco Konijnenburg, Chris Retif, and Duncan Verheij for technical support at AMOLF. We also thank collaborators and colleagues: Anna Akhmanova, Michel Bornens, Damian Brunner, Matt Footer, Tony Hyman, Tim Mitchison, Bela Mulder, Francois Nedelec, Tim Noetzel, Sam Reck-Peterson, Ted Salmon, Thomas Surrey, Catalin Tanase, and Ron Vale for help and discussions that were important for this work, and for making proteins or other reagents available to us. This work is part of the research program of the "Stichting voor Fundamenteel Onderzoek der Materie (FOM)", which is financially supported by the "Nederlandse organisatie voor Wetenschappelijk Onderzoek (NWO)".

References

1. Alberts, B & et al. (2002) *Molecular Biology of the Cell.* (Garland, New York).
2. Desai, A & Mitchison, T. (1997) *Annu. Rev. Cell Dev. Biol.* **13**, 83–117.
3. Grill, S & et al. (2001) *Nature* **409**, 630–633.
4. Tran, P & et al. (2001) *J. Cell Biol.* **153**, 397–411.
5. Inoue, S & Salmon, E. (1995) *Molecular Biology of the Cell* **6**, 1619–1640.
6. Dogterom, M & et al. (2005) *Curr. Op. Cell Biol.* **17**, 67–74.
7. Fygenson, D, Braun, E, & Libchaber, A. (1994) *Phys. Rev. E* **50**, 1579–1588.
8. Chretien, D, Fuller, S, & Karsenti, E. (1995) *J. Cell Biol.* **129**, 1311–1328.
9. Wang, H.-W & Nogales, E. (2005) *Nature* **435**, 911.
10. Wang, H.-W & et al. (2005) *Cell Cycle* **4**, 1157–1160.
11. VanBuren, V, Cassimeris, L, & Odde, D. (2005) *Biophys. J.* **89**, 2911–2926.
12. Janosi, I, Chretien, D, & Flyvbjerg, H. (2002) *Biophys. J.* **83**, 1317–1330.
13. Mandelkow, E, Mandelkow, E, & Milligan, R. (1991) *J. Cell Biol.* **114**, 977–991.
14. Howard, J & Hyman, A. (2003) *Nature* **422**, 753–758.
15. Akhmanova, A & Hoogenraad, C. (2005) *Curr. Op. Cell Biol.* **17**, 47.
16. Gundersen, G, Gomes, E, & Wen, Y. (2004) *Curr. Op. Cell Biol.* **16**, 106–122.
17. Hill, T. (1987) *Linear aggregation theory in cell biology.* (Springer., New York, Berlin, Heidelberg.).
18. Peskin, C, Odell, G, & Oster, G. (1993) *Biophys. J.* **65**, 316–324.
19. Mogilner, A & Oster, G. (1999) *Europ. Biophys. J.Biophys. Lett.* **28**, 235–242.
20. Doorn, G & et al. (2000) *Europ. Biophys. J.Biophys. Lett.* **29**, 2–6.
21. Stukalin, E & Kolomeisky, A. (2004) *J. Chem. Phys* **121**, 1097–1104.
22. Fygenson, D, Marko, J, & Libchaber, A. (1997) *Phys. Rev. Lett.* **79**, 4497–4500.
23. Dogterom, M & Yurke, B. (1997) *Science* **278**, 856–860.
24. Landau, L & Lifshitz, E. (1986) *Theory of elasticity.* (Pergamon, New York).
25. Janson, M & Dogterom, M. (2004) *Phys. Rev. Lett.* **92**, 248101–4.
26. Kerssemakers, J & et al. (2003) *Appl. Phys. Lett.* **83**, 4441.
27. Kerssemakers, J & et al. (2006) *Nature* **442**, 709–712.
28. Janson, M, Dood, M, & Dogterom, M. (2003) *J.Cell Biol.* **161**, 1029–1034.
29. Komarova, Y, Vorobjev, I, & Borisy, G. (2002) *J. Cell Sci.* **115**, 3527–3539.
30. Mimori-Kiyosue, Y & et al. (2005) *J. Cell Biol.* **168**, 141–153.
31. Biggins, S & Walczak, C. (2003) *Curr. Biol.* **13**, R449–R460.
32. Westermann, S & et al. (2006) *Nature* **440**, 565.
33. Grishchuk, E & et al. (2005) *Nature* **438**, 384–388.
34. Asbury, C & et al. (2006) *Proc. Natl. Acad. Sci. USA* **103**, 9873–9878.
35. Molodtsov, M & et al. (2005) *Proc. Natl. Acad. Sci. USA* **102**, 4353–4358.
36. Holy, T & et al. (1997) *Proc. Natl. Acad. Sci. USA* **94**, 6228–6231.
37. Dogterom, M & Yurke, B. (1998) *Phys. Rev. Lett.* **81**, 485–488.
38. Faivre-Moskalenko, C & Dogterom, M. (2002) *Proc. Natl. Acad. Sci. USA* **99**, 16788–16793.
39. Mata, J & Nurse, P. (1997) *Cell* **89**, 939–949.
40. Grill, S & et al. (2003) *Science* **301**, 518–521.
41. Brangwynne, C & et al. (2006) *J. Cell Biol.* **173**, 733–741.
42. Howard, J. (2006) *Phys. Biol.* **3**, 54–66.
43. Grill, S, Kruse, K, & Julicher, F. (2005) *Phys. Rev. Lett.* **94**, 108104.
44. Reck-Peterson, S & et al. (2006) *Cell* **126**, 335–348.

5

Mechanisms of Molecular Motor Action and Inaction

Sarah Rice

Northwestern University, Department of Cell and Molecular Biology, 303 E. Chicago Ave., Ward 8-007, Chicago, IL 60611 Phone: (312)503-5390 Fax: (312)503-7912 Email: s-rice@northwestern.edu

5.1 Preface

In the past few years, our understanding of molecular motor mechanisms has become sophisticated enough that animated models of kinesin and myosin have been made based on the experiments of many labs over several years [1]. A similarly detailed model of dynein is on the horizon. Now that we are gaining an understanding of how all three classes of molecular motors work, it has become possible to see parallels in their mechanisms of action.

This review begins with a summary of structural and mechanistic data on myosin, kinesin, and dynein, citing similarities between the motile mechanisms of all motors. In particular, the structural parallels between kinesin and myosin are striking. Deciphering the conformational changes that these motors undergo has been a task of matching x-ray crystal structures to the closest approximation of their true biochemical state, as determined from other experiments performed in the presence of their partner filaments.

All motors appear to have converged on certain core mechanisms. Myosin, kinesin, and dynein all have ATPase activity that is coupled to movement via a biased conformational change. That coupling occurs only when a motor is bound to its partner filament. This does two things. First, it makes the motor very efficient by ensuring that the expenditure of ATP energy occurs only while the motor is moving on the partner filament. Second, the activity of some motors may depend on their achieving a very specific conformation when they are bound to their partner filament. In the case of a processive molecular motor dimer, this conformation may be induced by the partner head. This type of gating mechanism can greatly enhance motor processivity.

What about the mechanisms of "inaction"? Are there common themes in motor regulation and in the mechanism by which inhibition is relieved by cargo binding? In contrast to their similar motile mechanisms, molecular motors are extremely divergent with respect to their regulation and cargo-binding mechanisms. In this review, the array of regulatory and cargo-binding mechanisms of motors is described in terms of a few examples. Because motors are poor enzymes unless they are bound to their partner filament, there are many potential regulatory mechanisms. Motors can be regulated by specific domains, regulatory proteins, phosphorylation, or small molecule drugs. Drugs are not physiological regulators. Regulation of smooth myosin is particularly striking, because it auto-regulates by adopting a conformation in which the two heads shut each other down by distinct mechanisms. Cargo movement by motors usually, but not always, depends on the attachment of adapter proteins or domains. Multiple adapter proteins can bind a single motor. For example, kinesin binds to either its own light chains or to the milton protein, depending on its cargo. Recent work on the myosin V tail domain has shown that multiple cargoes can bind a single adapter as well. There seems to be an example for every formal possibility of a regulatory or cargo-binding mechanism, and this vast array of mechanisms uniquely tailors each motor to its specific cellular function.

The situation appears simpler at the level of the entire cell, where a single chemical signal can instantly alter the course of thousands of motor cargoes in a purposeful and directed fashion. The best example of this is the movement of melanosomes, which can be induced to disperse throughout a pigment cell or aggregate in the center in response to a hormone. In particular, a different sort of "motor" may coordinate cargo attachment or regulation, engaging motors for unidirectional motion when and where they are needed. The intriguing problem of motor coordination makes it evident that despite the long lists of seemingly independent and unique mechanisms of regulation, inhibition, and cargo binding, the romantic days of building and testing models of molecular motors are far from over.

5.2 Making Molecular Movies

5.2.1 Myosin

The Swinging Lever Arm Hypothesis

The field of molecular motors has a rich tradition of making and testing models for movement. The most fundamental of these is the swinging lever arm model for muscle contraction, which remarkably has endured decades of intense scrutiny by myosin researchers [2, 3, 4, 5]. Its basic premise of a unidirectional conformational change of a motor coupled to partner filament binding applies to the mechanisms of all myosin superfamily members and has relevance to the mechanisms of the kinesin and dynein motors, which were discovered nearly 30 years after the first evidence for the hypothesis was published. Hugh Huxley has recently published a review of x-ray diffraction data that summarizes the evidence for the swinging lever arm hypothesis [6]. Importantly, this x-ray diffraction data established the connection between the ATPase activity of active muscle and the movement of myosin against actin, showing a conformational change (crossbridge tilting) of myosin on actin that is the basis of muscle contraction.

Kinetics and the Actin-binding Properties of Myosin

After the swinging crossbridge model was proposed, kinetic and structural experiments made it possible to create a moving picture of myosin based on that model. To do this, researchers had to determine the timing of the steps in myosin's ATPase cycle on actin. After observing that actin markedly increased the ATPase activity of myosin, Eisenberg and Moos applied a general mechanism of an enzyme (myosin), a substrate (ATP), and a modifier (actin) to this system [7]. This general framework formed the basis for creating kinetic models of all molecular motors. The kinetic cycle of myosin was soon thereafter described in much greater detail using pre-steady state kinetics [8]. This work showed that ATP binding very rapidly dissociates rigor-bound myosin from actin. After hydrolysis, myosin rebinds to actin. Phosphate release is the rate-limiting step for actin-activated ATPase, and is coupled to the rebinding of the myosin head on actin and the myosin powerstroke, the unidirectional conformational change that caused the crossbridge tilting described previously. The timing of myosin's movement on actin had been revealed.

Original Myosin Structures: Where is the Lever Arm?

Another key result in understanding myosin's mechanism was the first x-ray crystal structure of myosin subfragment 1, and its docking into electron micrographs of actomyosin in rigor [9, 10]. The myosin structure itself is a striking visual picture of an actin-binding ATPase machine with a lever arm structure, angled at 45° relative to actin (see Figure 5.1, left side). It was believed at the time that the first myosin structure was solved that subsequent structural data would soon confirm the different positions of the lever arm as observed in x-ray diffraction experiments on muscle fibers and reveal the structural states that corresponded to specific nucleotide states of the active site of myosin. X-ray crystallization of *Dictyostelium* myosin in the presence of at least 12 nucleotide analogs yielded extremely detailed data about the mechanism of nucleotide hydrolysis by myosin and showed a pre-powerstroke conformation of the nucleotide pocket [11, 12, 13]. However, there was a large discrepancy between these structures and crosslinking data that had shown that two cysteines at either end of the SH2 helix could be crosslinked by a very short length crosslinker, pPDM, in the absence of actin [14, 15]. These cysteines appeared separated in the x-ray crystal data at a distance far too great to allow crosslinking. Another difficulty in understanding myosin's mechanism was that the lever arm was missing from these x-ray crystal structures and its position relative to the actin filament in these structures was debated [16, 17]. The apparent position of the lever arm, judging from adjacent elements in x-ray crystal structures, contrasted with data from electron microscopy experiments showing a large, concerted change between the ADP·Pi-bound state and the ADP state of myosin on actin [18, 19].

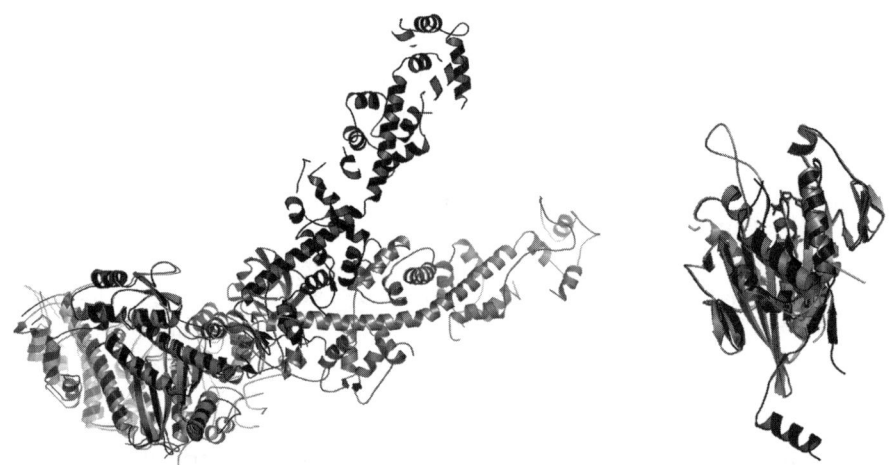

Figure 5.1. Two structures of scallop myosin (left) and human kinesin (right), showing different structural conformations of the same motor in the same nucleotide state. PDB accession numbers: 1B7T, scallop myosin ADP structure with lever arm almost parallel to actin (lighter) [20], 1S5G, scallop myosin ADP structure with lever arm ∼ 45° to actin (darker) [21], 1BG2, human kinesin with disordered neck linker (lighter) [22], 1MKJ, human kinesin with docked neck linker [23].

Recent Structures: A Breakthrough Interpretation

More recently, structures of smooth myosin and scallop myosin revealed lever arm conformations that were at 90° [24] and almost parallel to the actin filament [20]. Referencing the available crystallography, crosslinking, and electron microscopy data, the lever arm positions of these two structures were interpreted as a pre-powerstroke state and an actin-detached state, respectively. Transitions from the actin-detached conformation to the pre-powerstroke conformation and finally to the rigor conformation, would result in movement to the barbed (plus) end of actin. At the time, it seemed odd that the nucleotides present in the structures did not correspond to the structural conformations of myosin for those nucleotide states. ADP was present in the actin-detached structure, which the authors interpreted to be a weak-binding ATP-like state. Unlike other structures, the SH2 helix was melted into a conformation consistent with the observed crosslinking results on myosin in the absence of actin. A recent study has shown that ADP·BeFx (ADP·beryllium fluoride),

AMPPNP, and ADP all can be crystallized in the same actin-detached conformation, which remains virtually identical with the SH1 and SH2 cysteines crosslinked together [25] (see Figure 5.1). ADP-myosin can also be crystallized in a post-powerstroke conformation with its lever arm 45° to the actin filament [21]. All of this data indicates that myosin nucleotide states are not precisely coupled to structural states in x-ray crystal structures, which may be due to the absence of their partner filaments.

The revelation that nucleotides in x-ray crystal structures do not always correspond to the structural states that are predominantly occupied by the motor in solution was absolutely critical to our current understanding of myosin motors, and the story repeats itself with the kinesins in Section 5.2.2 of this review. This has been a key point in using data from x-ray crystal structures to understand the mechanisms of several other ATPases and GTPases as well, such as F1 ATPase [26], EF-Tu [27], the SERCA1a calcium pump [28], RecA [29], and actin [30].

5.2.2 Kinesin

Kinetics and Microtubule Binding Properties of Kinesin: Distinct from Myosin

When the microtubule-binding states and kinetic cycle of kinesin were revealed, they were completely different from myosin. Kinesin is weakly bound to microtubules in the ADP state but tightly bound in AMPPNP or ADP·AlF4 [31]. This was confirmed by pre-steady state kinetics experiments on kinesin monomers showing that kinesin·ADP releases ADP upon contact with microtubules, then remains tightly bound in the ATP state. The weak-binding ADP state is the rate-limiting enzymatic step for kinesin [32]. Given the differences between the kinesin and myosin kinetic cycles, and the huge difference in their molecular size, it seemed as if kinesin and myosin were completely distinct motors. The kinesin x-ray crystal structure, first solved in 1996 by Kull and colleagues showed that surprisingly and intriguingly they are very similar [22].

Structure: Kinesin is Myosin without a Lever

The first kinesin crystal structures showed that the ATP pocket and core structure of kinesin motors is quite homologous to myosin and to G proteins, and the implications of this similarity have been reviewed in-depth [33]. Both structures possess nucleotide-sensing elements that are homologous to each other and to the G-proteins. The central β-sheets of kinesin and myosin superimpose onto each other. Both motors have the switch-II, or relay helix element, that propagates a conformational change from the nucleotide-pocket and filament-binding regions to mechanical elements that exert the conformational change.

Given the core structural similarities between kinesin and myosin, it was surprising that there is no obvious lever arm structure in either monomer or dimer x-ray crystal structures of kinesin. In the structures of the rat kinesin monomer and dimer [34, 35], the core of the molecule is attached to a coiled-coil joining the two kinesin heads of the dimer by a 15-a.a. β-strand that is folded along the central β-sheet of kinesin, referred to as the neck linker region. This element is disordered in the original monomer structure solved by Kull, et al. (see Figure 5.1). Despite the intriguing structural similarities of kinesin and myosin, kinesin's mechanism could not be determined from x-ray crystal structure data alone.

Kinesin Mechanism: the Myosin Story Repeats Itself

The same type of observation just described for identifying the structural states of myosin's lever arm led to the current structural model of kinesin motility. An ADP-containing x-ray crystal structure with a docked neck linker [34], which is similar to the darker kinesin molecule in Figure 5.1, was interpreted as an ATP-like state. Cryo-electron microscopy as well as electron paramagnetic resonance (EPR) and fluorescence resonance energy transfer (FRET) spectroscopy data were consistent with the docked conformation of the neck linker conformation being predominantly present in the presence of ATP or ADP·Pi on microtubules [36]. In contrast, kinesin in the nucleotide-free state, or in the ADP state, showed an undocked conformation. This conformation was interpreted as disordered in the monomeric kinesin constructs used for this study, similar to the original kinesin structure shown in Figure 5.1 [22]. This interpretation led to a model for a conformational change that directs kinesin toward the plus end of the microtubule. Kinetic experiments had shown that upon ADP release, the motor binds to microtubules [32]. The transition of the neck linker element from the ADP-like conformation to the ATP-like conformation, while the motor is bound to microtubules, is a plus-end directed conformational change. Therefore, although kinesin lacks the rigid lever arm structure of myosin, motility of both motors is driven by an ATP-coupled conformational change of the motor on its partner filament. Remarkably, these conformational changes are propagated from the nucleotide- and filament-sensing elements to the mechanical elements of both motors in much the same way. In both motors, the conformational change is transmitted by an α-helix referred to as the "relay helix" , to either myosin's lever arm or kinesin's neck linker region [1].

5.2.3 Dynein

We are now on the verge of understanding the mechanism of dynein motility in detail, which means we will soon understand the basic mechanisms of all three classes of eukaryotic cytoskeletal motors. Looking at the structure of dynein, similarities to the kinesins and myosins seem almost inconceivable. Dynein is a megadalton complex AAA ATPase with no structural similarities to either of the other classes of motors. Nonetheless, mechanistic studies identified a lever arm that drives dynein's movement, and similar to myosin, dynein filament velocities *in vitro* are proportional to the length of this lever arm element [37, 38]. Single particle analysis of rotary-shadowed single dynein molecules in electron microscopy experiments have shown a significant movement of the dynein structure in the area that would normally join the two AAA domains of the dimer [39] that have been confirmed by more recent FRET spectroscopy experiments on the dynein AAA head and tail domains [40, 38].

The timing of dynein's ATPase cycle has recently been revealed by FRET spectroscopy and kinetics experiments. ADP release by the first AAA site (see Figure 5.2) is coupled to movement of the lever arm, unlike kinesin or myosin. Blocking hydrolysis at the third AAA site led to a state in which lever arm movement may be decoupled from ATPase activity [40]. The microtubule-binding domain of dynein is connected by a coiled-coil stalk to the AAA domain between the second and third sites. Conformational changes in the AAA ring of dynein resulting from ATPase activity may lead to changes in the coiled-coil register of the stalk leading to the microtubule-binding domain that coordinates ATP-dependent microtubule binding and release by dynein [41].

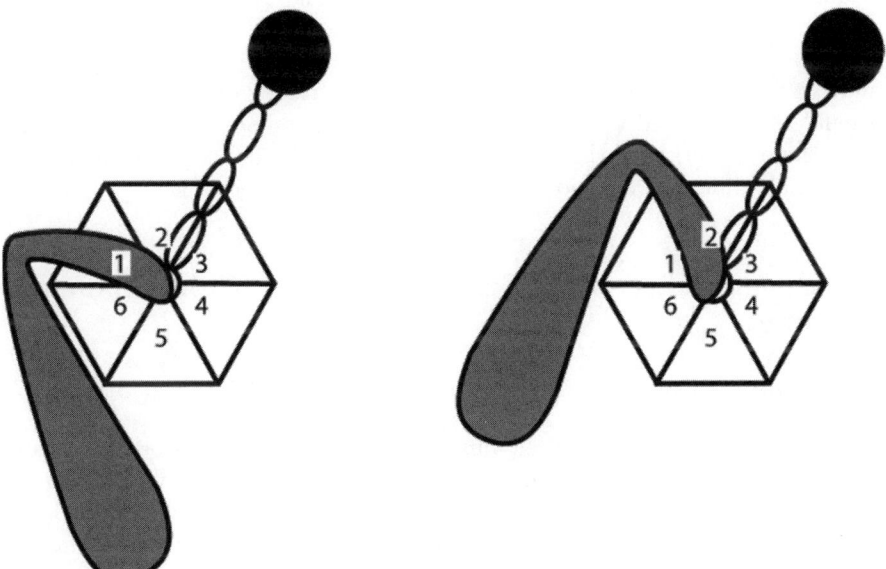

Figure 5.2. Dynein lever arm swing relative to the 6 AAA sites in the AAA domain. The microtubule-binding domain is the black dot, the stalk is shown as a coil connecting the microtubule-binding domain to the AAA head, and the tail, or lever arm, is depicted in gray. ATPase activity results in a swing of the tail.

5.2.4 Motor Movement without a Lever Arm?

A set of surprising results on all three types of molecular motors has revealed that motion of molecular motors along their partner filaments can be independent of the lever arm elements. This was shown in kinesin motility assays on a construct lacking the neck linker region that moved to the plus ends of microtubules at a velocity about 500-fold reduced from that of a wild-type kinesin control [42]. Similarly, a chimeric construct of the minus-end directed kinesin family member ncd with kinesin's plus-end directed neck linker, moved to the plus end at a velocity that was reduced about 100-fold from a wildtype ncd control [43]. Chimeras have been made of the converter domains and lever arms of myosin V and myosin VI, showing that the directionality of these motors does not depend on the converter domain or lever arm. These chimeric motors move at velocities that are reduced roughly 10-fold from the intact motors [44]. Recent work has determined that dynein that is anchored by its head domain such that it cannot use its lever arm is able to move slowly along microtubules towards the minus end [38]. One explanation for all of these results is that any biased conformational change of a motor on its partner filament will result in directed movement. This has been demonstrated for kinesin motors [45], which undergo a structural transition of their catalytic core on the microtubule that is independent of the neck linker. This movement could account for the motility of kinesin constructs lacking the neck linker element, and the same basic mechanism of a filament or nucleotide-induced "kick" of the motor domain on the partner filament may explain lever-independent movement of myosin and dynein motors as well. Results of these experiments should be interpreted with caution, because it is possible in all of these cases that the destruction of the lever arm element results in the motor adopting an entirely different (and possibly non-physiological) mechanism.

5.3 Reversibility of the Powerstroke and Tight Filament Binding

5.3.1 Motors Can Take Mechanical and Enzymatic Back Steps

Motors are defined by unidirectional, ATP-driven conformational changes, but interestingly, the mechanical conformational changes involved appear to be at least somewhat reversible. All three types of motors appear to have mechanically uncoupled states. For myosin, this state occurs when the motor is not bound to its partner filament [25]. A similar argument can be made for kinesin based on crystal structures showing different conformations in solution [22, 35], and spectroscopic data on kinesin in solution that did not reveal significant conformational changes in different nucleotide states [36]. The mechanical coupling of dynein to its ATPase activity may depend specifically on the activity of its third AAA repeat [40]. Even more interestingly, dynein appears to have bidirectional movement *in vitro* that may be the result of a reversal in its coupling of ATPase activity to microtubule binding [46], and dynein may also have an unproductive mode of binding to microtubules that is released by the tension that is created when multiple motors move a cargo at one time [47]. Motors from all three classes have been observed to take backward mechanical steps [48, 49, 50, 51]. In the case of kinesin, a detailed quantitative study of the dependence of kinesin force-velocity curves on ATP concentration was consistent with a small change in free energy creating kinesin's forward bias [49]. EPR spectroscopic data showed this as well, although this may have been due to the use of nucleotide analogs and monomeric kinesin constructs [52]. More convincingly, docking of the kinesin neck linker can be induced in solution by adding sulfate ions, and the free energy change associated with that docking was estimated by EPR spectroscopy to be small ($\sim$ 4 kJ/mol) [23]. However, comparing rates of ATP synthesis and ATP hydrolysis for kinesin has shown that a significant amount of free energy from ATP is expended as the neck linker docks, in apparent contrast to this data [53]. This is discussed in more detail next.

Rates of ATP synthesis from ADP and phosphate have been measured for all three classes of molecular motors, and these data argue for a change in the enzymatic coupling of motors when they bind to their partner filaments. Both myosin and dynein have significant rates of ATP synthesis in solution [54, 55, 56] that are slowed by interaction with the partner filament. In contrast, kinesin's rate of ATP synthesis is markedly increased by its interaction with microtubules [53]. Myosin and dynein thus appear to transition from a weakly coupled, solution state to a strongly coupled, active state on their partner filaments. Kinesin is a very poor enzyme for both the hydrolysis and the synthesis of ATP in the absence of microtubules, and it activates, enabling both reactions, upon microtubule binding. Recent high resolution cryo-electron microscopy data on kinesin family members has revealed large structural rearrangements of the central β-sheet when the motor is docked to microtubules [57, 58]. It may be that for kinesin, most of the free energy available to perform work when it is bound to the microtubule is used to engage the motor by shifting its position on the microtubule. A small change in free energy may then accompany neck linker docking. This scheme may improve the efficiency of molecular motors in converting ATP energy into mechanical work.

5.3.2 Uncoupling and Disengaging Makes Motors Efficient

Molecular motors may use most of their ATP energy in transitioning from an uncoupled or inactive state to a coupled, active state with a mechanical bias, rather than in generating the mechanical bias itself. Motors use very little ATP in the absence of their partner filament because either there is little change in free energy between various conformational states or because they are enzymatically inactive. They remain uncoupled or inactive in solution because of the large energetic barrier in achieving the coupled, active state. Once it achieves the coupled state on its partner filament, a motor need not have a very large change in free energy associated with its powerstroke to have highly biased, efficient forward movement [52], see Figure 5.3. The scenario of Figure 5.3 is certainly oversimplified, but it illustrates that using only a small investment of free energy, a motor can be heavily biased in one direction. The remainder of the available free energy that does work can be interpreted as a free energy of binding to the microtubule [49]. A large investment of free energy at that stage binds a motor tightly to its partner filament and enables significant changes in the motor's enzymatic mechanism in the filament-bound state so that motors can be truly "off" when they are not bound.

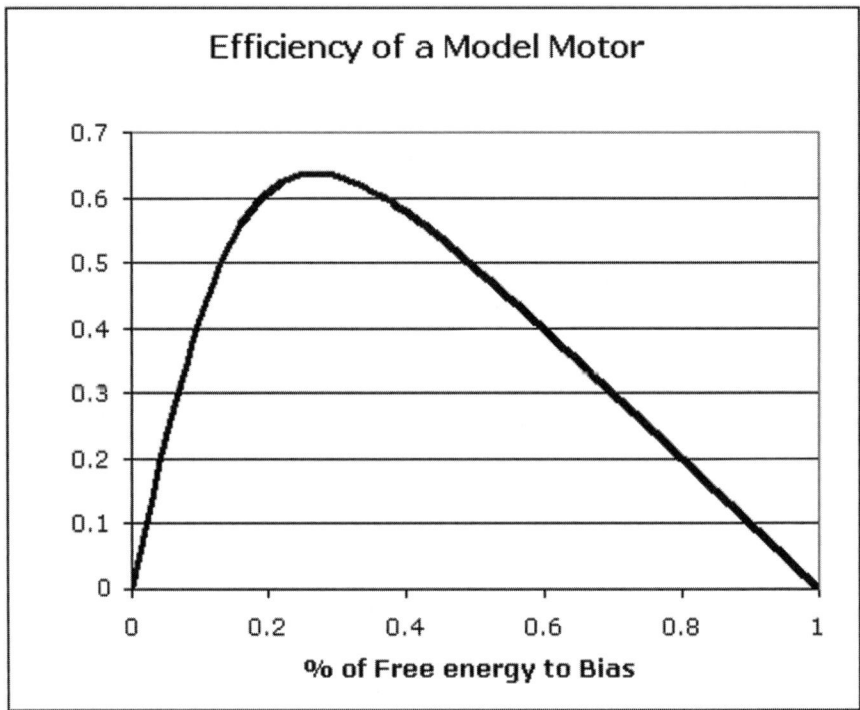

Figure 5.3. A simple model for molecular motor bias and efficiency of movement. In this model, it is assumed that ΔG from ATP hydrolysis goes either to generate a bias or to perform work. Equations (5.1)-(5.3) characterize the model, first described in [52].

Figure 5.3 is based on the following equations:

$$\Delta G_{\text{ATP}} = \Delta G_{\text{Bias}} + \Delta G_{\text{Work}} \tag{5.1}$$

$$\Delta G_{\text{Bias}} = -RT \ln \left(\frac{\text{forward steps}}{\text{back steps}} \right) \qquad (5.2)$$

$$\text{Efficiency} = \Delta G_{\text{Work}} \left(\text{forward steps} - \text{back steps}\right). \qquad (5.3)$$

5.4 The Processive Motors: The Importance of Gating

A striking behavior of several kinesin, myosin, and dynein family members that was not predicted by the swinging crossbridge hypothesis is processivity, which is the ability of a single motor molecule to take several ATP-coupled steps on its partner filament before dissociating. A critical property of highly processive motors is gating. When gating occurs in a molecular motor dimer, the gated head is enzymatically trapped at specific points in the chemomechanical cycle of coupled ATPase activity and conformational changes. The trapped head cannot advance a step in its chemomechanical cycle until the other head has done so. Gating of the two motor heads has been shown to be the major determinant of processivity for kinesin, and gating mechanisms have been observed for several other motor types as well.

5.4.1 Are Processive Motors Always Gated?

It is possible for molecular motors to be processive simply because the two tethered motor heads in a dimer both spend the majority of their ATPase cycles tightly bound to their partner filament [59]. However, the duty cycle (relative amount of time in the ATPase cycle that is spent tightly bound to the partner filament) of single motor heads correlates poorly with motor processivity in general. Kinesin takes hundreds of steps without falling off of the microtubule [60], but its processivity would be extremely low if it were not gated. ADP-kinesin is not tightly bound to microtubules [31] and ADP release is the rate-limiting step in kinesin's enzymatic cycle [32]. The kinesin family member ncd, meanwhile, spends the majority of its ATPase cycle tightly bound to microtubules [61], but ncd is categorically nonprocessive: it never steps twice on the microtubule without falling off [62]. The processive myosins V and VI both spend the majority of their ATPase cycles bound to actin. Therefore, they are predicted to be processive without gating mechanisms [59], but data on both of these motors has shown that they, too, are gated [59, 63, 64]. Data from several different techniques is used to demonstrate the gating mechanisms of molecular motors. The data for kinesin, myosin V, myosin VI, and ncd are discussed in Section 5.4.2.

5.4.2 Gating Mechanisms

Kinesin

The first clear demonstration of gating for a molecular motor was made by David Hackney in 1994, performing ADP release experiments on kinesin dimers [65]. While all ADP was released from a solution of kinesin monomers upon addition of microtubules, only half of the ADP was released from a solution of dimers. Thus, one head released ADP and bound tightly to micro-tubules in an apo-state, while the other was somehow prevented from doing so. Subsequent experiments found additionally that ATP binding to the apo-head triggers ADP release by the other head [66]. More recently, gating has been demonstrated for the "second" step of kinesin, which is the release of the rear head that occurs before the front head hydrolyzes ATP and detaches from the microtubule. ATPase kinetic studies have been performed on a kinesin mutant that decoupled the phosphate release and microtubule detachment steps (E164A) [67]. The E164A mutant bound to microtubules and under-went a single round of ATP hydrolysis. Phosphate release occurred, but the head remained tightly bound to the microtubule after phosphate release. ATP binding by the second front head was prevented. This work provided evidence that phosphate release occurs while the head is still bound to the microtubule and that the front head must wait until the rear head detaches to bind ATP. Mechanical measurements using optical trapping arrived at a similar conclu-sion. BeFx was added to kinesin molecules that were stepping processively in an optical trap [68]. Kinesin pauses its stepping when ADP·BeFx binds to one of its heads, and steps backwards once before releasing BeFx and ADP from the leading head, binding ATP, and resuming stepping. A possible in-terpretation is that ADP·BeFx, when bound to either head, enables a single forward step of the other head, consistent with previous results [66, 36]. After this forward step, BeFx cannot release until the lead head takes a backward step, placing the BeFx head in front. This result shows that the affinity of the front head of a kinesin dimer for ADP·BeFx is less than that of the rear head. Thus, the front head of a kinesin dimer may not bind ATP tightly until the rear head goes into a weak-binding state. The gating mechanism described by these two papers may help to prevent the front head from binding and hydrolyzing ATP to enter a weak-binding state at the same time as the rear head.

Myosin VI

A kinetic comparison of myosin VI monomers and dimers suggests that a gating mechanism enhances its processivity. The phosphate release rate in a myosin VI dimer is substantially less than that of the monomer. The lead head of myosin VI is prevented from going into a weak binding state until the trailing head releases ADP [69, 63]. Because the lead head waits for the trailing head to release ADP, the trailing head is prevented from entering a weakly bound ATP or ADP·Pi state at the same time as the front head.

Myosin V

Mechanical measurements on myosin V using optical trapping have demon-strated that it is also gated. Purcell and colleagues [64] simulated the intramolecular strain of the two heads of a myosin V dimer by subjecting single myosin V heads to either forward or backward loads of 2 pN. The backward load causes myosin V heads to detach more slowly from the actin filament independent of ATP concentration, while forward loads result in detachment rates that are consistent with unloaded ATPase assays. These results are consistent with backward loads slowing the weak-to-strong transition of myosin V·ADP [64]. Thus, the leading head of myosin V may be gated by internal strain in the molecule.

5.4.3 Gating and Step Size

Gating would explain one other seemingly perplexing property of kinesin, myosin V, and myosin VI. All of these motors seem structurally capable of taking many different-sized steps on their partner filaments, based on a simple comparison of motor geometries to filament geometries. Despite this, kinesin and myosin V both have fairly tightly distributed, load-independent step sizes [48, 49, 70, 71]. The step sizes of myosin VI and dynein are more variable than those of kinesin and myosin V, but it appears structurally possible that these motors could have far more variability in their step size than is measured [50, 51]. The gating mechanisms of these motors may result in internal strain strictly limiting the number of enzymatically active positions of motor heads on their partner filaments and thus limiting the sizes of steps these motors take. While this is an attractive idea, it is difficult to measure internal strain in molecular motors.

5.4.4 Gating in Nonprocessive Motors?

Gating mechanisms can make motors nonprocessive as well as processive. Single molecules of the kinesin family member, ncd cannot take multiple mechanical steps on the microtubule before dissociating [62]. Structural and mechanistic studies on ncd revealed that the unbound head of ncd is never positioned such that it can bind to microtubules at the same time as the bound head, and the velocity and ATPase activity of a single-headed ncd heterodimer is the same as wild-type ncd [72]. Thus, the unbound head cannot bind the microtubule until the bound head has let go. It is unknown what biological function is served by this conformationally gated non-processivity of ncd, or the gating mechanisms of the processive motors either, for that matter. It is interesting to speculate, however, that because enzymatic mechanisms of motors are exquisitely sensitive to their conformation, they may be controlled or regulated very effectively in subtle ways.

5.5 Mechanisms of Motor "Inaction"

Similarities between the mechanical mechanisms of kinesin and myosin were not apparent at first, but when researchers became aware of them, both motor mechanisms were better understood. Can kinesin researchers gain insight on mechanisms of "inaction" from available data on myosin motors in the same way that they did for the mechanisms of action? Perhaps not. Myosin super-family proteins have highly varied mechanisms of *in vivo* regulation by specific domains or proteins despite having virtually the same mechanical mechanisms. A striking example of this is smooth myosin, which self-regulates using differ-ent mechanisms on each of its two heads. The actin-binding cleft of one head of smooth myosin appears to bind the converter domain of the other head, shutting both heads down [73]. In Section 5.4 of this chapter it was argued that motors must bind a partner filament in a specific geometry to be active ATPases. This constraint on motor activity means that molecular motors can be shut down through regulation in several different ways. Similarly, small molecule inhibitors that shut down motors in nonphysiological ways can have multiple modes of inhibition of motor activity. Both in the case of regulation and inhibition by small molecule drugs, the active site of a motor can be blocked, partner filament binding can be blocked, or a conformation of the motor that is necessary to activate the ATPase activity can be blocked. All of these types of mechanisms exist. Table 5.1 contains a number of regulatory and inhibitory mechanisms for motors, just to demonstrate the variation in them.

5.6 Cargo-Binding Mechanisms

With motor-cargo interactions, as in the case of motor regulation, there are numerous formally possible mechanisms and it seems that every combinatorial possibility for a motor-cargo interaction has been exploited. Motors have light chains that can themselves bind to cargo [85] or connect to other adapter proteins that then connect to cargoes [86, 87]. In some cases, cargoes can bind light chains or even heavy chains of motors directly [88, 89]. The same motor heavy chain can bind to different light chains [90] or different light chain isoforms [85, 91], or bind to different cargoes directly [92], and the same cargo can bind to at least three different motors [93].

There is not much structural or mechanistic data available on motor-cargo interactions, but the recently solved x-ray crystal structure of the myosin V tail shows how it can recognize two different cargoes, vacuoles and secretory vesicles. Interestingly, the two cargoes are recognized through two completely distinct molecular surfaces on the tail [92]. Although it is likely that bind-ing of myosin V to one organelle sterically prevents binding to another, this structural work entertains the formal possibility of two cargoes simultaneously binding to one motor.

Table 5.1. Regulation and inhibition mechanisms for motors.

Regulator/Inhibitor	Motor	Mechanism	Reference
Physiological Regulators			
10S Conformation	Smooth Myosin	Autoregulated state, the converter of one head jams into the actin binding cleft of the other	[73]
Tail	Myosin V	May bind somewhere near the upper lobe, slows ADP release in a Ca^{++}-dependent manner	[74, 75, 76]
Tail	Kinesin	Prevents microtubule-stimulated ADP release, structural basis unknown	[77]
Troponin/Tropomyosin	Skeletal Myosin	Competes for binding site on actin	[78]
ADP	Myosin VI	Extremely high ADP affinity causes the motor to spend most of its time in stall under cellular conditions	[79]
Calmodulin-like Regulatory Domain	Kinesin-like Calmodulin Binding Protein	Upon binding of Ca^{++}, calmodulin-like domain folds into microtubule-binding region, competing with microtuble binding	[80]
Non-Physiological Small-molecule Inhibitors			
Blebbistatin	Myosin II	Allosterically inhibits Pi release	[81]
Monastrol	Eg5 (kinesin family)	Allosterically blocks microtubule-stimulated ADP release	[82]
Rose Bengal	Kinesin	Competitive inhibitor of microtubule binding	[83]
Adociasulfate-2	Kinesin	Competitive inhibitor of microtubule binding	[84]

5.7 Motor Coordination in Cells: The Return of Simple Models?

Is there any room for the reductionist thinking that motor mechanism researchers are accustomed to in dealing with the problem of how cargo gets to its destination within cells? Despite the endless combinatorial array of motor regulatory and cargo-binding mechanisms, several experiments indicate that the answer is yes. While regulatory and cargo-binding mechanisms are complex and unique to each motor, they can be controlled by simple chemical signals on the cellular level. This has been clearly demonstrated in pigment cells. Treatment of these cells with a single chemical signal (such as adrenalin and caffeine, in the case of fish pigment cells) can induce full-scale aggregation or dispersion of pigment granules [94].

Kinesin, dynein, and myosin V have all been shown to move the same organelles in this system, and in fact they are all attached at once [93]. The attachment of myosin V can be regulated by Calmodulin kinase II, but kinesin and dynein do not readily dissociate from these organelles, once attached [95]. Do these two oppositely directed motors work against each other in a tug-of-war, or do they somehow coordinate their movements? Several experiments indicate that their movements may be coordinated. Movement of lipid droplets connected to both kinesin and dynein is inhibited in both directions by mutations that slow dynein's movement only, indicating that dynein assists kinesin's processivity [96]. The mechanism of cooperation between the two motors may involve dynactin, which has been shown to have a single binding site for both kinesin II and dynein in *Xenopus* melanophores [97].

Surprisingly, even at the level of single motor steps, kinesin and dynein motors on single organelles appear to cooperate. Kural and colleagues tracked the movement of peroxisomes *in vivo* with a precision of a few nanometers. They observed organelles with both kinesin and dynein attached, which made several discrete 8 nm steps in one direction, then reversed, and took several more steps. One cargo could reverse and take several unidirectional, 8 nm steps several times [98]. Purely random connections between motor and cargo would not predict this behavior because opposing motions of the two motors would lead to the observation of intermediate step sizes. This result demonstrates that not only do kinesin and dynein appear to cooperate, but their coordination is extremely rapid, such that an organelle can switch directions in milliseconds without any apparent pause in movement or tug-of-war between the two motors.

The mechanism of such a coordinator would require energy to switch instantly between kinesin and dynein motor movement, which perhaps could be harnessed "parasitically" from the ATPase activity of the two motors it controls. Alternatively, the coordinator could have ATPase or GTPase activity of its own, which would enable several mechanisms of external control. Regardless of how this part of the story turns out, it will be interesting to see how the vast array of cargo-binding and regulatory mechanisms can yield fairly simple and sensitive cellular responses to particular cues.

References

1. Vale, R & Milligan, R. (2000) *Science* **288**, 88–95.
2. Huxley, A & Niedergerke, R. (1954) *Nature* **173**, 971–3.
3. Huxley, H & Hanson, J. (1954) *Nature* **173**, 973–6.
4. Huxley, A. (1957) *Prog. Biophys. Biophys. Chem.* **7**, 255–318.
5. Huxley, H. (1969) *Science* **164(886)**, 1356–65.
6. Huxley, H. (2004) *Eur. J. Biochem.* **271**, 1403–15.
7. Eisenberg, E & Moos, C. (1968) *Biochem.* **7**, 1486–9.
8. Lymn, R & Taylor, E. (1971) *Biochem.* **10**, 4617–24.
9. Rayment, I, Holden, H, & et al. (1993) *Science* **261**, 58–65.
10. Rayment, I, Rypniewski, W, & et al. (1993) *Science* **261**, 50–8.
11. Fisher, A, Smith, C, & et al. (1995) *Biochem.* **34**, 8960–72.
12. Smith, C & Rayment, I. (1996) *Biochem.* **35**, 5404–17.
13. Gulick, A, Bauer, C, & et al. (1997) *Biochem.* **36**, 11619–28.
14. Miller, L, Coppedge, J, & et al. (1982) *Biochem. Biophys. Res. Commun.* **106**, 117–22.
15. Nitao, L & Reisler, E. (1998) *Biochem.* **37**, 16704–10.
16. Gulick, A & Rayment, I. (1997) *Bioessays* **19**, 561–9.
17. Holmes, K. (1997) *Curr. Biol.* **7**, R112–8.
18. Jontes, J, Wilson-Kubalek, E, & et al. (1995) *Nature* **378**, 751–3.
19. Whittaker, M, Wilson-Kubalek, E, & et al. (1995) *Neuron* **378**, 748–51.
20. Houdusse, A, Kalabokis, V, & et al. (1999) *Cell* **97**, 459–70.
21. Risal, D, Gourinath, S, & et al. (2004) *Proc. Natl. Acad. Sci. USA* **101**, 8930–5.
22. Kull, F, Sablin, E, & et al. (1996) *Nature* **380**, 550–5.
23. Sindelar, C, Budny, M, & et al. (2002) *Nat. Struct. Biol.* **9**, 844–8.
24. Dominguez, R, Freyzon, Y, & et al. (1998) *Cell* **94**, 559–71.
25. Himmel, D, Gourinath, S, & et al. (2002) *Proc. Natl. Acad. Sci. USA* **99**, 12645–50.
26. Chen, C, Saxena, A, & et al. (2006) *J. Biol. Chem.* **281**, 13777–83.
27. Andersen, G, Nissen, P, & et al. (2003) *Trends Biochem. Sci.* **28**, 434–41.
28. Toyoshima, C, Nomura, H, & et al. (2003) *FEBS Lett* **555**, 106–10.
29. Xing, X & Bell, C. (2004) *Biochem.* **43**, 16142–52.
30. Klenchin, V, Khaitlina, S, & et al. (2006) *J. Mol. Biol.* **362**, 140–50.
31. Romberg, L & Vale, R. (1993) *Nature* **361**, 168–70.
32. Ma, Y.-Z & Taylor, E. (1995) *Biochem.* **34**, 13242–51.
33. Vale, R. (1996) *J. Cell. Biol.* **135**, 291–302.
34. Kozielski, F, Sack, S, & et al. (1997) *Cell* **91**, 985–94.
35. Sack, S, Muller, J, & et al. (1997) *Biochem.* **36**, 16155–16165.
36. Rice, S, Lin, A. W, & et al. (1999) *Nature* **402**, 778–783.
37. Uyeda, T, Abramson, P, & et al. (1996) *Proc. Natl. Acad. Sci. USA* **93**, 4459–64.
38. Shima, T, Kon, T, & et al. (2006) *Proc. Natl. Acad. Sci. USA* **103**, 17736–40.
39. Burgess, S, Walker, M, & et al. (2003) *Nature* **421**, 715–8.
40. Kon, T, Mogami, T, & et al. (2005) *Nat. Struct. Mol. Biol.* **12**, 513–9.
41. Gibbons, I, Garbarino, J, & et al. (2005) *J. Biol. Chem.* **280**, 23960–5.
42. Case, R, Rice, S, & et al. (2000) *Curr. Biol.* **10**, 157–60.
43. Case, R, Pierce, D, & et al. (1997) *Cell* **90**, 959–66.
44. Homma, K, Yoshimura, M, & et al. (2001) *Nature* **412**, 831–4.
45. Kikkawa, M, Sablin, E, & et al. (2001) *Nature* **411**, 439–45.

46. Ross, J, Wallace, K, & et al. (2006) *Nat. Cell. Biol.* **8**, 562–70.
47. Mallik, R, Petrov, D, & et al. (2005) *Curr. Biol.* **15**, 2075–85.
48. Rief, M, Rock, R, & et al. (2000) *Proc. Natl. Acad. Sci. USA* **97**, 9482–6.
49. Schnitzer, M, Visscher, K, & et al. (2000) *Nature Cell. Biol.* **2**, 718–723.
50. Rock, R, Rice, S, & et al. (2001) *Proc. Natl. Acad. Sci. USA* **98**, 13655–13659.
51. Reck-Peterson, S, Yildiz, A, & et al. (2006) *Cell* **126**, 335–48.
52. Rice, S, Cui, Y, & et al. (2003) *Biophys. J.* **84**, 1844–54.
53. Hackney, D. (2005) *Proc. Natl. Acad. Sci. USA* **102**, 18338–43.
54. Sleep, J, Hackney, D, & et al. (1980) *J. Biol. Chem.* **255**, 4094–9.
55. Holzbaur, E & Johnson, K. (1986) *Biochem.* **25**, 428–34.
56. Holzbaur, E & Johnson, K. (1986) *Biochem.* **25**, 428–34.
57. Hirose, K, Akimaru, E, & et al. (2006) *Mol. Cell* **23**, 913–23.
58. Kikkawa, M & Hirokawa, N. (2006) *Embo J.* **25**, 4187–94.
59. De La Cruz, E & Ostap, E. (2004) *Curr. Opin. Cell. Biol.* **16**, 61–7.
60. Vale, R, Funatsu, T, & et al. (1996) *Nature* **380**, 451–453.
61. Pechatnikova, E & Taylor, E. (1997) *J. Biol. Chem.* **272**, 30735–40.
62. Foster, K & Gilbert, S. P. (2000) *Biochem.* **39**, 1784–91.
63. Robblee, J, Olivares, A, & et al. (2004) *J. Biol. Chem.* **279**, 38608–17.
64. Purcell, T, Sweeney, H, & et al. (2005) *Proc. Natl. Acad. Sci. USA* **102**, 13873–8.
65. Hackney, D. (1994) *Proc. Natl. Acad. Sci. USA* **91**, 6865–6869.
66. Ma, Y.-Z & Taylor, E. (1997) *J. Biol. Chem.* **272**, 724–730.
67. Klumpp, L, Hoenger, A, & et al. (2004) *Proc. Natl. Acad. Sci. USA* **101**, 3444–9.
68. Guydosh, N & Block, S. (2006) *Proc. Natl. Acad. Sci. USA* **103**, 8054–9.
69. De La Cruz, E, Ostap, E, & et al. (2001) *J. Biol. Chem.* **276**), 32373–81.
70. Yildiz, A, Forkey, J, & et al. (2003) *Science* **300**, 2061–5.
71. Yildiz, A, Tomishige, M, & et al. (2004) *Science* **303**, 676–8.
72. Endres, N, Yoshioka, C, & et al. (2006) *Nature* **439**, 875–8.
73. Liu, J, Wendt, T, & et al. (2003) *J. Mol. Biol.* **329**, 963–72.
74. Liu, J, Taylor, D, & et al. (2006) *Nature* **442**, 208–11.
75. Olivares, A, Chang, W, & et al. (2006) *J. Biol. Chem.* **281**, 31326–36.
76. Thirumurugan, K, Sakamoto, T, & et al. (2006) *Nature* **442**, 212–5.
77. Hackney, D & Stock, M. (2000) *Nature Cell. Biol.* **2**, 257–260.
78. Phillips, G. J, Fillers, J, & et al. (1986) *J. Mol. Biol.* **192**, 111–31.
79. Altman, D, Sweeney, H, & et al. (2004) *Cell* **116**, 737–49.
80. Vinogradova, M, Reddy, V, & et al. (2004) *J. Biol. Chem.* **279**, 23504–9.
81. Allingham, J, Smith, R, & et al. (2005) *Nat. Struct. Mol. Biol.* **12**, 378–9.
82. Yan, Y, Sardana, V, & et al. (2004) *J. Mol. Biol.* **335**, 547–54.
83. Hopkins, S, Vale, R. D, & et al. (2000) *Biochem.* **39**, 2805–14.
84. Sakowicz, R, Berdelis, M, & et al. (1998) *Science* **280**, 292–5.
85. Glater, E, Megeath, L, & et al. (2006) *J. Cell. Biol.* **173**, 545–57.
86. Verhey, K, Meyer, D, & et al. (2001) *J. Cell Biol.* **152**, 959–970.
87. Fukuda, M, Kuroda, T, & et al. (2002) *J. Biol. Chem.* **277**, 12432–6.
88. Kanai, Y, Dohmae, N, & et al. (2004) *Neuron* **43**, 513–25.
89. Ling, S, Fahrner, P, & et al. (2004) *Proc. Natl. Acad. Sci. USA* **101**, 17428–33.
90. Gyoeva, F, Bybikova, E, & et al. (2000) *J. Cell. Sci.* **113**, 2047–54.
91. Wozniak, M & Allan, V. J. (2006) *Embo J.* **25**, 5457–68.
92. Pashkova, N, Jin, Y, & et al. (2006) *Embo J.* **25**, 693–700.
93. Rogers, S & Gelfand, V. (1998) *Curr. Biol.* **8**, 161–4.
94. Nascimento, A, Roland, J, & et al. (2003) *Annu. Rev. Cell. Dev. Biol.* **19**, 469–91.

95. Rogers, S, Tint, I, & et al. (1997) *Proc. Natl. Acad. Sci. USA* **94**, 3720–5.
96. Gross, S, Welte, M, & et al. (2002) *J. Cell. Biol.* **156**, 715–24.
97. Deacon, S, Serpinskaya, A, & et al. (2003) *J. Cell. Biol.* **160**, 297–301.
98. Kural, C, Kim, H, & et al. (2005) *Science* **308**, 1469–72.

6

Molecular Mechanism of *Mycoplasma* Gliding - A Novel Cell Motility System

Makoto Miyata

Department of Biology, Graduate School of Science, Osaka City University
Sumiyoshi-ku, Osaka 558-8585 JAPAN, PRESTO, JST, JAPAN Tel:
+81-6-6605-3157; Fax: +81-6-6605-3158; E-mail:
miyata@sci.osaka-cu.ac.jp

Summary. More than ten *Mycoplasma* species from two independent groups bind to glass surfaces and exhibit movement while maintaining their binding, a process known as gliding motility. They form a small membrane protrusion at a cell pole and move in the direction of the protrusion. Genomic sequencing and analysis has revealed that the motility mechanism must differ from those of other bacteria and other motor protein systems. *M. mobile* glides at $2.0-4.5\mu m/s$. Its gliding machinery is composed of three very large proteins and one regular-sized protein around the cell membrane. Four hundred and fifty units of this machinery are located around the base of the protrusion, an area designated the "neck" and supported by a unique cytoskeletal structure. The gliding occurs based on the energy supplied from ATP hydrolysis on a solid surface coated by sialic acid. Gli349, the plausible leg protein, has a music-note shape that is 100nm in length and a suggested flexible 50nm protrusion. Our working model suggests that the flexible legs cyclically bind and release the sialic acid, and pull the cell body via conformational changes triggered by change in tension. *M. pneumoniae* glides at $0.3-1.0\mu m/s$. The protrusion, known as the attachment organelle, is a large structure into which more than 11 proteins are integrated. Recent studies by electron microscopy have revealed a detailed structure, comprised of i) a surface structure, ii) segmented paired plates, iii) a terminal button, iv) a wheel, and v) a translucent area. Structural studies suggest an inchworm model for gliding, whereas a power stroke model similar to that of *M. mobile* can explain the inhibitory effects caused by an antibody against the plausible leg protein, P1 adhesin.

6.1 Introduction

Mollicutes, including *Mycoplasmas, Spiroplasmas, Achoreplasmas*, and some others, constitute a class of parasitic or commensal bacteria featuring reduced genome sizes (560 to 2300 kbp) [1]. Phylogenetically, they belong to the high-AT branch of gram-positive bacteria, which also includes *Chlostridium* and *Bacillus* [2]. Unlike the cells of other bacterial groups, mollicute cells totally lack a peptidoglycan layer and are covered with membrane-anchored proteins, including antigenic variants [1]. Several *Mycoplasma* species form membrane protrusions. On solid surfaces, these species exhibit gliding motility in the direction of the protrusion; this motility is believed to be involved in the pathogenicity of *Mycoplasmas* [3, 4, 5]. Interestingly, *Mycoplasmas* have no surface flagella or pili, and their genomes contain no genes related to known bacterial motility. In addition, no homologs of conventional motor proteins that are common in eukaryotic motility have been found [6, 5]. This means that the mechanisms by which they glide are unknown. We have studied this mystery for almost ten years, mainly in relation to the fastest species, *Mycoplasma mobile* (abbreviated below as *M. mobile*), and suggest a working model for the mechanism, which may be applied to *Mycoplasma pneumoniae* (abbreviated below as *M. pneumoniae*).

6.2 *Mycoplasma* Gliding

6.2.1 How Do They Glide?

M. mobile, a pathogen of fresh water fish, was isolated from a sick gill organ of a tench (an aquarium fish). It showed prominent gliding activity and gave us the chance to study the gliding mechanism. Figure 6.1 shows the gliding motility of *M. mobile*. Color was added to frames captured in a 4s video at 0.1 s intervals, and the frames were integrated into a single image [7]. It is difficult to convey how rapidly these organisms move using a static image. To gain an appreciation of their speed, visit *http://www.sci.osaka-cu.ac.jp/ miyata/myco1.htm* and watch the film. In one second, *Mycoplasmas* travel a distance of $2.0 - 4.5\mu$m, or 3-7 times their length [8]. As shown in Figure 6.2, the cell is flask-shaped. The protrusion has been called a "head" [9, 10]. Gliding always occurs in the direction of the head, without halting or reverse movement. The gliding movement occurs only on solid surfaces - *Mycoplasmas* show Brownian motion when they detach from a solid surface. It is easy to observe the gliding of *M. mobile*, because all cells are always moving if they are alive.

M. pneumoniae, a cause of human "walking pneumonia", also has a membrane protrusion at one pole of the cell and exhibits gliding motility. The motility of this species is dependent on the growth conditions and the stage of the culture, and can be affected by damage caused by centrifugation during the collection of cells. Its gliding speed reaches $0.3 - 1\mu$m/s, much slower than that of *M. mobile* [12, 13, 14, 15] (see Figure 6.3). The movement occurs in the direction of the conical protrusion formed at one cell pole, as observed in *M. mobile* (see Figure 6.4). This conical structure is called the "attachment organelle" or the "tip structure" [16].

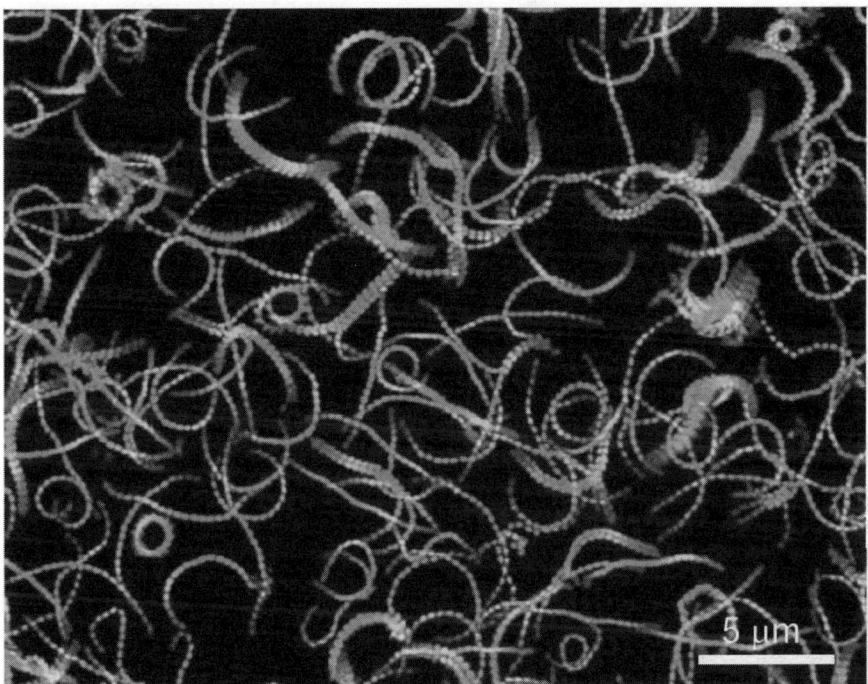

Figure 6.1. Gliding of *M. mobile*. Selected frames from 4s of footage were differently colored from purple to red and integrated into one image [7, 11]. The traces in the image show gliding speed, direction of gliding, and time relations with other traces. Reprinted from [11]. Copyright 2005 National Academy of Sciences, U.S.A. (See color insert.)

6.2.2 Phylogenetic Relationship of Gliding *Mycoplasmas* with Other Surface Motile Organisms

The gliding motility of *Mycoplasmas* may remind researchers working on eukaryotic motility of the motility of an amoeba or nematode sperm, because they move on solid surfaces such as glass and plastics. However, *Mycoplasma* gliding is unlikely to share structures or mechanisms with eukaryotic motility, because *Mycoplasmas* are phylogenetically very distant from eukaryotes.

What about the relationship with the surface motility of other bacteria? The surface motility of bacteria was first discovered more than 130 years ago and has been reported for a wide variety of species [17]. The motility of most of these species has been lumped together as "gliding motility", and the mechanisms responsible have remained unclear. However, recent studies using new experimental techniques have led to the classification of gliding motility into three groups, excluding *Mycoplasmas* (see Figure 6.5) [18, 17, 19, 20, 21]. The first, pili motility, includes the twitching motility of *Pseudomonas aeruginosa*, the social (gliding) motility of *Myxococcus xhansus*, and the gliding motility of some species of cyanobacteria [18, 20]. These bacteria extend Type 4 pili

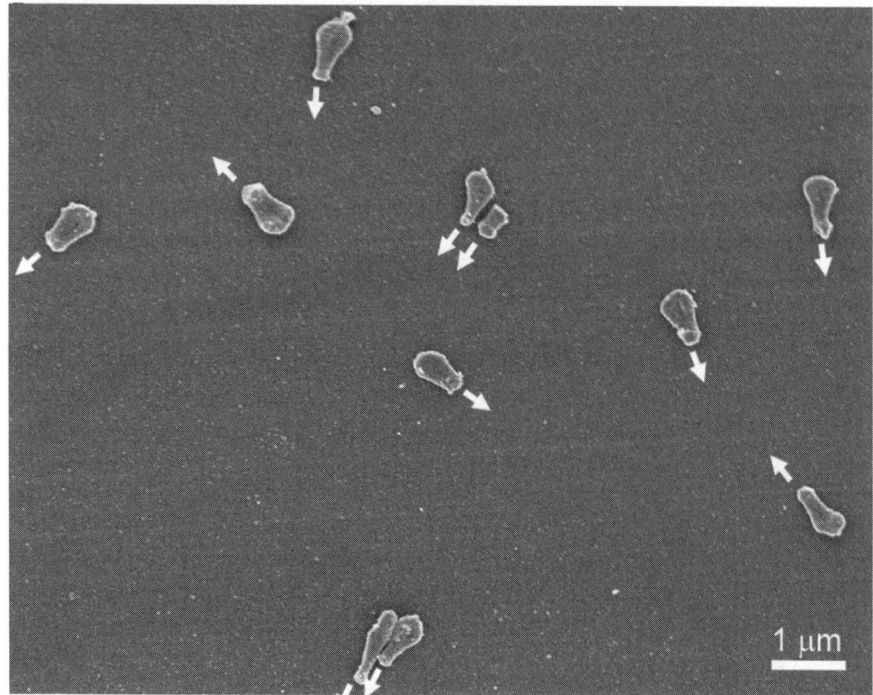

1 μm

Figure 6.2. Scanning electron micrograph of *M. mobile*. The cells are gliding in the direction indicated by arrows.

at a cell pole, attach to a solid surface by the end of the pili, and then retract the pili while maintaining the attachment. Thus, the cells move by retraction relative to the surface. The second group of gliding motilities, slime motility, includes the gliding motility of cyanobacteria living in a connected form as a filament, and the adventurous (gliding) motility of *Myxococcus xhansus* [19, 21]. Here the bacteria secrete polysaccharide through slime nozzles at a cell pole and move through a reaction against the hydration of the secreted polymer. However, another hypothesis was suggested recently [22]. The third group comprises the gliding of the *Cytophaga-Flavobacterium*-group, the mechanism of which has not yet been well clarified [17, 23]. Possible mechanisms include the flow of a secreted polysaccharide on the cell surface or the lateral movements of outer membrane components. To date, 12 species of *Mycoplasmas* in more than 200 species have been reported to glide: *M. mobile*, *M. pulmonis*, *M. agassizii*, *M. testudineum*, *M. pneumoniae*, *M. testudimis*, *M. amphoriforme*, *M. pirum*, *M. gallisepticum*, *M. imitans*, *M. genitalium*, and *M. penetrans* [24, 25, 4]. Phylogenetically, they are classified into two groups; the first four species belong to the hominis group represented by *M. mobile*, and the latter eight belong to the *pneumoniae* group [26, 2]. Those groups are distantly located in the phylogenetic tree; no homologs of proteins

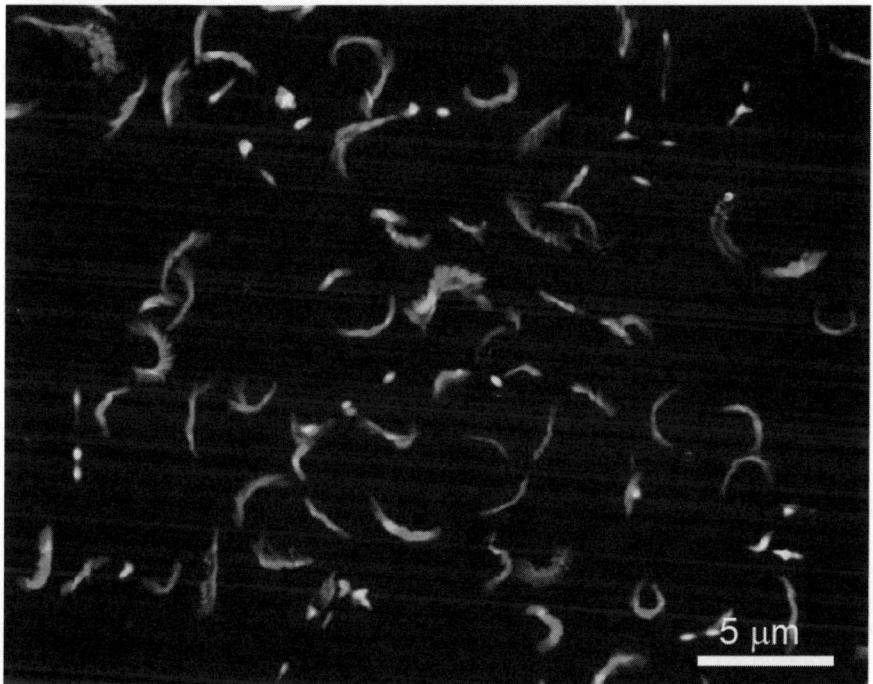

Figure 6.3. Gliding of *M. pneumoniae*. Selected frames from a 10s video were treated in the same way as those in Figure 6.1 [7, 15]. (See color insert.)

involved in the gliding mechanism of *M. mobile* are present in the genome of *M. pneumoniae* [6, 27, 28]. Therefore, some researchers have suggested that these groups glide with different mechanisms [6, 29]. What is the definition of "different" mechanism? Does it mean that they originated from unrelated systems? I guess that twice of establishment of surface motility is not easy in the relatively short history of *Mycoplasma* evolution. At the same time, if the gliding motility is determinant for natural selection of *Mycoplasmas*, even such extensive evolution might be possible. The answer may be given when the details of gliding mechanisms are elucidated for both subgroups.

6.2.3 Purpose and Pathogenicity

Why do *Mycoplasma* species glide? So far, eight different mechanisms of active bacterial motility have been reported: flagella swimming, synechococcus swimming, pili motility, slime motility, *Cytophaga* gliding, *Spiroplasma* swimming, and *Mycoplasma* gliding. Other motilities than those of *Mycoplasma* and *Spiroplasma* are all known to be controlled by "two-component systems", usually exhibiting chemotaxis [31]. Generally, bacteria move to access nutrients and escape from wastes and predators. However, *Mycoplasma* gliding

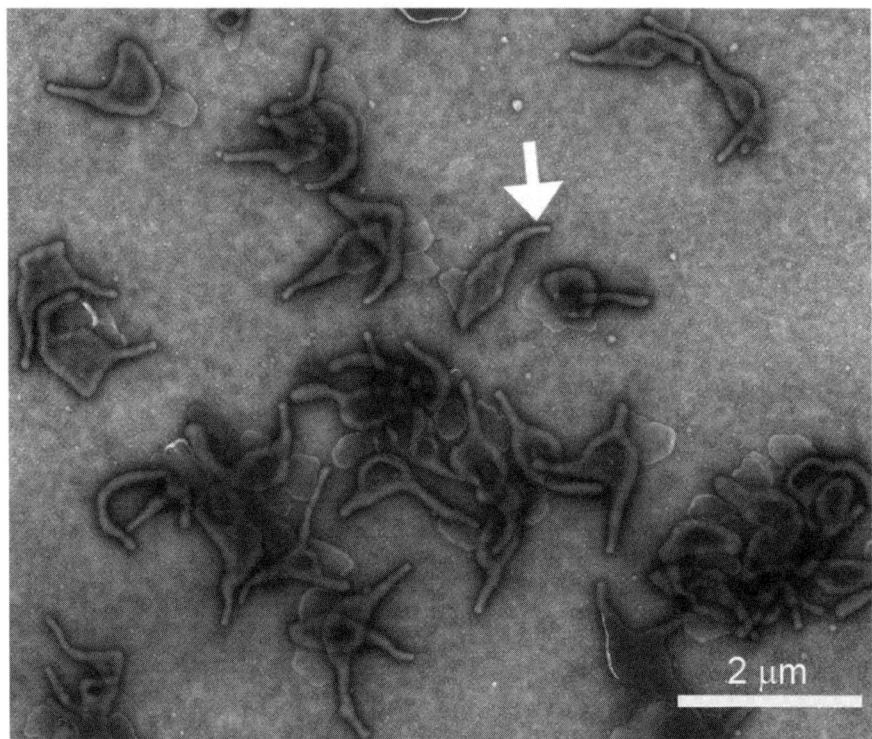

Figure 6.4. Electron micrograph of negatively stained *M. pneumoniae*. The attachment organelle of a cell is pointed by an arrow.

does not show any obvious chemotaxis [4, 32, 5]. Granted, this failure to detect chemotaxis may be caused by the experimental conditions, because the environments of *Mycoplasmas* in nature are distinct from those of other motile bacteria, which may be reflected to the remarkable reduction of genome sizes in *Mycoplasma* species [1]. Thus, we cannot rule out the possibility of chemotaxis in *Mycoplasmas*. Still, if they have chemotaxis, it is caused by a novel mechanism different from that of other bacteria, just as gliding motility in *Mycoplasmas* must be caused by different mechanisms because no homologs of two-component system genes can be found in their genomes [33, 34, 35, 6, 36, 37]. Furthermore, the physical conditions of *Mycoplasmas* in their hosts require motion for the sake of motion; even random movements are sufficient. Generally, gliding *Mycoplasmas* bind to animal tissues tightly and cluster. If they stay in the same position without movement, they will use up the nutrients and pollute their environment with their wastes. Random movements will enable them to get better conditions for propagation. In the case of *M. mobile*, the cells show rheotaxis, by which the organism moves upstream of liquid flow as usually observed in fish [38]. *M. mobile* responds to a flow of $100 - 300 \mu m/s$ with a time lag of 1-3 s and exhibits a change of direction and

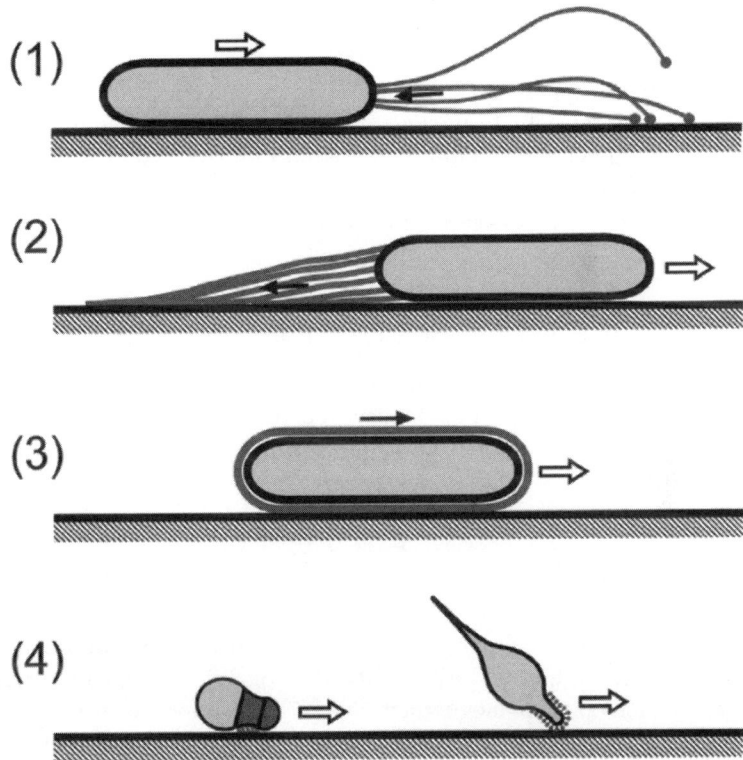

Figure 6.5. Mechanisms for the various gliding motilities of bacteria. (1) Pili motility, (2) Slime motility, (3) *Flavobacterium* motility, and (4) *Mycoplasma* gliding are schematically presented. Reprinted from [30]. (Copyright 2005 Kyoritsu Shuppan.)

28% acceleration of speed. *M. mobile* is a pathogen isolated from the gills of a fresh water fish [9, 10, 39, 40, 41]; thus, this response may help to prevent the organism from being swept out of the gills. The change in the gliding direction could also be explained by physical mechanisms, specifically the subcellular position of the adhesion protein (see Section 6.4.2) [42, 43, 27] and the fact that gliding always occurs in the direction of the head. The head of the *Mycoplasma* may be turned around by the force of the liquid flow, resulting in upstream movement. This phenomenon can be observed more clearly by attaching a bead to the tail of a *Mycoplasma* placed in a liquid flow [44]. Simple physical explanations, however, cannot account for the acceleration seen [5].

Cytadherence of *Mycoplasmas*, linked to gliding motility, is well-known to be involved in parasiticity and pathogenicity [16, 1]. When a *Mycoplasma* strain loses cytadherence, it is easily removed from the tissue by the host animal. It is less clear how gliding affects or contributes to pathogenicity. Many *Mycoplasmas* confirmed for gliding motility are pathogenic [4, 1, 39, 40, 41].

However, there have been few studies focusing on the relationship between gliding and pathogenicity; it is difficult to separate the effects of cytadherence from those of gliding because no cytadherence-positive and gliding-negative strains have been isolated. D. Krause's group at Georgia University constructed such a strain of *M. pneumoniae*, and examined the infection by the strain in cultured cells that had differentiated into tracheal cells. The results showed that gliding motility is essential to translocation of *Mycoplasmas* to the cell surface after they attach to cilia tip [45, 46].

6.3 Studies on Mechanism of *M. mobile* Gliding

M. mobile, isolated from lesions on the gills of freshwater fish, does not have direct clinical or industrial importance. Therefore, only a few people were working on it up to 1997, and knowledge of its genetic system remains scarce even today. However, its remarkable gliding properties are attracting an increasing number of researchers.

6.3.1 Gliding Mutants

To identify the molecules involved in the gliding mechanism, we isolated mutants defective in gliding [47]. Generally, the colony morphology of microorganisms depends on their moving activity [17]. Therefore, to isolate mutants defective in gliding, we used a variety of agar concentrations in the culture media. *M. mobile* formed a typical "fried-egg" shape colony in media containing greater than 0.5% agar. However, the colony shape changed in media containing less than 0.5% agar. Under such conditions, the colonies tended to be composed of small clusters; in 0.1% agar, they were reduced to a stack of small clusters. Presumably, *Mycoplasmas* occasionally glide on the filamentous network of agar and form small blocks of colonies. We exposed *M. mobile* to ultraviolet irradiation and screened for nongliding mutants by examining colony morphology in 0.1% agar. We obtained ten mutants with altered gliding and binding characteristics. One group, which consisted of five strains, did not bind to glass, and consequently did not glide. Many of their cells had irregular shapes. All of these mutants had mutations in the open reading frames (ORFs) of the very large proteins that have been shown to be involved in the gliding mechanisms. Another group, consisting of four strains, had a 20% greater gliding speed and a shorter cell axis than that of the wild-type strain. We have not identified the mutation points of this group, nor can we explain why these gliding strains show a colony morphology similar to that of nonmotile mutants. Perhaps, the colony morphology may also be related to some chemotactic activity that has not yet been discovered.

6.3.2 Antibodies Inhibiting Gliding Motility

We adopted another strategy to identify the proteins involved in the mechanism. Because some gliding reactions occur at the interface of the *Mycoplasma* cell and a solid surface, *Mycoplasma* should expose at least some parts of its gliding machinery. Therefore, the reaction is expected to be inhibited by the binding of antibodies from outside the cell. Mice were immunized with whole *M. mobile* cells or the TritonX-100 extracted structure, and hybridoma were constructed from the spleen cells [42, 48]. The hybridoma were screened for the production of an antibody that specifically inhibits gliding motility. We isolated several inhibitory antibodies which could be classified into two groups. The first group removed the gliding cells from the glass, while the second group stopped the gliding and kept the cells on the solid surface.

6.3.3 Identification of Gliding Proteins

We initially identified two identical gliding proteins using nongliding and nonbinding mutants, as well as inhibitory monoclonal antibodies [42, 48, 27]. Briefly, the target proteins of the inhibitory antibodies were found to be lost in two nonbinding mutants based on immunoblotting analysis. The corresponding protein at the 220 kDa position in SDS-PAGE was excised from the gel and its internal amino acid sequences were determined. The DNA fragment coding the protein was amplified by a polymerase chain reaction (PCR) using degenerated oligonucleotide DNA, and sequenced. Sequencing of an approximately 30kb region, including the flanking regions showed that four ORFs are coded tandemly on the genome (see Figure 6.6). Three of the four ORFs were predicted to code proteins of extraordinary molecular mass, namely 123k, 349k, and 521k. Because the very high molecular weights of these proteins are an obvious feature, we named these gliding genes *gli123*, *gli349*, and *gil521*. We found that the target protein of the second group of antibodies is coded by *gli521*. The features of the proteins are summarized in Table 6.1. Detailed analyses of protein profiles of three other mutants revealed that a protein band of about 123 kDa is missing in one of the nonbinding mutants [28]. We made a polyclonal antibody against protein fragments from *gli123* expressed in *Escherichia coli*, and confirmed that the protein missing in the mutant is coded by *gli123*. Sequencing the 30kb region of genomes of the nonbinding and nongliding mutants revealed that each of the three mutants has a nonsense mutation at one of the ORFs; another mutant has a single amino acid substitution from serine to leucine at the 2770th amino acid of *gli349*, and the other has an insertion of four amino acids in *gli521*. The amino acid sequences of these three proteins did not have any significant similarities with any proteins of known function, and did not contain any motifs suggesting protein functions. Moreover, no similarity was found with the amino acid sequence of the P1 adhesin of *M. pneumoniae*, with the exception that neither it, nor any of the three proteins, contain a cysteine residue. Orthologs were found in a similar arrangement in the genome of a *Mycoplasma pulmonis*, a gliding *Mycoplasma* that is closely related to *M. mobile* [6, 27, 28], but not in any other species. The nonexistence of obvious orthologs in other than closely related species may suggest variation in the structures of gliding machinery. It may also suggest that the number of proteins involved in *Mycoplasma* gliding is quite small compared to that of other bacterial motility systems such as flagella and pili, possibly because a large number of proteins and interactions would hamper evolutionary change in the components.

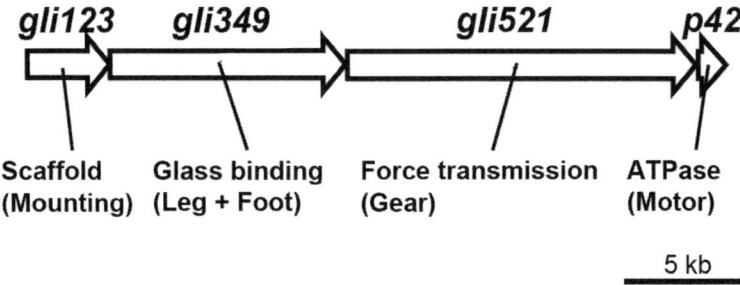

Figure 6.6. Gene locus of *M. mobile* coding gliding proteins. Four proteins involved in the gliding mechanism are coded tandemly on the genome. The typical role of each protein is presented in the parentheses.

6.3.4 Where is the Machinery?

To localize the gliding machinery we attached a small bead to the surface of the *Mycoplasma* and traced its movement during gliding [43]. The bead was carried by the *Mycoplasma*, remaining bound to the same position on the cell, which did not rotate on its axis. After prolonged culture, the heads of *M. mobile* tend to elongate. When cultured for an additional 24 hours after cell numbers have reached their maximum, some of the cells that were still gliding had an elongated head. Their movement suggested that the head dragged the remainder of the organism (here referred to as the "body") and sometimes the body became stuck to something on the glass and the head was stretched. This suggests that the force for gliding is generated around the head.

Localization of the gliding machinery was achieved by immunofluorescence microscopy with a procedure modified to fit to *Mycoplasmas* [42, 49, 27]. The results from all antibodies against Gli123, Gli349, and Gli521 proteins showed that the gliding machinery localizes at the base of the membrane protrusion (see Figure 6.7). As the protrusion has already been called the "head", we

Table 6.1. Features of gliding proteins.

Protein	Gli123	Gli349	Gli521	P42
ORF code	MMOB1020	MMOB1030	MMOB1040	MMOB1050
Amino acid residues	1,128	3,181	4,684 (4,728)^a	356
Predicted MW	123,318	348,516	515,863 (520,568)	42,003
pI	5.1	4.9	5.2	9.6
Amino acid content	Cys(0%) Asn(14.8%)	Cys(0%) Asn(12.0%)	Cys(0%) Asn(13.5%)	Cys (0%)
Transmembrane segment	near N-terminus	near N-terminus	near C-terminus^b	none
Motifs		none		
Subcellular localization		neck		ND^c
Number on a cell	450	450	450	ND
Deficiency of mutant		non-binding, non-gliding		ND
Function	machinery formation	glass binding	force transmission	ATPase

^{a}: Edman degradation analysis of the amino-terminal sequence has suggested cleavage of this protein between residues 43 and 44.

^{b}: A second transmembrane segment is predicted near the amino-terminus in the peptide cleaved from the mature Gli521 protein.

^{c}: Not determined.

named the gliding machinery part the "neck". We labeled Gli349 and Gli521 molecules on fixed and negatively stained cells with gold particles conjugated to the antibody, and examined the cells three-dimensionally using electron microscopy (EM). Our conclusion was that these molecules are distributed evenly around the neck [48, 27]. We also obtained three classes of uninhibitory monoclonal antibodies which specifically recognize surface proteins. Immunofluorescence microscopy using these antibodies revealed that the surface proteins are localized to the head, neck, body, and head-body [42]. The targets of these antibodies, MvspN, MvspO, MvspI, and MvspK proteins, are coded as gene clusters composed of 16 homologs on the genome [6] (where Mvsp stands for mobile variable surface protein). These proteins are suggested to be responsible for antigenic variation, based on the following facts: i) Antibodies against these proteins were preferentially raised, when whole *Mycoplasma* cells were injected into mice; ii) The amino acid sequences feature certain short repeats identical to those observed in *Mycoplasma* antigenic proteins; and iii) The amino acid sequences suggest membrane anchoring via the N-terminal transmembrane (TM) segment, or the lipid attached to the N-terminus. These observations show that the surface is comprised of three parts, and the neck is specialized for gliding and cytadherence.

6.3.5 Architecture Around the Gliding Machinery

What does this machinery look like? The careful observation of a negatively stained cell surface under EM showed that the neck surface is covered with

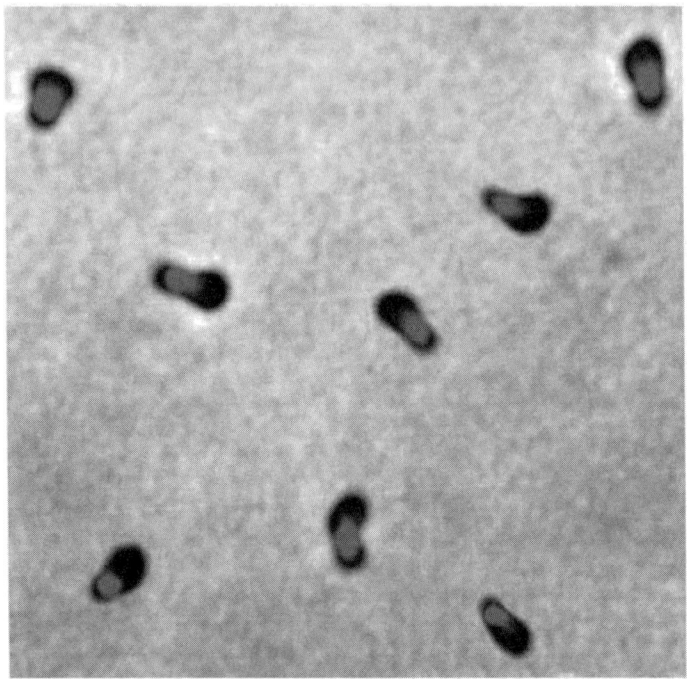

1μm

Figure 6.7. Subcellular localization of gliding proteins in *M. mobile*. The Gli349 protein is labeled red by immunofluorescence microscopy. Gli123 and Gli521 are also localized at the same position.

many thin filaments. Considering that gliding proteins are localized exclusively at the neck, the filaments should be a part of the machinery. As the molecular shape of isolated Gli349 is filamentous (see Section 6.3.6) and because Gli349 is highly accessible to the antibody from the outside, the filamentous structure on the surface may be composed of this protein. When cells were treated and extracted by TritonX-100, a jellyfish-like structure emerged, where a latticed solid "bell" part is connected with dozens of flexible "tentacles". These filaments are attached with ring-like particles about 20nm in diameter (see upper part of Figure 6.9). The filaments in the jellyfish structure were obviously disturbed in mutants lacking each of gliding proteins, suggesting that this protein is involved in the structure or the formation process of jellyfish. Recently, actin and tubulin homologs (MreB and FtsZ, respectively) have been found to be conserved widely in bacteria [50]. Generally, MreB is involved in cell shape maintenance and chromosome segregation, and FtsZ is responsible for cytokinesis. However, there are no genes coding these proteins on the genome of *M. mobile* [6]. Hence, the jellyfish structure should be composed of cytoskeletal proteins specific to *Mycoplasmas*. To analyze the interaction between a cell and a solid surface, *Mycoplasmas* gliding on a mica plate were chemically fixed, embedded into resin, sectioned, and observed by EM. Structures connecting the cell and the mica plate were faintly visible, but the distance between them was estimated to be around 25nm.

Visualization of the connecting structure was achieved by employing "rapid freeze/fracture/deep etch/replica" EM, which may be one of the best methods for visualizing fragile and fast-moving structures [51]. We fast-froze *M. mobile* cells gliding on glass in a matter of milliseconds, prepared a replica, and examined the replica using EM. In the fracturing process, the knife generally tends to travel along the interface between the outer and inner leaflets of a cell membrane. Thus, gliding *Mycoplasmas* may remain on the glass (as shown in Figure 6.8A) with the structure that might be responsible for gliding. We looked for and found such structures, i.e., structures responsible for gliding, as shown in Figure 6.8B. In this figure, the outer leaflet of the lower cell membrane can be seen at the top. Leg-like structures about 50nm long appear to be sticking out from the neck, extending to the solid surface, and attaching to the surface at their distal end. The angle of the leg structure to the cell axis varied widely. These observations suggest that these structures might pull the cell forward by repeatedly binding to, and releasing from, the glass surface. Based on the predicted function, appearance, and dimensions, we assume that the leg is mainly composed of the Gli349 protein. The cell architecture is schematically summarized in Figure 6.9.

6.3.6 Gli349 Protein

The Gli349 protein is composed of 3181 amino acids and features a transmembrane segment near the N-terminus [27]. Because this protein was identified first among the four proteins involved in gliding, it is rather well characterized. A monoclonal antibody against Gli349 reduces gliding speed and finally removes the cells from the glass. Therefore, we believe that Gli349 is responsible for glass binding during gliding and should convey movement. Gli349 forms a broad band in SDS-PAGE (polyacrylamide gel electrophoresis) gel, and is sensitive to protease treatment, suggesting the flexible character of the molecule [52].

The protein was isolated from *Mycoplasma* cells and its structure was analyzed [53]. Gel filtration analysis showed that the isolated Gli349 protein is monomeric. Rotary shadowing EM revealed that the molecules are about 100nm long and resemble the symbol for an eighth note in music (see Figure 6.10A). They contain an oval part 14nm long at a pole, which has been named the "foot". From this foot extend three rods, in tandem, of 43, 20, and 20 nm, in that order (see Figure 6.10C). The hinge connecting the first and second rods is flexible, while the next hinge has a distinct preference in its angle, about 90 degrees. Analyses using a monoclonal antibody suggested that the foot corresponds to the C-terminal region. To determine the molecular shape

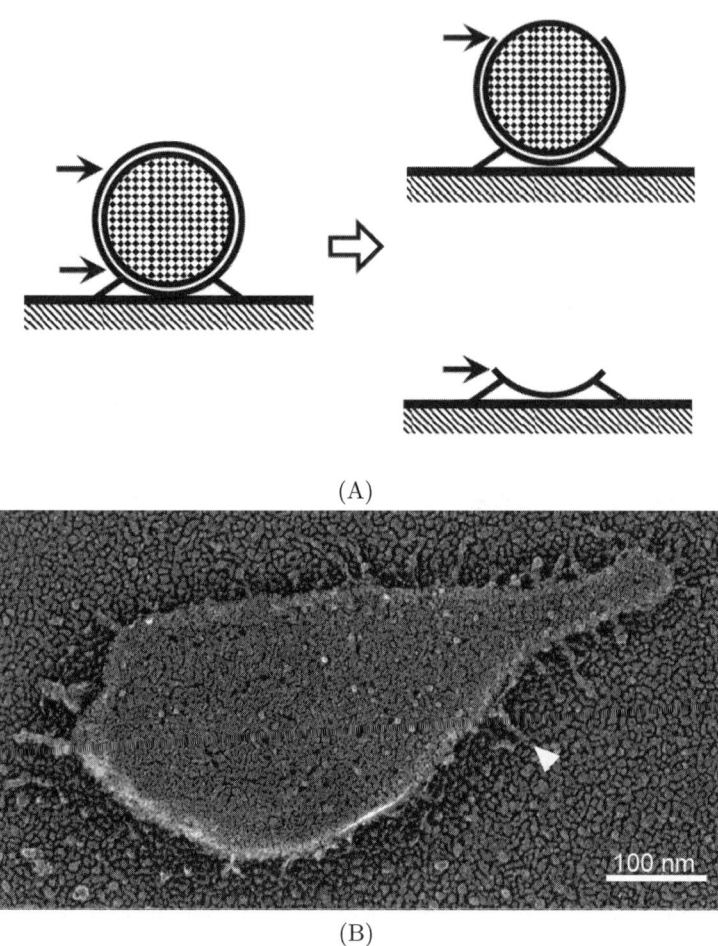

(A)

(B)

Figure 6.8. Freeze-fracture EM of *M. mobile*. (A) Schematic diagram of freeze-fracture faces. Left: *Mycoplasma* cell attached to the glass surface prior to fracture. Outer and inner leaflets of the cell membrane, cytoplasm, glass and legs at the lower surface of a cell are represented. Fracture of cells mostly occurred at two points, shown by the black arrows. Right top: (P-face) Outer view of inner leaflet of the cell membrane. Right bottom: (E-face) Inner view of outer leaflet of the membrane. (B) A typical image of the E-face of a gliding *Mycoplasma*. The head is at the right. The outer leaflet of the membrane on the lower side of the cell is viewed from the top. The upper side of the cell, the cytoplasm, and the inner leaflet of the membrane on the lower surface of the cell have been removed by fracturing. The leg-like structure is indicated by the triangle.

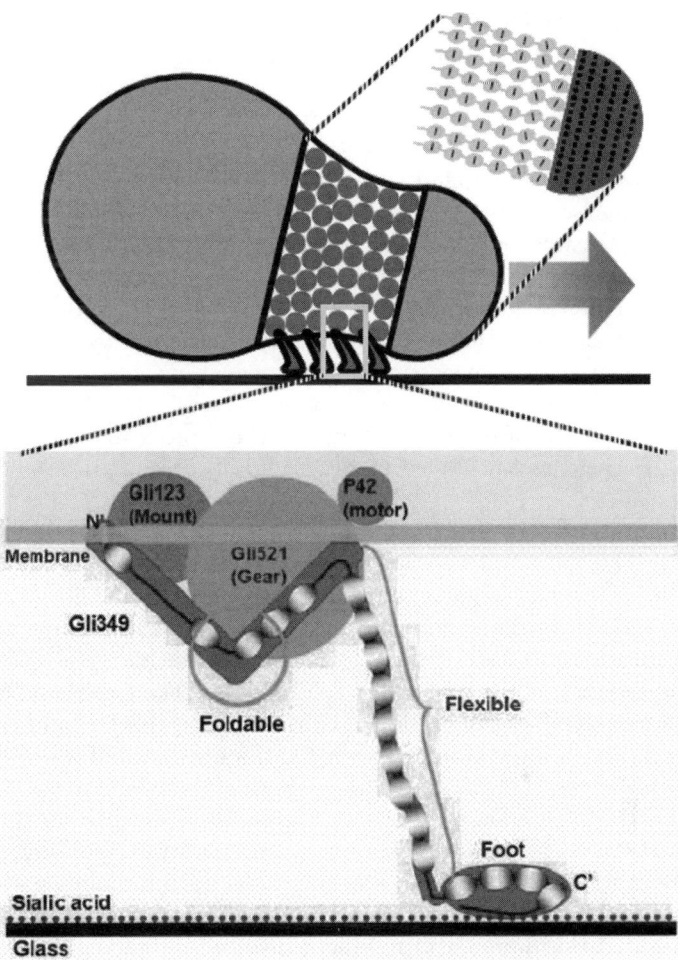

Figure 6.9. Cell architecture of *M. mobile*. Four proteins involved in gliding mechanisms are localized near the cell membrane, and aligned at the cell neck forming 450 units per cell. Gli349, the plausible leg protein, sticks out from the C-terminal region; its distal end can bind to sialic acid fixed on the solid surface. The gliding units are supported from inside the cell by the jellyfish-like structure recently identified, as presented in the top illustration. The jellyfish structure is composed of latticed solid "head" part and flexible strings attached with particles about 20nm in diameter. (See color insert.)

in water, atomic force microscopy (AFM) was applied (see Figure 6.10D). The molecules on mica showed a morphology in which a large globule about 25nm in diameter is connected to a smaller one with a filament. Because the images obtained from AFM generally reflect the stiffness of the molecular parts, the images suggested that the connecting filament is very flexible. We assigned the parts of the schematic molecular shape found by the shape and dimension data of EM, to the images found by AFM.

Detailed analyses of the amino acid sequence were carried out with the sequence of a plausible ortholog of *M. pulmonis*, MYPU2110 [52]. Gli349 and MYPU2110 were found to contain, respectively, 18 and 22 repeats of about 100 amino acid residues each (see Figure 6.10C). No sequence homologous to any of the repeats was found and no compatible fold structure was found among known protein structures, suggesting that the repeat in these proteins is novel and takes a new folding structure. Proteolytic analyses of Gli349 revealed that cleavage positions are often located between the repeats, implying that the regions connecting the repeats are unstructured, flexible, and exposed to the environment [52].

The information about the Gli349 molecule is summarized in Figure 6.10C [53, 52, 27]. The short rod end is suggested to be the N-terminus, where a transmembrane segment is predicted. Two shorter rods should lie near the cell membrane, and the longer flexible rod sticks out from the membrane when the molecule is bonded to a solid surface. This part may lie along the cell surface when the molecule is not bound to the solid surface. The foot is composed of the domain near the C-terminus domain and is responsible for binding to the solid surfaces. We isolated two monoclonal antibodies, each of which removes gliding *Mycoplasmas* from the solid surface. Both of them bind to the positions near the flexible hinge marked *theta*$_2$ in Figure 6.10C, suggesting that these domains perform conformational changes. A mutant strain containing Gli349 with a substitution from serine to leucine at the 2770th amino acid cannot bind to solid surfaces, and consequently cannot glide, suggesting that the structure around this amino acid residue is involved in binding.

6.3.7 Gli521 Protein

Gli521 is coded as an ORF of 4728 amino acids featuring transmembrane segments at both ends, of which the N-terminus one is processed [48]. This protein was identified as the target of an inhibitory antibody that removes gliding *Mycoplasmas* from glass.

When the antibodies against this protein are added to gliding *Mycoplasmas*, some *Mycoplasmas* are removed from the glass but some remain. The remaining ratio increases to near 100% with the concentration of antibodies used. Therefore, we believe that the role of this protein is the motor or the gear-transmitting force from the motor to the leg. Identification of ATPase activity in P42, coded downstream of Gli521 on the genome, suggests that the role of Gli521 is the transmission of force from the motor to the leg, i.e., a

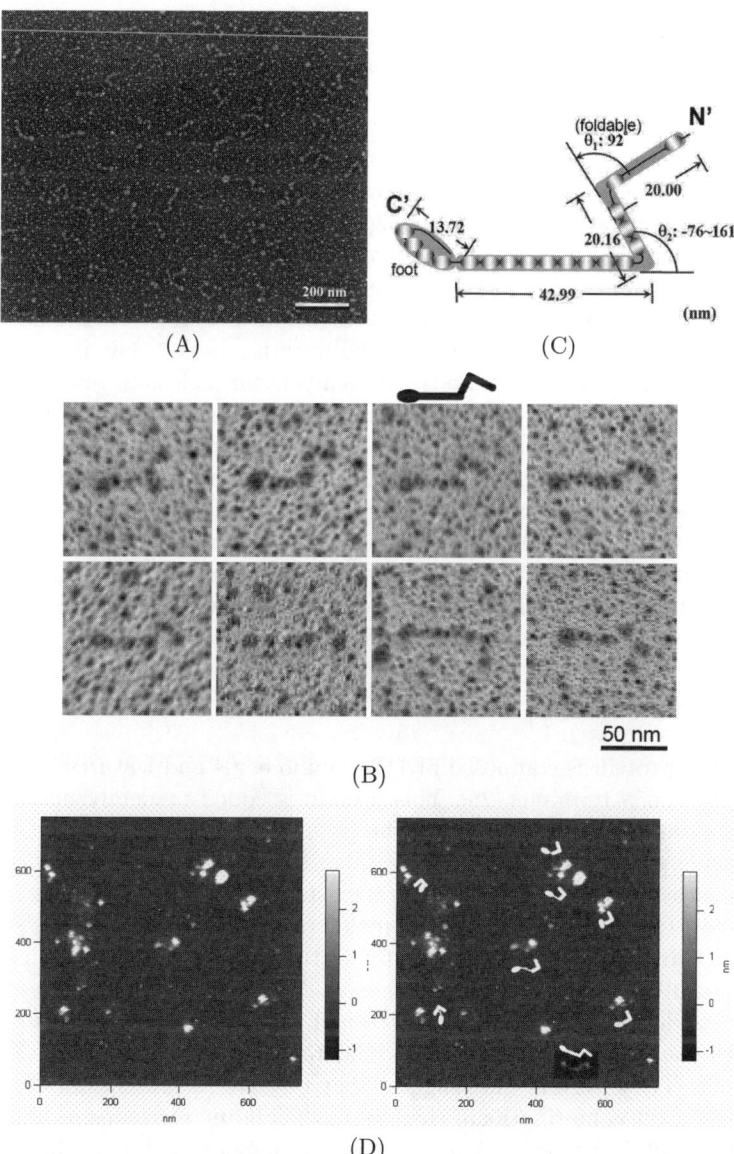

Figure 6.10. Molecular shapes of the Gli349 protein. (A) Field image of rotary shadowing EM of an isolated protein. (B) Typical molecular images. The images feature a globular end designated as the "foot" and two hinges. A schematic diagram of the molecule is presented at the top. (C) Summary of statistical analyses of EM images. (D) AFM image in a liquid condition. Molecules in 10mM ammonium acetate were fixed to a N-hydroxysuccinimide (NHS)-activated substrate and scanned by AsylumReseach MFP-3D AFM equipped with a probe, RC150VB (OLYMPUS). In the right panel, an image from a dry condition is presented as an inset. Each part of the AFM images is assigned to a part of the molecular shape summarized from the rotary shadowing EM, as indicated in the right panel.

"gear" (see Section 6.3.11). Because the mutant lacking Gli521 does not bind to solid surfaces and shows a reduced amount of Gli349, another presumable role of Gli521 is to provide scaffolding for Gli349.

The Gli521 protein was isolated from *Mycoplasma* culture and characterized. The protein was suggested to behave basically as a monomer. Rotary shadowing EM revealed that this protein molecule is triad-shaped, like clathrin [54], and that its arms are 130nm long. Although Gli521 does not show any similarities with clathrin, it may also form basket-like structures, and support the cell shape and other gliding proteins. Gli521 also shows a tendency to form two-dimensional sheets with spheres 13nm in diameter. The structure of a sheet is obviously different from the putative basket, but its formation may be caused by the interactions between protein molecules responsible for the formation of the basket. Partial digestion by protease treatments showed that this protein is divided into three domains. Each of these domains may correspond to the arms of the triad. TM segments are predicted from amino acid sequences at the N- and C-terminal domains, and the N-terminal TM is actually cleaved by processing between residues 43 and 44 [48]. The binding sites of inhibitory antibodies were mapped in the region spanning from amino acids 3736 to 4020, suggesting that this region faces to the outside and obeys conformational changes in the gliding mechanism.

6.3.8 Gli123 Protein

The Gli123 protein is composed of 1128 amino acids and features a TM segment near the N-terminus [28]. This protein is coded tandemly upstream of the Gli349 and Gli521 proteins on the genome (see Figure 6.6). In a strain lacking this protein, the localization and stability of Gli349 and Gli521 proteins are significantly disturbed. Gli123 exists at the same cellular position as Gli349 and Gli521, at an equivalent molar ratio. These observations suggest that this protein provides binding sites in the cell for other gliding proteins and forms a complex of gliding machinery. We made anti-Gli123 antibodies from a few dozen animals, but failed to make antibodies that inhibit gliding, although the produced antibodies bound to the molecule from the outside of the cell. These results may suggest that the Gli123 protein does not perform significant conformational changes in the gliding mechanism, unlike the other two gliding proteins. Alternatively, the moving part is occluded from the outside.

6.3.9 Interaction Between Gliding Proteins

Their localization at the same position in the cell, the identity of their molecular ratio, and the tandem gene organization suggest that the gliding proteins bind one another and form a complex. However, they do not show obvious interactions in purification processes or in experiments to detect molecular interactions, such as surface plasmon resonance (SPR). Their interaction may

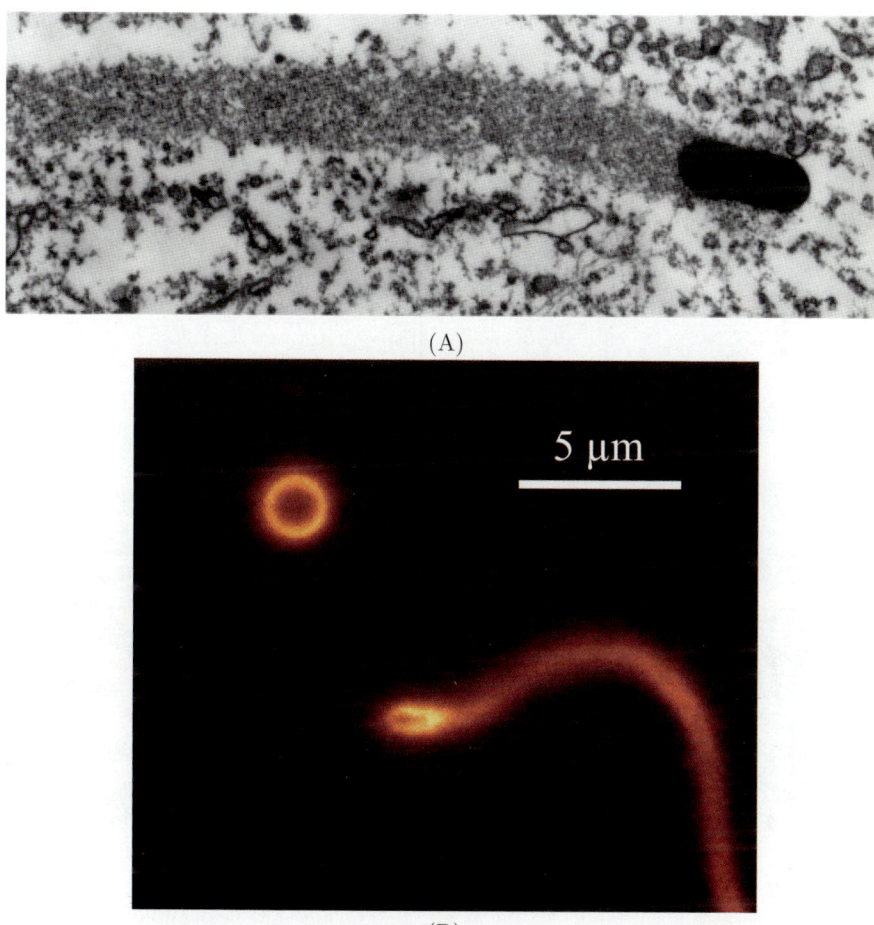

(A)

(B)

Figure 1.1. Example of wild type *Listeria* and its homogeneous comet as seen by electron microscopy (A) and confocal microscopy (B). Note the actin layer in front of the bacterium is missing in (A) while the fluorescence image (B) clearly shows that there is a gel. Figure (A) is reprinted from [9]. ©(1992), with permission from Elsevier, Figure (B) courtesy of Vincent Noireaux.

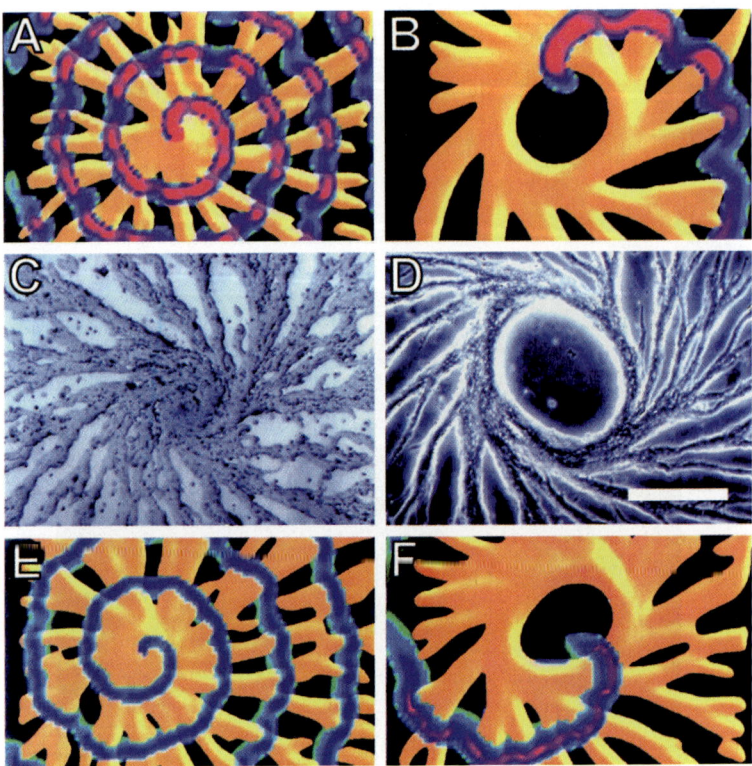

Figure 3.2. Close-up view of the aggregation center in experiments (C, D) and a model (A,B; also E,F) due to Vasiev et al. [2]. The hole has been enhanced in going from A to B and C to D by lowering the excitability of the system, accomplished experimentally by adding caffeine. (courtesy of C. Weijer).

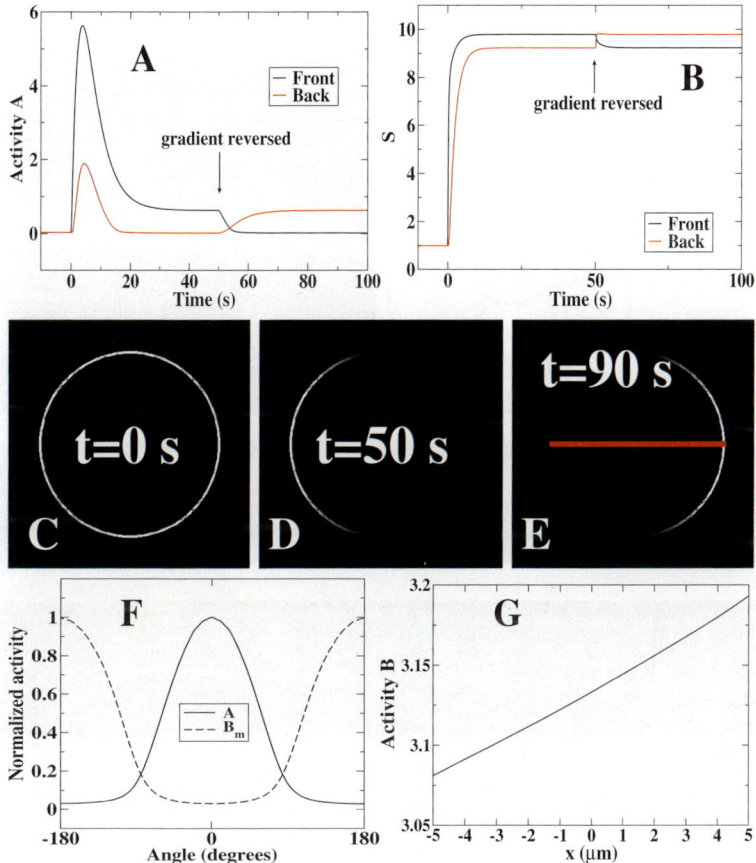

Figure 3.13. A: The value of A as a function of time at the front of the cell (black line) and the back of the cell (red line) as a response to a gradient reversal experiment. B: The external concentration (S) before and after gradient reversal at t=50 s. C-E: Value of A along the perimeter of the cell in a linear gray scale at different times. F: Values of A and B_m, normalized by their maximum value, along the perimeter at t=90 s. G: Value of B as a function of space measured along the symmetry axis of the cell, parallel to the gradient (drawn as a red line in E).

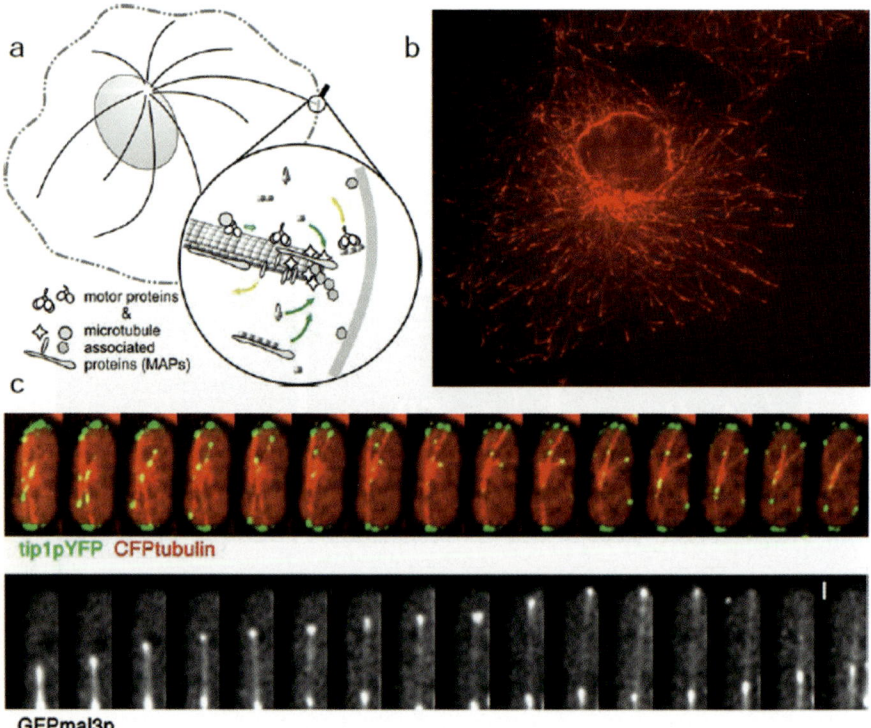

Figure 4.3. Microtubule end-binding proteins. Specialized proteins interact with the dynamic ends of microtubules *in vivo* to control microtubule dynamics, deliver cargo, or mediate interactions with the cell cortex. (a) Artist impression of the protein complexity at microtubule ends in cells. (b) Fluorescent antibody staining of EB1, a catastrophe suppressing end-binding protein in mammalian cells. Image of COS-7 cell courtesy of Dr. Anna Akhmanova. (c) Life imaging using fluorescent protein (FP) labeling of two different end binding proteins, tip1p and mal3p, in interphase fission yeast cells. Images taken every 10.4 sec (top) and 4 sec (bottom); courtesy of Dr. Damian Brunner. When microtubules reach the cell ends, Mal3p disappears before catastrophes occur.

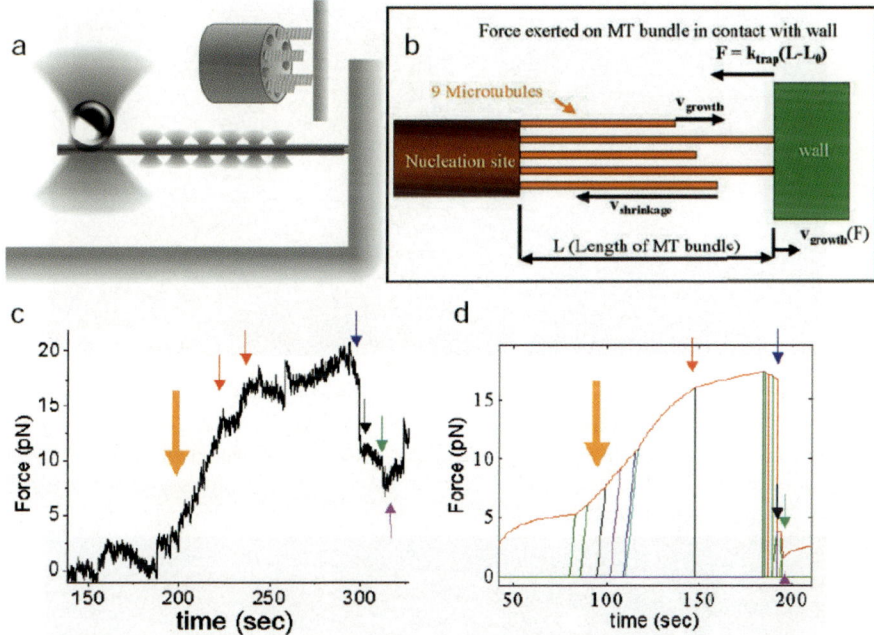

Figure 4.10. Microtubules can coordinate their dynamics under force. (a) Bead-axoneme construct in a keyhole optical trap. Axonemes can nucleate up to 9 microtubules simultaneously (see enlargement) when the tubulin concentration and temperature are high enough. (b) Parameters used to simulate growth of a microtubule bundle in the optical tweezers experiment. (c) Experimental result for force generated by a growing microtubule bundle. Arrows indicate events that likely correspond to events observed in the simulations. (d) Simulation result. Yellow arrows indicate arrivals of new microtubules, red arrows indicate catastrophes of individual microtubules without shrinkage of the bundle, blue and green arrows indicate catastrophes of the entire bundle and black and pink arrows indicate rescues of the bundle due to the arrival of new microtubules.

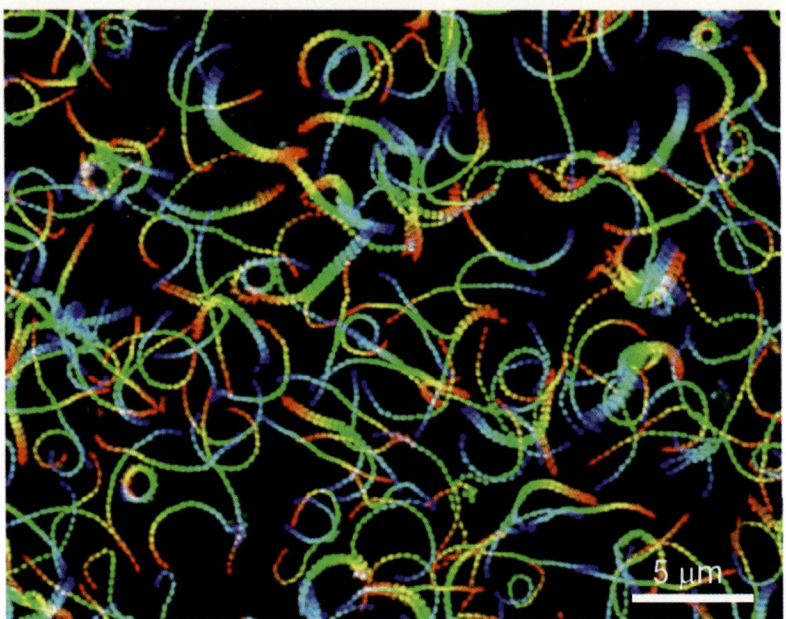

Figure 6.1. Gliding of *M. mobile*. Selected frames from 4s of footage were differently colored from purple to red and integrated into one image [7, 11]. The traces in the image show gliding speed, direction of gliding, and time relations with other traces. Reprinted from [11]. Copyright 2005 National Academy of Sciences, U.S.A.

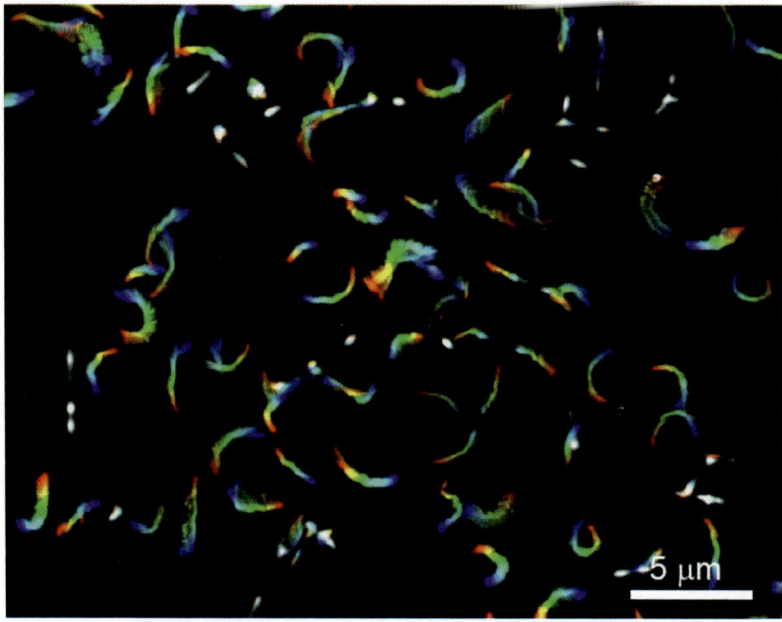

Figure 6.3. Gliding of *M. pneumoniae*. Selected frames from a 10s video were treated in the same way as those in Figure 6.1 [7, 15].

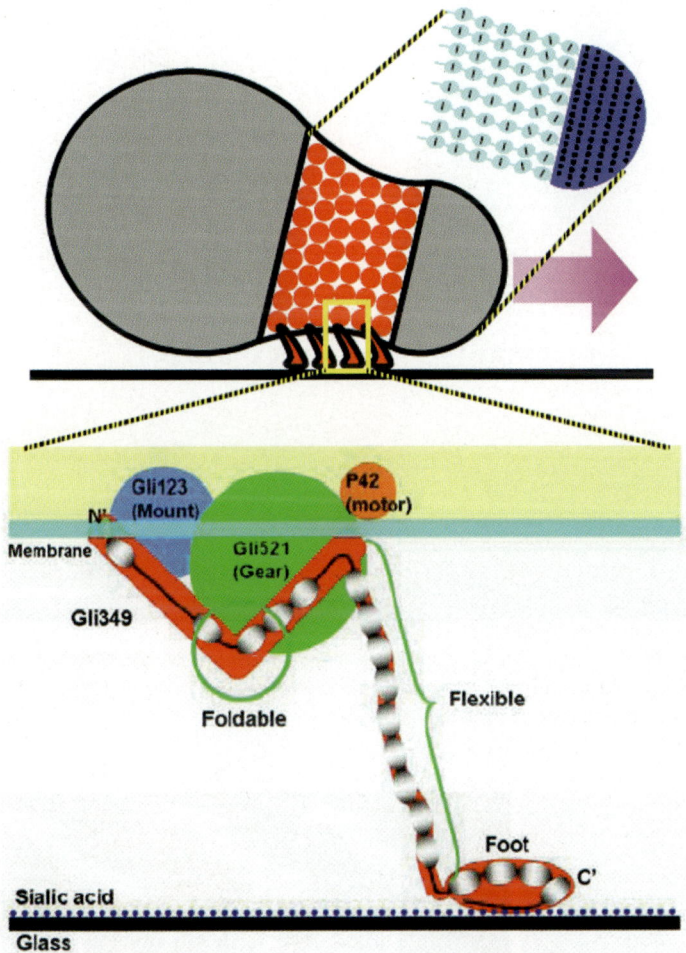

Figure 6.9. Cell architecture of *M. mobile*. Four proteins involved in gliding mechanisms are localized near the cell membrane, and aligned at the cell neck forming 450 units per cell. Gli349, the plausible leg protein, sticks out from the C-terminal region; its distal end can bind to sialic acid fixed on the solid surface. The gliding units are supported from inside the cell by the jellyfish-like structure recently identified, as presented in the top illustration. The jellyfish structure is composed of latticed solid "head" part and flexible strings attached with particles about 20nm in diameter.

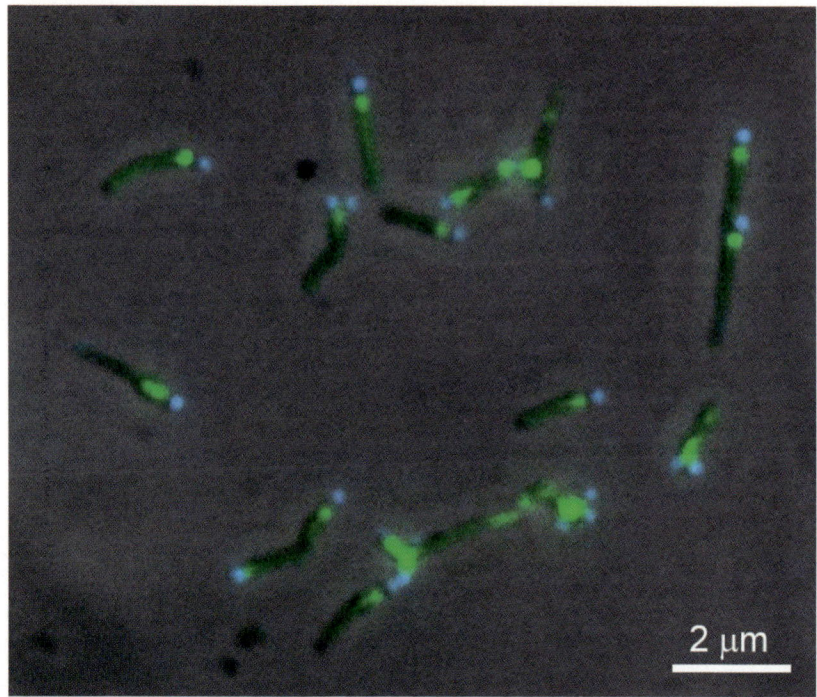

Figure 6.16. Localization of cytadherence proteins in *M. pneumoniae* cells. P65 and HMW3 are labeled by CFP (cyan) and YFP (green) fusions, respectively. Phase contrast and fluorescent microscopic images were independently taken and merged. Courtesy of Dr. Tsuyoshi Kenri [74].

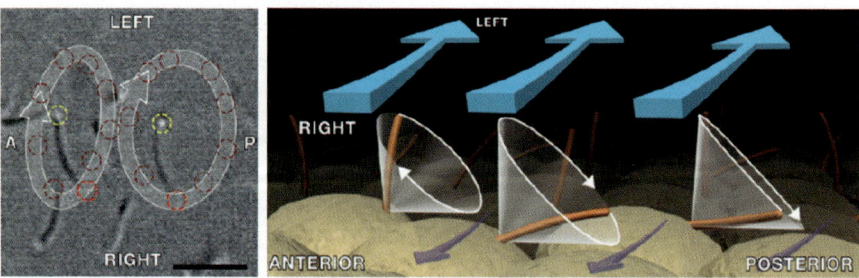

Figure 8.7. The trajectory described by the tip of nodal monocilia is elliptic with the axis of rotation tilted toward the posterior side. The larger the distance between tip and nodal surface, the stronger the fluid layer is pushed into the direction of ciliar motion. The ciliar motion thus gives rise to a fast flow just above the tip from right to left, and to a backward flow close to the nodal surface from left to right. The backward flow is much slower due to the viscous resistance of the wall. Figure is reprinted from [13]. Copyright (2006), with permission from Elsevier.

require other factors, such as the cell membrane. The antibody against Gli349 easily inhibits binding of *Mycoplasmas* to solid surfaces such as glass and erythrocytes. We isolated two mutants resistant to this antibody and found that they have two amino acid substitutions in the N-terminal region of Gli521. One substitution is from proline to arginine at the 476th amino acid, and the other is from serine to arginine at the 859th amino acid. These results may suggest that the Gli521 molecule interacts with Gli349 at the domain including these sites.

6.3.10 Sialic Acid as Scaffold

The gliding motility of *Mycoplasmas* can be easily observed on glass surfaces. However, in nature, *Mycoplasmas* live in animal tissues, suggesting that their intrinsic binding target is the surface structure of animal cells or an extracellular matrix. Jaffe et al. found that coating glass with serum is sufficient to allow *M. mobile* to bind to and glide on glass, and that this coating is sensitive to protease treatment [55]. These observations suggest that protein components in serum share a structure with those on animal tissues and mediate the glass binding essential for gliding. However, identification of the binding targets derived from serum is not easy, because it contains so many components. Previously, the net binding of *M. pneumoniae* monitored by radiolabeling was reported to depend on N-acetylneuraminic acid (the predominant sialic acid) [56]. Sialic acids are comprised of a sugar together with a glycerol and an amino group (see Figure 6.11), and are widely distributed as the tip structure of polysaccharides attached to proteins, lipids and sugars in cell membranes. To identify the direct binding target we examined the factors that affect the binding of *Mycoplasma pneumoniae* to solid surfaces and concluded that N-acetylneuraminyllactose (sialyllactose) attached to a protein can mediate glass binding, on the basis of the following four lines of evidence: i) glass binding was inhibited by N-acetylneuraminidase; ii) glass binding was inhibited by N-acetylneuraminyllactose; iii) binding occurred on glass pretreated with bovine serum albumin attached to N-acetylneuraminyllactose; and iv) gliding speed depended on the density of N-acetylneuraminyllactose on glass [57].

Because, as discussed above, the C-terminal domain of Gli349 is likely to be related to binding to solid surfaces, this domain might share a structure with known sialic acid binding proteins. Therefore, we compared the amino acid sequence of Gli349, especially focusing on the C-terminal region, with those of sialic acid binding proteins. However, no features related to sialic acid binding proteins were identified. This attempt also failed for the whole sequence of the P1 adhesin, the leg protein of *M. pneumoniae*. These facts may suggest that the binding proteins of these two species are novel types of sialic acid binding proteins, otherwise other proteins associated with the leg proteins are responsible for binding to sialic acid.

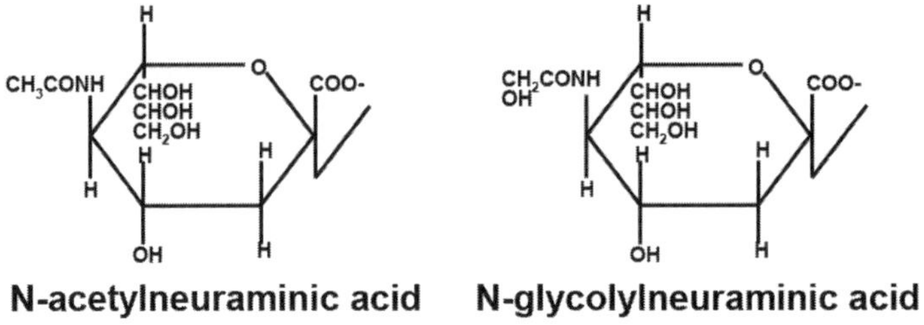

N-acetylneuraminic acid   N-glycolylneuraminic acid

Figure 6.11. Structures of representative sialic acids.

6.3.11 Energy Source

Information about direct energy sources is essential for elucidation of biomotility systems. One can effectively obtain this information by analyzing the inhibitory effects of various drugs on motility. We examined the effects of various drugs on *M. mobile* gliding and found that the addition of arsenate reduces gliding speed with a decrease in cellular ATP concentration. However, ionophores had little or no effect on gliding speed, even if they diminished the membrane potential. These observations suggest that the gliding mechanism is coupled to ATP hydrolysis [55].

In a study of the motility of eukaryotic axonemes such as sperm and protozoa, Gibbons et al. [58] constructed a useful system, called the "Triton model", where the cell membrane is permeabilized and the intracellular reactions can be manipulated externally. This system demonstrated sliding between microtubules, ATP as the energy source, dynein as the motor protein, and so on. Therefore, we constructed this kind of model for *M. mobile*, probed the internal reactions, and concluded that the gliding mechanism is driven by the energy of ATP [11]. Treatment with Triton X-100 immediately stopped the gliding and converted the cells to permeabilized "ghosts". Surprisingly, when ATP was added exogenously, 85% of the ghosts were reactivated, gliding at speeds similar to those of living cells. The reactivation activity and inhibition by various nucleotides and ATP analogs, as well as their kinetic parameters, showed that the machinery is driven by the hydrolysis of ATP to ADP plus phosphate, caused by an unknown ATPase.

P42 is a 356-amino-acid protein encoded downstream of Gli521 (see Figure 6.6). This protein was found to have ATPase activity although the amino acid sequence does not have any motifs suggesting such activities [59]. The preferences of nucleotides and ATP analogs are similar to those of the unknown ATPase presumed to be the motor of gliding ghosts. These observations suggest that gliding is driven by the ATPase activity of P42.

6.3.12 Gliding on Small Etched Patterns

To obtain information about the movement of the legs, we made a series of small steps, ranging from 50 to 1000nm by lithography, which is generally used to make silicon and DNA chips, and observed the behavior of *M. mobile* cells at each step [7]. The cells were able to climb over the 50nm step normally, but the number of cells able to scale the step decreased as steps became higher. The critical height was around 400nm, which is higher than the cell itself. As far as stepping down, when the cells came to a cliff edge, they tended to turn and go along the edge. We also etched a small groove with a tapered end. *M. mobile* glided to the end of the groove and became stuck for a moment, but did not reverse its direction. If gliding occurs by the repeated movement of rigid sticky legs, reversing the direction of a cell's movement when gliding through such a groove might happen by chance when unusual binding of the structures occurs. Why does *M. mobile* never reverse? As suggested from the molecular shape of the Gli349 protein, the legs have flexible parts that pull rather than push the cell body of the *Mycoplasma*. When *Mycoplasma* was observed gliding on a red blood cell (RBC) by interference microscopy, the RBC membrane seemed to be pulled to the direction of *Mycoplasma* [60]. The pulling character of the legs may also be suggested from this observation.

6.3.13 Mechanical Characteristics of Gliding and Binding

Temperature dependence was analyzed for glass binding and gliding speed. The gliding speed changed linearly as a function of temperature from 4 to 40°C (see Figure 6.12A) [44], as observed for conventional motor proteins and the bacterial flagella motor [61, 62, 63]. This suggests that the protein movements depend on the heat wobble of water. Similarly, the number of *Mycoplasma* cells bound to glass decreased linearly with temperature (see Figure 6.12B). This would also suggest that the stability of the bond between sialic acid and the leg protein depends on the heat wobble of water.

The mechanical characteristics of motility are a crucial piece of information about the mechanism. We therefore measured the force velocity relationship by applying a load to a bead attached to the tail of a *M. mobile* cell, using one of the two methods shown in Figure 6.13A [44]. The first method employs a controlled liquid flow. Because *M. mobile* is reoriented upstream when a flow of liquid is applied, the load applied to *M. mobile* (or rather to the bead) can be calculated from Stoke's law. Another method, the "optical tweezer" (laser trap) method, utilizes the fact that a strong beam applied to an object with a mismatched refractive index traps the object at the center of the beam. The lateral force generated on the object (the bead carried by a *Mycoplasma* in this case) can be calculated from the bias between the centers of the laser and the object. Because the liquid flow washed all gliding cells from the glass surface, we could not measure the maximal (stall) force using the first method. Therefore, the optical tweezer method was used to measure the maximal (stall)

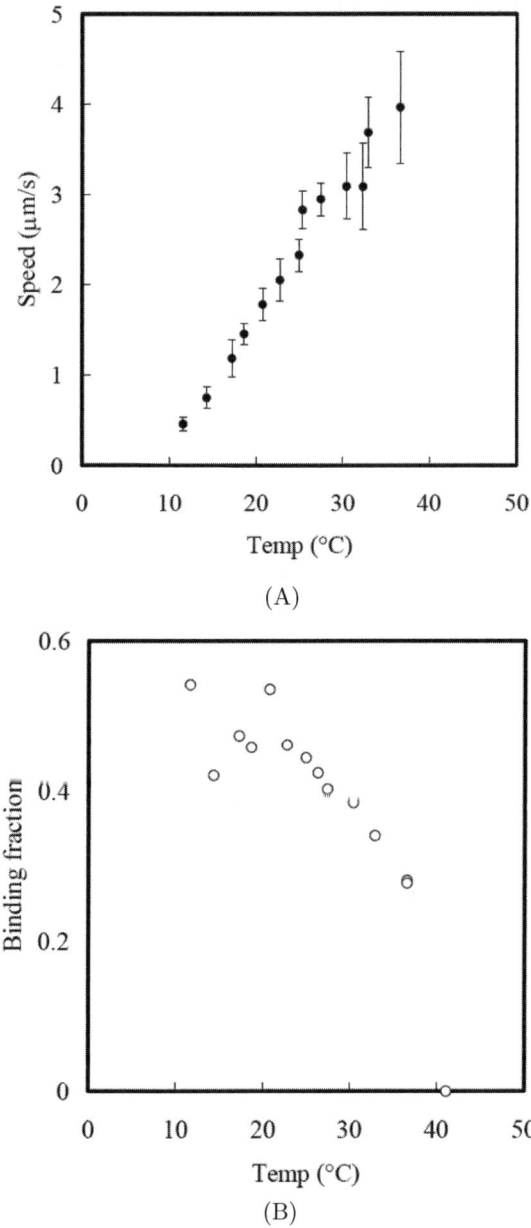

Figure 6.12. Temperature dependence of binding and gliding. (A) The fraction of the number of bound cells to the number of total cells in the microscopic field. About 40 cells were analyzed for each temperature. (B) Gliding speeds averaged from 10 cells are presented as a function of temperature. (A) is reprinted from [44]. Copyright (2002), with permission from The American Society for Microbiology.

force. We obtained the force and velocity relationship for *M. mobile* gliding (see Figure 6.13B) and found that i) *M. mobile* can continuously change speed, responding to loads at various speeds of flow, suggesting that the gliding can respond to a wide variety of loads, as may occur in mucous on the surface of host tissues; ii) the maximal (stall) force was 27 piconewtons (pN), or roughly 1,800 times the force [15 femtonewtons (fN)] required to thrust a *Mycoplasma* cell [64, 65]; and iii) the maximal force did not depend on the temperature but the speed did, suggesting that force generation and stroke are different steps.

The number of Gli349 molecules on the cell surface was quantified by titrating the antibody binding to the cell surface [27]. We estimated that there are around 450 Gli349 molecules on a cell. Assuming that a quarter of them are involved in gliding at any given moment and participate in generating the maximal force detected, the force exerted through each leg would be 0.24 pN, which is much smaller than the value of conventional motor proteins [66, 67].

6.3.14 Features of Mechanism and Centipede Model

The mechanism may depend on repeated binding based on energy from ATP. It has these features in common with conventional motor proteins such as myosin and kinesin. However, *Mycoplasma* gliding has the following unique features: i) the machinery is half exposed; ii) the ATPase and binding site are positioned about 50nm apart; iii) the "rail", fixed sialic acid, has no polarity; iv) the stall force is as small as 0.24pN per unit, if we assume that one-quarter of all gliding units are engaged at any moment; and v) the mechanism probably does not have a pushing step because the longest rod of the Gli349 molecule is flexible.

Considering these features and other results obtained so far, we suggest a "centipede (power stroke) model", a working model to explain the mechanism [68, 5, 48, 27]. There, each gliding unit is in a mechanical cycle, which is composed of a series of states [(a-f) in Figure 6.14] and transition steps: i) stroke, ii) movement, iii) release, iv) return, v) initial binding, and vi) tight binding. The major assumption here is that the tension exerted to the leg causes a progression of steps, as indicated by the lightning dashes in the figure. In Stage (a), the leg (Gli349) tightly binds to sialic acid on the solid surface. The force exerted to the leg triggers a conformational change in the upper part of the leg, causing the stroke (i). This trigger is assumed. The unit is waiting for a new molecule of ATP at Stage (a), because the ghosts stop gliding when ATP is depleted from permeabilization by triton. At Stage (b), higher tension is exerted to the leg and actual movement (ii) occurs. Considering that the leg length about 50nm, and the gliding speed is 2.5 microns per second, this step should occur within 10ms at the slowest. The stall by mechanical force should occur at this step. At Stage (c), tension applied to the leg decreases and this change triggers the leg to release sialic acid at step (iii). The decrease in tension also causes return (iv) of the upper part of the leg to the initial

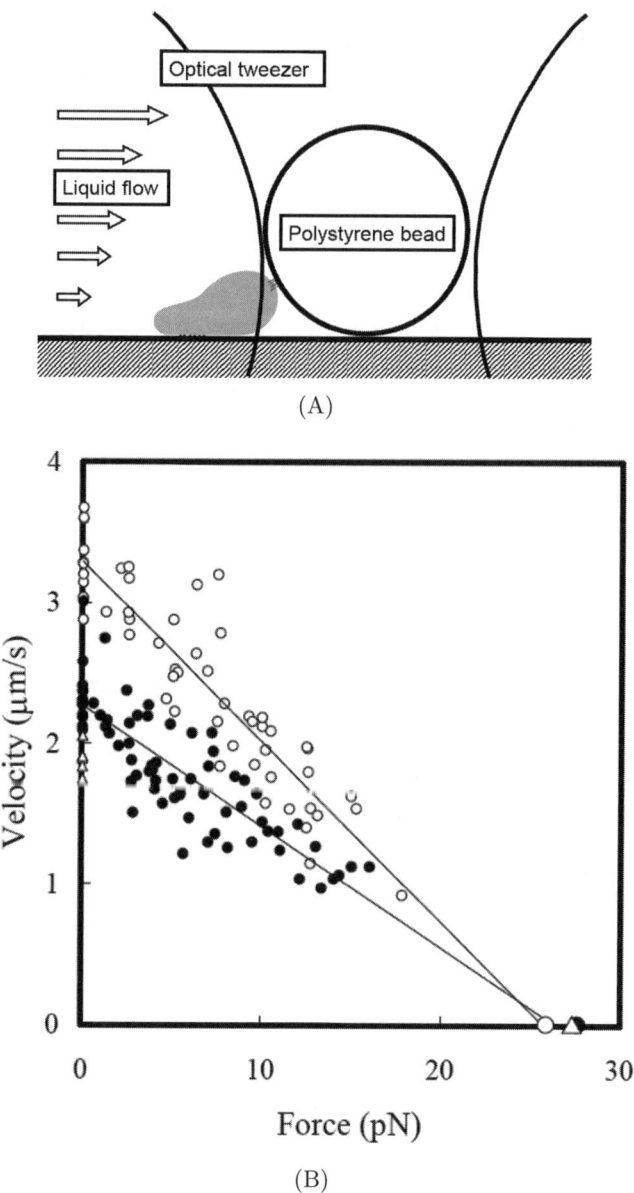

(A)

(B)

Figure 6.13. Force velocity relationship. (A) Experimental design. Gliding *Mycoplasmas* were attached with a bead, and loaded by either of two methods, controlled liquid flow or optical tweezer. (B) Force velocity relationships under three different temperatures. Data for 17.5, 22.5, and 27.5°C are presented by triangles, closed circles, and open circles, respectively. Stall forces were measured by optical tweezer and other data were from flow experiments. Figures reprinted from [44]. Copyright (2002), with permission from The American Society for Microbiology.

conformation, as Stage (e). Then, the unit returns to the tightly binding state via an initial binding state (f). An antibody against Gli349 removes the gliding *Mycoplasmas* from glass but also reduces the gliding speed. As the maximum (stall) force is 1,800 times larger than that required to pull a *Mycoplasma* cell with the observed gliding speed (see Section 6.3.13), a decrease in the number of working legs by antibody alone cannot explain the decrease in the gliding speed. Thus, a drag force caused by the antibody must be considered, possibly derived from the loosely bound state, i.e., the initial binding state (f). Removal of a cell by this antibody occurs via a reversal of step (v).

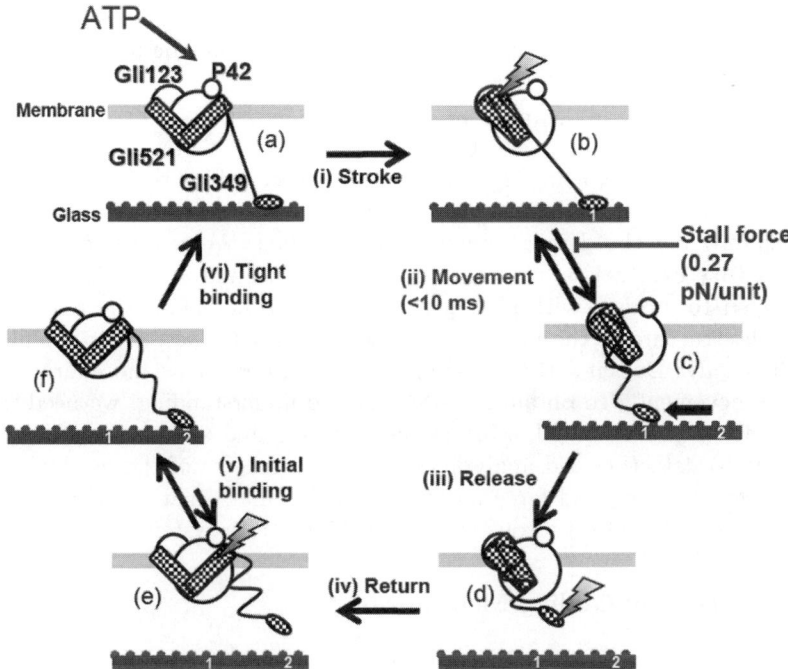

Figure 6.14. Centipede model for gliding mechanism. The *Mycoplasma* cell membrane is presented by a gray line. Four gliding proteins are assembled around the cell membrane to form a unit of gliding machinery. The leg protein, Gli349, repeats the sequence of binding, movement, and release with sialic acid on the solid surface, and then pulls the cell body forward.

We have estimated that the gliding unit consists of around 450 Gli349 molecules on each cell using the anti-Gli349 antibody [27, 11]. If this is correct, the units can be expected to cooperate in pulling the cell forward. The interaction between molecules is enhanced at the micro scale, and the legs are very flexible. Hence, the movements are likely to cooperate physically. So far, we have isolated three inhibitory antibodies, and identified mutation points affecting gliding and binding [42, 47, 48, 27, 28]. Mapping of working points of these effectors on the gliding proteins has shown that the effectors are well explained by this working model.

This model has been suggested based on observations of *M. mobile*. Obviously, this model can be applied to *Mycoplasma* species closely related to *M. mobile*, such as *M. pulmonis*, whose genomes contain easily detectable orthologs of gliding proteins. But what about nonrelated species, such as those belonging to the *pneumoniae* group? Because of the poor gliding activity, it has not been easy to conduct such extensive analyses on other gliding *Mycoplasmas*, but recently we obtained similar results for *M. pneumoniae* gliding using an anti-P1 antibody [48]. The antibody reduced the gliding speed of *M. pneumoniae* and eventually removed the cells from the glass surface. This suggests that the mechanism of *Mycoplasma* gliding may be universal.

6.3.15 Toward a Complete Model

Although we can explain all of our current observations with the mechanical cycle described in Section 6.3.14, the model has not been substantiated adequately and the details remain to be clarified. We recently identified a jellyfish-like cytoskeletal structure inside the head and neck. Although it is suggested to interact with gliding proteins we do not know the role of this jellyfish structure in the mechanism. Does it support the gliding machinery mechanically? Does it sort the gliding proteins? Does it have more direct roles in the movement? To obtain a more concrete understanding, we need more information about the following issues: i) whole and atomic images of machinery, ii) detection and analysis of actual movements of the machinery by fluorescent labeling and force measurement, iii) mathematic inspection of a working model, and iv) reconstruction of the motility system.

6.3.16 Origin of Gliding Mechanism

Because no other proteins as yet have been found to be related to gliding proteins, we cannot know the origins of the gliding mechanism. Generally, *Mycoplasmas*, which parasitize animal tissues, have advanced systems for antigenic variations. Mvsps are suggested to be involved in the antigenic variations, as discussed in Section 6.3.4 [6, 42]. MvspI (the largest Mvsp at 220 kDa) is an abundant surface protein clustering at regions other than the neck. We isolated this protein, and characterized and observed it using EM with rotary shadowing. The protein forms a homodimer, where the C-terminal flexible filamentous parts stick out from the N-terminal globular part. The outline of this protein is reminiscent of the whole shape of the gliding unit as presented by Figure 6.15 [53]. The amino acid sequences of Mvsps also feature repeats of about 90 amino acids, although the sequence is not common with those of Gli349. These observations may suggest that the gliding proteins and the antigenic proteins may share an ancestor protein.

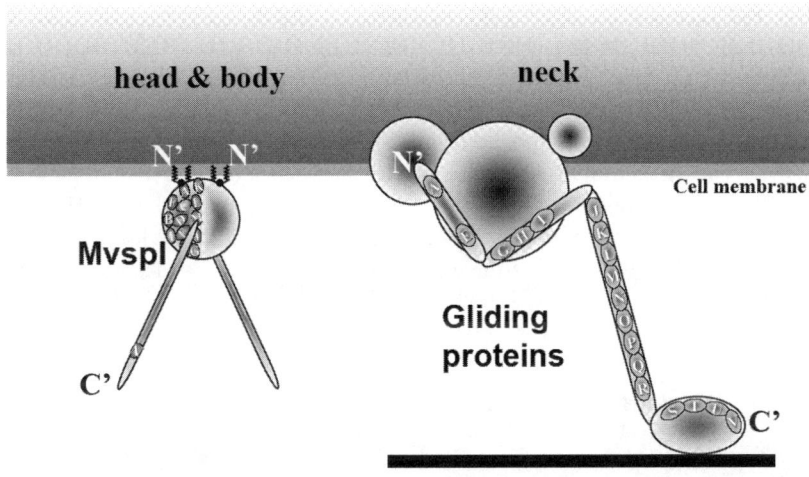

Figure 6.15. Morphological similarities between MvspI and gliding machinery. The MvspI protein forms a dimer and is anchored to the membrane at other parts than the cell neck via lipids attached to the N-terminal. The protein molecule features an N-terminal globular part, C-terminal filamentous part, and repeat sequences of about 90 amino acids. This protein shares features found in the gliding unit: a globular part near the membrane, a filamentous part, and repeat sequences.

6.3.17 Harnessing *Mycoplasma* Motion

Some researchers working on biomotility systems dream that their discoveries will eventually be used as the bases of artificial motors. Hiratsuka et al. examined in detail the behaviors of *M. mobile* on micropatterns made by lithography, and found that *Mycoplasmas* preferentially move along walls. Based on this feature, they made an arabesque pattern and controlled the direction of *Mycoplasma* gliding. The directions of 90% of the gliding *Mycoplasmas* were controlled in repetitive broken circles connected by lines [7]. Then, the motion of gliding *Mycoplasmas* was harnessed by some additional artifices. *Mycoplasmas* that were controlled in terms of their gliding direction in a pattern were introduced into a circle covered by a rotor 20μm in diameter, and attached to the rotor via biotin labeling on the *Mycoplasma* cell surface. The rotor moved at 2.0 rpm [69], becoming the first example of a micromechanical device to integrate inorganic materials with living bacteria.

6.4 Studies on *M. pneumoniae* Gliding

The study of *M. pneumoniae* gliding has progressed to a different point, and along a different path from that of *M. mobile*. The study of *M. pneumoniae* began in the early 1970s because this organism had long been known as a human pathogen. The machinery features an organized structure, and the gliding of *M. pneumoniae* is relatively slow compared to that of *M. mobile*, and is interrupted by pauses. Furthermore, unlike *M. mobile*, some genetic analyses have been possible based on the genome information available since 1996. At one pole of the *M. pneumoniae* cell, a cone-shaped membrane protrusion, known as the attachment organelle, is formed as shown in Figure 6.4 [70, 16, 5, 71]. When the interaction between the organism and host tissues is examined, this structure always binds to the tissue [72, 73]. Moreover, many proteins essential for cytadherence and gliding are clustered around it (see Figure 6.16). Therefore, the attachment organelle should be responsible for binding and gliding.

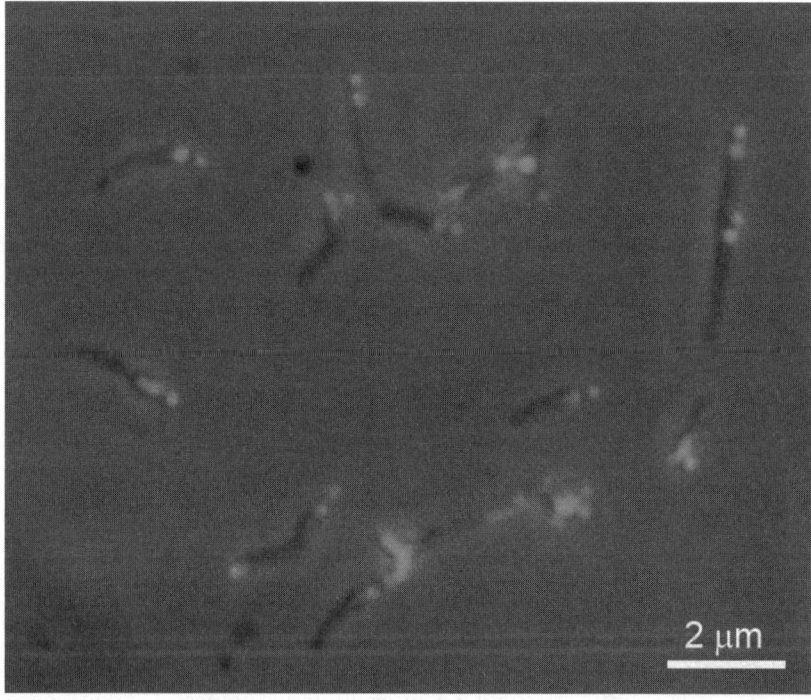

Figure 6.16. Localization of cytadherence proteins in *M. pneumoniae* cells. P65 and HMW3 are labeled by CFP (cyan) and YFP (green) fusions, respectively. Phase contrast and fluorescent microscopic images were independently taken and merged. Courtesy of Dr. Tsuyoshi Kenri [74]. (See color insert.)

6.4.1 Ultrastructure of the Attachment Organelle

Although the structural features of this organelle were originally found in the early 1970s [75, 76, 77, 73], detailed study of the structure has not been well-focused. Recently, some EM study on the structure, including electron cryotomography (ECT), has achieved some progress [78, 79, 71, 80, 81]. The structure of the attachment organelle is apparently composed of five parts (see Figure 6.17): i) the surface structure (nap), ii) segmented paired plates (rod core), iii) the distal end of the rod (terminal button), iv) the proximal end of the rod (bowl or wheel), and v) the translucent area.

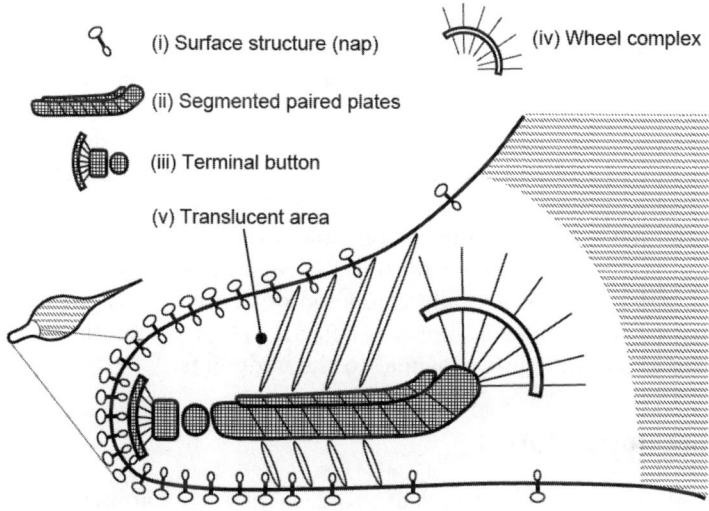

Figure 6.17. Architecture of attachment organelle of *M. pneumoniae*. The structures can be divided into five parts. Each part has been suggested to include the cytadherence proteins.

Surface Structure (i)

Negative-staining EM revealed a surface structure, called a "nap", at the tip [82, 4, 83]. Similar structures can also be found on other gliding species of the *M. pneumoniae* group, including *M. genitalium, M. pulmonis,* and *M. gallisepticum.* EM has also been used to demonstrate filamentous elements between *M. pneumoniae* and the host tissue to which it is bound, but the relationship between this structure and the nap is unknown [72, 73]. Recent studies using ECT also revealed clustering of membrane proteins on the surface of the attachment organelle [79, 81]. The units are about 16nm in height, composed of a proximal stalk part about 10nm in length, and a distal globular part about 4 − 8nm thick and 8nm wide [81]. This structure is reported to span the cell membrane with an intracellular domain extending 24nm into the cytoplasm with a distal wider part. Considering that gliding occurs at the interface between the cell membrane and a solid surface, this structure may play the role of the foot, as hypothesized for *M. mobile*.

Segmented Paired Plates (ii)

In sectional images of the attachment organelle, the center has an electron-dense core surrounded by a translucent area [82, 84, 85, 73]. After extraction with Triton X-100 or Triton X-114, an Triton-insoluble shell, which appears to be a cytoskeletal structure, remains [76, 77, 80]. It is composed of a relatively thick "rod" and a filamentous network forming a basket-like structure. It was thought that the rod, i.e., the electron-dense core in sectional images, supports the attachment organelle and that the basket supports the remainder of the cell [5, 71]. As the basket cannot be found after treatment of the Triton-insoluble shell by DNase and RNase (our unpublished data), this structure may be partly or thoroughly composed of nucleic acids. The isolated rods are apparently divided into two parts; one is a striated rod and the other is the distal end structure (terminal button). A recent study using ECT revealed that this striated rod is actually composed of thicker and thinner segmented plates paired with about a 7nm gap [79, 81]. The rods are flexible, and are often bent about 150° just proximal to the midpoints.

Terminal Button (iii)

The distal end of the electron-dense core in the sectional images is enlarged and attached to the cell membrane; this end is called the "terminal button" [86, 87, 16, 71, 73]. These features can also be observed in the negatively stained image of isolated rods, although they are sometimes attached with membrane pieces, suggesting a complex structure including the polar cell membrane [78, 80]. ECT studies clarified that the terminal button can be divided into three major parts; the distal part is attached to the inner layer of peripheral membrane proteins [79, 81].

Wheel Complex (iv)

Hegermann et al. suggested a striking model, based on the observations of cryosections by EM [78]. In that model, the rod is attached at its proximal end to a "wheel complex" with fibrils, which connect the rod and the periphery of the cell. A similar structure, called the "bowl", has been found at the proximal end of rod by ECT, although the fibrils are not easily seen.

Translucent Area (v)

The electron-dense core is surrounded by an electron-lucent area, from which dense complexes are excluded [73]. Hegermann et al. examined the structure of this area by treating chemically fixed cells with Triton X-100, and suggested that spoke structures connect the rod and the cell periphery. Seybert et al. reported that the proximal end of the rod is attached to the membrane based on ECT findings. Henderson and Jensen [79], however, proposed that this part may be filled with fluid having properties distinct from those of cytosol, and may play a role in transmitting the force and movement generated around the rod.

6.4.2 Proteins Related to Cytadherence

As cytadherence is a pathogenic determinant, the adhesion protein and ten other proteins essential for the process have been identified [88, 16, 71]. All of them examined so far have been localized in the attachment organelle. Therefore, localization at the attachment organelle is now one of criteria for identifying cytadherence proteins [13, 49, 89]. The cytadherence-related proteins are also generally essential for gliding, because *Mycoplasmas* can glide only when they are bound to solid surfaces. Recent progress in the techniques of genetic rearrangement in *M. pneumoniae* have accelerated identification of cytadherence proteins [90, 91, 13, 92]. These proteins can be classified into five groups, based on subcellular localization and roles, as follows (see Figure 6.17)[1]:

i) **P1 adhesin (MPN141):** a 170-kDa protein with transmembrane segments, responsible for binding to solid surfaces such as host cells, glass, and plastics provably via sialic acid or sulfated glycolipids [94, 56], and also for gliding motility [15].

ii) **Proteins that support P1 adhesin with physical interactions:** P90, P40, (These two proteins are derived from an ORF, MPN142) etc. [95, 96, 49, 89]. These proteins are localized within a rather wide area surrounding the attachment organelle, like P1 adhesin. The complex formed by P1 and these proteins may be observed as the nap structure under EM.

iii) **Proteins localizing at the mid part of the rod:** HMW1 (MPN447) and HMW2 (MPN310). These proteins are essential to the rod formation [92, 97, 89]. The segmented paired plates of the rod may be composed of these proteins.

iv) **Proteins localizing at the proximal end of the rod, suggested to be components of the wheel:** These include P41 (MPN311), P24 (MPN312), and P200 (MPN567)

v) **Proteins localizing near the distal end of the attachment organelle:** suggested to be components of the terminal button, such as P65

---

[1] ORF names will be noted as well as the conventional protein names [93].

(MPN309), P30 (MPN453), and HMW3 (MPN452) [13, 98, 49, 89, 99]. [45, 74, 13].

Most proteins in these groups except P200 are coded in three loci on the genome, i.e. i) P1 operon coding P1 adhesin, P90, P40, and one ORF; ii) crl locus coding P65, HMW2, P41, and P24; and iii) HMW operon coding P30, HMW3, HMW1 and other 6 ORFs [87]. The proteins in each group are likely to function at the same time, as often observed for many bacterial systems. Conversely, the ORFs in these loci could have a role in cytadherence. HMW1, HMW3, P65, and P200 proteins share a feature in SDS-PAGE, by which they migrate with much slower speed than that predicted from their molecular weights. This feature is caused by the obvious bias in amino acid contents, found in the domains, named "acidic proline rich (APR)", comprising more than half of these proteins [88]. This feature in the amino acid sequence may suggest that these proteins exist in conditions specific for the attachment organelle, and any difference in the condition from the usual cytosol is reflected by the translucent appearance of the organelle under EM.

The *M. pneumoniae* genome has the gene for the FtsZ protein, the bacterial tubulin homolog, but not the gene for MreB (the bacterial actin homolog) both of which are widely distributed in bacteria and function as cytoskeletal proteins [93, 35]. FtsZ is expressed in *M. pneumoniae* cells [100], but obvious localization at the attachment organelle was not suggested (our unpublished data), showing that this protein is not involved in the attachment organelle.

6.4.3 Search for Proteins Involved in Gliding Mechanism, But Not in Cytadherence

As discussed above, a number of proteins have been identified as being involved in cytadherence. However, these identifications do not provide direct answers about the gliding mechanism. Therefore, researchers have decided to isolate mutants specifically deficient in gliding motility. *Mycoplasmas* were mutated by transposons, and each clone was examined for spreading of colonies (as originally done in the studies of *M. mobile* gliding mutants) and also examined for binding and normal growth rate. Pich et al. used *Mycoplasma genitalium* (closely related to *M. pneumoniae*) and identified two proteins that fit the criteria [29]. Hasselbring et al. identified 11 proteins of *M. pneumoniae* and finally focused on three proteins [12]. These results did not immediately resolve the mystery of gliding, but they did provide some suggestions and tools for future studies:

i) Mutations in most of the identified proteins did not produce complete defects in motility. Even when they reduced the motility severely, some activity remained. The direct involvement of these genes in the gliding mechanism is unclear, because if they are essential for gliding, some of transposon-insertion mutants should be defective in gliding. Alternatively, the genes specific for gliding motility are essential also for *Mycoplasma* growth.

ii) mg386, identified from *M. genitalium*, is an ortholog of P200 of *M. pneumoniae* [29]. P200 is known to localize at the wheel position in *M. pneumoniae* (see above) [45, 74]. Analyses of an occasionally isolated mutant of *M. pneumoniae* suggested participation of this protein in the gliding mechanism [45]. The wheel and this protein may play a more direct role in gliding than the other protein components.

iii) *M. pneumoniae* cells lacking the MPN387 protein cannot glide at all, suggesting a direct involvement of this protein in the mechanism [12].

iv) In *M. pneumoniae* mutant cells lacking P41 proteins, the attachment organelle detaches from the cell body and continues to glide [101]. This observation is consistent with the roles of these proteins predicted from the subcellular localization. The isolated piece of attachment organelle might be a clue in defining and analyzing the machinery necessary for the mechanism, including the conditions produced by factors other than macromolecules themselves [102].

6.4.4 Suggested Gliding Mechanisms

The ends of the rod in the attachment organelle are bound to the peripheries, and the rod is flexible with a bend. The inchworm model has therefore been suggested [79, 81]. The attachment organelle appears rather stiff during gliding in optical microscopy, although other cell parts look flexible. Therefore, I am not willing to support the inchworm model. Another model similar to inchworm is "contraction-extension" model, where the rod repeats cyclic contraction and extension with association and dissociation of legs with the solid surface [103]. This controversy would be settled by time-resolved and detailed structures around the attachment organelle, including the legs.

We have found that when monoclonal antibodies against the P1 adhesin were added to the medium they reduced the gliding speed and finally dislodged the gliding cells from the glass surface. This phenomenon can be explained by one of the following three scenarios, based on the assumption of a power stroke of the P1 adhesin molecule, as hypothesized for *M. mobile* [5, 48, 15, 27]. The first is that the binding of the antibody reduces the rate of release of P1 molecules from the glass, resulting in the generation of a drag force, and also prohibits rebinding after the release, as discussed for *M. mobile*. The second scenario is that only a fraction of P1 molecules are in the propelling cycle, while others are in a state of static binding, keeping the cells on the glass and also causing a drag force in normal gliding. In this case, the binding of the antibody causes a decrease in the number of P1 molecules in the cycle, resulting in a shortage of the propelling force required for a cell to travel at the normal speed. In the third scenario, a fraction of P1 molecules are in the propelling cycle, as proposed in the second scenario, but the drag force is not large enough to balance the propelling force exerted through a short duration, and the speed of the cell depends on the sum of its stroke durations. In this case, the decrease in the number of P1 molecules in the cycle directly reduces the cell's gliding speed.

6.4.5 Formation of the Attachment Organelle

Examination of the cells of species with a polarized morphology reveals that most cells have a protrusion at one pole (the attachment organelle in *M. pneumoniae* and the head in *M. mobile*). This suggests that the formation of the protrusion is coupled with the growth and division of a cell [13, 104, 105, 106, 49]. Bredt examined cell division in *M. pneumoniae* and suggested that a nascent attachment organelle was formed adjacent to the original one [3]. We have labeled the P1 adhesin with a fluorescent dye and grouped different images of the cells [49]. The DNA replication stage of each cell was also estimated by quantifying the amount of DNA in the cell, using DAPI, a fluorescent dye that binds to DNA, because *Mycoplasmas* are known to replicate their DNA throughout the replication period [105, 106, 107, 108]. The results showed that a nascent attachment organelle was formed adjacent to the original one, and that it then moved to the opposite pole along the lateral face, and that cell division then occurred. Hasselbring et al. persisted with long-term observation of the behaviors of attachment organelles in live propagating cells, and obtained the following conclusions (see Figure 6.18) [109]: i) The attachment organelle during cell division cycle behaves with the outline suggested from fixed cells. A nascent organelle is formed adjacent to the original one, and moves laterally to the other pole. However, the next nascent organelle tends to emerge before the end of cytokinesis. Therefore, the cell images are varied. ii) An EM study suggested that the components of the original attachment organelle divide into two pieces and form two nascent ones [78]. However, quantification of a protein component in the formation process denied this hypothesis. Nascent organelles are assembled near the original ones from *de novo* protein components. iii) Cytokinesis is caused by the force generated by gliding motility, which is coupled with the cell division cycle.

6.5 Prospect

The gliding of *Mycoplasmas* was first reported in *M. pulmonis* in 1946 [110] and the prominent gliding activity of *M. mobile* was reported in 1984 [10]. The gliding is intriguing to watch, and attracts many researchers. However, the mechanisms of *Mycoplasma* gliding have not been studied extensively until recently, probably because they had not been identified as truly unique systems. Now, this feature is clear, and increasing numbers of studies about *Mycoplasma* gliding are presented every year. This situation should provide distinct progress in the near future. However, *Mycoplasma* gliding is not yet

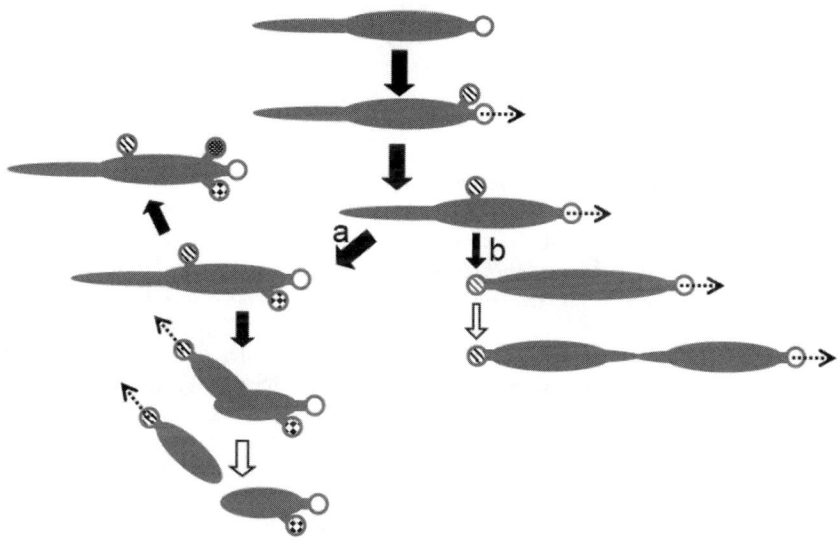

Figure 6.18. Cell division scheme of *M. pneumoniae* suggested from direct observation of living cells [109]. The attachment organelles are presented by small circles. The dashed arrows indicate movement of the attachment organelles, and open arrows indicate cytokinesis. Solid arrows indicate steps in the cell cycle, with arrow size reflecting relative frequency. In most cases, the multiple formation of a terminal organelle occurs before cytokinesis, as presented by step (a). Figure is taken from [109]. Copyright 2005 National Academy of Sciences, U.S.A.

a field crowded with researchers. Thus, we can expect faster progress in this field if more researchers from various fields begin studying the problem. Some researchers may not wish to study phenomena that are not widely distributed in living organisms. However, I suspect that novel general concepts about the behavior and physics of living machines can be discovered in the field of *Mycoplasma* gliding.

Acknowledgements

I am very grateful to those collaborators who shared these exciting times with me, and also to those colleagues who provided valuable comments and encouraged our studies. I thank Dr. Tshuyoshi Kenri, Naoko Uchida, Adan-Kubo Jun, Daisuke Nakane, and Takahiro Nonaka for providing figures and unpublished data. This research has been supported by grants from the Ministry of Education, Science, Sports, Culture and Technology of Japan, and by a grant from the Institution for Fermentation in Osaka.

References

1. Razin, S, Yogev, D, & Naot., Y. (1998.) *Microbiol. Mol. Biol. Rev.* **62**, 1094–1156.
2. Weisburg, W, Tully, J, Rose, D, Petzel, J, Oyaizu, H, Yang, D, Mandelco, L, Sechrest, J, Lawrence, T, Etten, J. V, Maniloff, J, & Woese., C. (1989.) *J. Bacteriol.* **171**, 6455–6467.
3. Bredt, W. (1968.) *Pathol. Microbiol.* **32**, 321–326.
4. Kirchhoff, H. (1992.) *Mycoplasmas- Molecular Biology and Pathogenesis.* (American Society for Microbiology, Washington, D.C.), pp. 289–306.
5. Miyata, M. (2005.) *In: Mycoplasmas: pathogenesis, molecular biology, and emerging strategies for control.* (Horizon Biosciene, Norfolk, U.K.), pp. 137–163.
6. Jaffe, J, Stange-Thomann, N, Smith, C, DeCaprio, D, Fisher, S, Butler, J, Calvo, S, Elkins, T, FitzGerald, M, Hafez, N, Kodira, C, Major, J, Wang, S, Wilkinson, J, Nicol, R, Nusbaum, C, Birren, B, Berg, H, & Church., G. (2004.) *Genome Research* **14**, 1447–1461.
7. Hiratsuka, Y, Miyata, M, & Uyeda., T. (2005.) *Biochem. Biophys. Res. Commun.* **331**, 318–324.
8. Rosengarten, R & Kirchhoff., H. (1987.) *J. Bacteriol.* **169**, 1891–1898.
9. Kirchhoff, H, Beyene, P, Fischer, M, Flossdorf, J, Heitmann, J, Khattab, B, Lopatta, D, Rosengarten, R, Seidel, G, & Yousef., C. (1987.) *Int. J. Syst. Bacteriol.* **37**, 192–197.
10. Kirchhoff, H & Rosengarten., R. (1984.) *J. Gen. Microbiol.* **130**, 2439–2445.
11. Uenoyama, A & Miyata., M. (2005.) *Proc. Natl. Acad. Sci. USA* **102**, 12754–12758.
12. Hasselbring, B, Page, C, Sheppard, E, & Krause., D. (2006.) *J. Bacteriol.* **188**, 6335–6345.
13. Kenri, T, Seto, S, Horino, A, Sasaki, Y, Sasaki, T, & Miyata., M. (2004.) *J. Bacteriol.* **186**, 6944–6955.
14. Radestock, U & Bredt., W. (1977.) *J. Bacteriol.* **129**, 1495–1501.
15. Seto, S, Kenri, T, Tomiyama, T, & Miyata., M. (2005.) *J. Bacteriol.* **187**, 1875–1877.
16. Krause, D & Balish., M. (2004.) *Mol. Microbiol.* **51**, 917–924.
17. McBride, M. (2001.) *Annu. Rev. Microbiol.* **55**, 49–75.
18. Mattick, J. (2002.) *Annu. Rev. Microbiol.* **56**, 289–314.
19. Mogilner, A & Oster., G. (2003.) *Curr. Biol.* **13**, R721–733.
20. Sogaard-Andersen, L. (2004.) *Curr. Opin. Microbiol.* **7**, 587–593.
21. Wolgemuth, C & Oster., G. (2004.) *J. Mol. Microbiol. Biotechnol.* **7**, 72–7.
22. Mignot, T, Shaevitz, J, Hartzell, P, & Zusman., D. (2007.) *Science* **315**, 853–856.
23. McBride, M. (2004.) *J. Mol. Microbiol. Biotechnol.* **7**, 63–71.
24. Hatchel, J, Balish, R, Duley, M, & Balish., M. (2006.) *Microbiology* **152**, 2181–2189.
25. Jurkovic, D, Friedberg, A. J, Hatchel, J. M, Relich, R. F, & Balish., M. F. (2006.) *Adherence and gliding motility at low temperatures by tortoise-associated mycoplasmas.* p. 22.

26. Vasconcelos, A, Ferreira, H, Bizarro, C, Bonatto, S, Carvalho, M, Pinto, P, Almeida, D, Almeida, L, Almeida, R, Alves-Filho, L, Assuncao, E, Azevedo, V, Bogo, M, Brigido, M, Brocchi, M, Burity, H, Camargo, A, Camargo, S, Carepo, M, Carraro, D, de Mattos Cascardo, J, Castro, L, Cavalcanti, G, Chemale, G, Collevatti, R, Cunha, C, Dallagiovanna, B, Dambros, B, Dellagostin, O, Falcao, C, Fantinatti-Garboggini, F, Felipe, M, Fiorentin, L, Franco, G, Freitas, N, Frias, D, Grangeiro, T, Grisard, E, Guimaraes, C, Hungria, M, Jardim, S, Krieger, M, Laurino, J, Lima, M, Lopes, M, Loreto, E, Madeira, H, Manfio, G, Maranhao, A, Martinkovics, C, Medeiros, S, Moreira, M, Neiva, M, Ramalho-Neto, C, Nicolas, M, Oliveira, S, Paixao, R, Pedrosa, F, Pena, S, Pereira, M, Pereira-Ferrari, L, Piffer, I, Pinto, L, Potrich, D, Salim, A, Santos, F, Schmitt, R, Schneider, M, Schrank, A, Schrank, I, Schuck, A, Seuanez, H, Silva, D, Silva, R, Silva, S, Soares, C, Souza, K, Souza, R, Staats, C, Steffens, M, Teixeira, S, Urmenyi, T, Vainstein, M, Zuccherato, L, Simpson, A, & Zaha., A. (2005.) *J. Bacteriol.* **187**, 5568–5577.
27. Uenoyama, A, Kusumoto, A, & Miyata., M. (2004.) *J. Bacteriol.* **186**, 1537–1545.
28. Uenoyama, A & Miyata., M. (2005.) *J. Bacteriol.* **187**, 5578–5584.
29. Pich, O, Burgos, R, Ferrer-Navarro, M, Querol, E, & Pinol., J. (2006.) *Mol. Microbiol.* **60**, 1509–1519.
30. Miyata. (2005.) *Tanpakushitsu Kakusan Koso* **50**, 239–245.
31. Bourret, R, Charon, N, Stock, A, & West., A. (2002.) *J. Bacteriol.* **184**, 1–17.
32. Kirchhoff, H, Boldt, U, Rosengarten, R, & Klein-Struckmeier., A. (1987.) *Curr. Microbiol.* **15**, 57–60.
33. Chambaud, I, Heilig, R, Ferris, S, Barbe, V, Samson, D, Galisson, F, Moszer, I, Dybvig, K, Wroblewski, H, Viari, A, Rocha, E, & Blanchard., A. (2001.) *Nucleic. Acids Res.* **29**, 2145–2153.
34. Fraser, C, Gocayne, J, White, O, Adams, M, Clayton, R, Fleischmann, R, Bult, C, Kerlavage, A, Sutton, G, Kelley, J, Fritchman, R, Weidman, J, Small, K, Sandusky, M, Fuhrmann, J, Nguyen, D, Utterback, R, Saudek, D, Phillips, C, Merrick, J, Tomb, J.-F, Dougherty, B, Bott, K, Hu, P.-C, Lucier, T, Peterson, S, Smith, H, III, C. H, & Venter., J. (1995.) *Science* **270**, 397–403.
35. Himmelreich, R, Hilbert, H, Plagens, H, Pirkl, E, Li, B.-C, & Herrmann., R. (1996.) *Nucleic Acids Res.* **24**, 4420–4449.
36. Papazisi, L, Gorton, T. S, Kutish, G, Markham, P. F, Browning, G. F, Nguyen, D. K, Swartzell, S, Madan, A, Mahairas, G, & Geary., S. J. (2003.) *Microbiology* **149**, 2307–2316.
37. Sasaki, Y, Ishikawa, J, Yamashita, A, Oshima, K, Kenri, T, Furuya, K, Yoshino, C, Horino, A, Shiba, T, Sasaki, T, & Hattori., M. (2002.) *Nucleic Acids Res.* **30**, 5293–5300.
38. Rosengarten, R, Fisher, M, Kirchhoff, H, Kerlen, G, & Seack., K.-H. (1988.) *Curr. Microbiol.* **16**, 253–257.
39. Stadtlander, C & Kirchhoff., H. (1990.) *Vet. Microbiol.* **21**, 339–343.
40. Stadtlander, C & Kirchhoff., H. (1995.) *Br. Vet. J.* **151**, 89–100.
41. Stadtlander, C, Lotz, W, Korting, W, & Kirchhoff., H. (1995.) *J. Comp. Pathol.* **112**, 351–359.
42. Kusumoto, A, Seto, S, Jaffe, J, & Miyata., M. (2004.) *Microbiology* **150**, 4001–4008.

43. Miyata, M & Uenoyama., A. (2002.) *FEMS Microbiol. Lett.* **215**, 285–289.
44. Miyata, M, Ryu, W, & Berg., H. (2002.) *J. Bacteriol.* **184**, 1827–1831.
45. Jordan, J, Chang, H, Balish, M, Holt, L, Bose, S, Hasselbring, B, 3rd, R. W, Krunkosky, T, & Krause., D. (2007.) *Infect. Immun.* **75**, 518–522.
46. Krunkosky, T, Jordan, J, Chambers, E, & Krause., D. (2007.) *Microb. Pathog.* **42**, 98–103.
47. Miyata, M, Yamamoto, H, Shimizu, T, Uenoyama, A, Citti, C, & Rosengarten., R. (2000.) *Microbiology* **146**, 1311–1320.
48. Seto, S, Uenoyama, A, & Miyata., M. (2005.) *J. Bacteriol.* **187**, 3502–3510.
49. Seto, S, Layh-Schmitt, G, Kenri, T, & Miyata., M. (2001.) *J. Bacteriol.* **183**, 1621–1630.
50. Gitai, Z. (2005.) *Cell* **120**, 577–586.
51. Miyata, M & Petersen., J. (2004.) *J. Bacteriol.* **186**, 4382–4386.
52. Metsugi, S, Uenoyama, A, Adan-Kubo, J, Miyata, M, Yura, K, Kono, H, & Go., N. (2005.) *Biophysics* **1**, 33–43.
53. Adan-Kubo, J, Uenoyama, A, Arata, T, & Miyata., M. (2006.) *J. Bacteriol.* **188**, 2821–2828.
54. Edeling, M, Smith, C, & Owen., D. (2006.) *Nat. Rev. Mol. Cell Biol.* **7**, 32–44.
55. Jaffe, J, Miyata, M, & Berg., H. (2004.) *J. Bacteriol.* **186**, 4254–4261.
56. Roberts, D, Olson, L, Barile, M, Ginsburg, V, & Krivan., H. (1989.) *J. Biol. Chem.* **264**, 9289–9293.
57. Nagai, R & Miyata., M. (2006.) *J. Bacteriol.* **188**, 6469–6475.
58. Gibbons, B & Gibbons., I. (1972.) *J. Cell. Biol.* **54**, 75–97.
59. Ohtani, N & Miyata., M. (2006.) *Biochem. J.* **403**, 71–77.
60. Fischer, M, Kirchhoff, H, Rosengarten, R, Kerlen, G, & Seack., K.-H. (1987.) *FEMS Microbiol. Lett.* **40**, 321–324.
61. Kawaguchi, K & Ishiwata., S. (2000.) *Biochem. Biophys. Res. Commun.* **272**, 895–899.
62. Kawai, M, Kawaguchi, K, Saito, M, & Ishiwata., S. (2000.) *Biophys. J.* **78**, 3112–3119.
63. Meister, M, Lowe, G, & Berg., H. (1987.) *Cell* **49**, 643–650.
64. Rosengarten, R & Kirchhoff., H. (1988.) *Curr. Microbiol.* **16**, 247–252.
65. Rosengarten, R, Klein-Struckmeier, A, & Kirchhoff., H. (1988.) *J. Bacteriol.* **170**, 989–990.
66. Finer, J, Simmons, R, & Spudich., J. (1994.) *Nature* **368**, 113–119.
67. Svoboda, K & Block., S. (1994.) *Cell* **77**, 773–784.
68. Charon, N. (2005.) *Proc. Natl. Acad. Sci. USA* **102**, 13713–13714.
69. Hiratsuka, Y, Miyata, M, Tada, T, & Uyeda., T. (2006.) *Proc. Natl. Acad. Sci. USA* **103**, 13618–13623.
70. Balish, M & Krause., D. (2006.) *J. Mol. Microbiol. Biotechnol.* **11**, 244–255.
71. Miyata, M & Ogaki., H. (2006.) *J. Mol. Microbiol. Biotechnol.* **11**, 256–264.
72. Razin, S & Jacobs., E. (1992.) *J. Gen. Microbiol.* **138**, 407–422.
73. Wilson, M & Collier., A. (1976.) *J. Bacteriol.* **125**, 332–339.
74. Kenri, T, Seto, S, Horino, A, Arakawa, Y, Sasaki, T, & Miyata., M. (2006.) *Mapping of localization sites of cytadherence-related and cytoskeletal proteins of Mycoplasma pneumoniae by fluorescent protein tagging.* p. 137.
75. Biberfeld, G & Biberfeld., P. (1970.) *J. Bacteriol.* **102**, 855–861.
76. Göbel, U, Speth, V, & Bredt., W. (1981.) *J. Cell Biol.* **91**, 537–543.
77. Meng, K & Pfister., R. (1980.) *J. Bacteriol.* **144**, 390–399.

78. Hegermann, J, Herrmann, R, & Mayer., F. (2002.) *Naturwissenschaften* **89**, 453–458.
79. Henderson, G & Jensen., G. (2006.) *Mol. Microbiol.* **60**, 376–385.
80. Regula, J, Boguth, G, Görg, A, Hegermann, J, Mayer, F, Frank, R, & Herrmann., R. (2001.) *Microbiology* **147**, 1045–1057.
81. Seybert, A, Herrmann, R, & Frangakis., A. (2006.) *J. Struct. Biol.* **156**, 342–354.
82. Hu, P, Cole, R. M, Huang, Y, Graham, J, Gardner, D, Collier, A, & Jr., W. C. (1982.) *Science* **216**, 313–315.
83. Kirchhoff, H, Rosengarten, R, Lotz, W, Fischer, M, & Lopatta., D. (1984.) *Israel J. Med. Sci.* **20**, 848–853.
84. Shimizu, T & Miyata., M. (2002.) *Curr. Microbiol.* **44**, 431–434.
85. Wall, F, Pfister, R, & Somerson., N. (1983.) *J. Bacteriol.* **154**, 924–929.
86. Balish, M. (2006.) *Front. Biosci.* **11**, 2017–2027.
87. Krause, D. (1996.) *Mol. Microbiol.* **20**, 247–253.
88. Balish, M & Krause., D. (2002.) *Molecular Biology and Pathogenicity of Mycoplasmas.*, eds. Herrmann, R & Razin, S. (Kluwer Academic / Plenum Publishers, London.), pp. 491–518.
89. Seto, S & Miyata., M. (2003.) *J. Bacteriol.* **185**, 1082–1091.
90. Balish, M, Santurri, R, Ricci, A, Lee, K, & Krause., D. (2003.) *Mol. Microbiol.* **47**, 49–60.
91. Hedreyda, C, Lee, K, & Krause., D. (1993.) *Plasmid* **30**, 170–175.
92. Krause, D, Proft, T, Hedreyda, C. T, Hilbert, H, Plagens, H, & Herrmann., R. (1997.) *J. Bacteriol.* **179**, 2668–2677.
93. Dandekar, T, Huynen, M, Regula, J, Ueberle, B, Zimmermann, C, Andrade, M, Doerks, T, Sanchez-Pulido, L, Snel, B, Suyama, M, Yuan, Y, Herrmann, R, & Bork., P. (2000.) *Nucleic Acids Res.* **28**, 3278–3288.
94. Krivan, H, Olson, L, Barile, M, Ginsburg, V, & Roberts., D. (1989.) *J. Biol. Chem.* **264**, 9283–9288.
95. Layh-Schmitt, G & Herrmann., R. (1994.) *Infect. Immun.* **62**, 974–979.
96. Layh-Schmitt, G, Podtelejnikov, A, & Mann., M. (2000.) *Microbiology* **146**, 741–747.
97. Layh-Schmitt, G, Hilbert, H, & Pirkl., E. (1995.) *J. Bacteriol.* **177**, 843–846.
98. Romero-Arroyo, C, Jordan, J, Peacock, S, Willby, M, Farmer, M, & Krause., D. (1999.) *J. Bacteriol.* **181**, 1079–1087.
99. Stevens, M & Krause., D. (1992.) *J. Bacteriol.* **174**, 4265–4274.
100. Jaffe, J, Berg, H, & Church., G. (2004.) *Proteomics* **4**, 59–77.
101. Hasselbring, B & Krause., D. (2007.) *Mol. Microbiol.* **63**, 44–53.
102. Jensen, G. (2007.) *Mol. Microbiol.* **63**, 4–6.
103. Wolgemuth, C, Igoshin, O, & Oster., G. (2003.) *Biophys. J.* **85**, 828–842.
104. Krause, D. (1998.) *Trends Microbiol.* **6**, 15–18.
105. Miyata, M. (2002.) *Molecular Biology and Pathogenicity of Mycoplasmas.*, eds. Herrmann, R & Razin, S. (Kluwer Academic / Plenum Publishers, London.), pp. 117–130.
106. Miyata, M & Seto., S. (1999.) *Biochimie* **81**, 873–878.
107. Seto, S & Miyata., M. (1998.) *J. Bacteriol.* **180**, 256–264.
108. Seto, S & Miyata., M. (1999.) *J. Bacteriol.* **181**, 6073–6080.
109. B.M., H, Jordan, J, Krause, R, & Krause., D. (2006.) *Proc. Natl. Acad. Sci. USA* **103**, 16478–16483.
110. Andrewes, C & Welch., F. (1946.) *J. Pathol. Bacteriol.* **58**, 578–580.

7

Hydrodynamics and Rheology of Active Polar Filaments

Tanniemola B. Liverpool[1] and M. Cristina Marchetti[2]

[1] Department of Mathematics, University of Bristol, University Walk, Bristol BS8 1TW, UK t.liverpool@bristol.ac.uk
[2] Physics Department, Syracuse University, Syracuse, NY 13244, USA mcm@phy.syr.edu

Summary. The cytoskeleton provides eukaryotic cells with mechanical support and helps them perform their biological functions. It is a network of semiflexible polar protein filaments and many accessory proteins that bind to these filaments, regulate their assembly, link them to organelles and continuously remodel the network. Here we review recent theoretical work that aims to describe the cytoskeleton as a polar continuum driven out of equilibrium by internal chemical reactions. This work uses methods from soft condensed matter physics and has led to the formulation of a general framework for the description of the structure and rheology of active suspension of polar filaments and molecular motors.

7.1 Introduction

Cells are living soft matter. They are composed of a variety of soft materials, such as lipid membranes, polymers, and colloidal aggregates, often constrained to a reduced spatial dimensionality and geometry. It is then reasonable to expect that the dynamics and interactions of these constituents that control cell function take place on the same time and energy scales as those of synthetic soft materials. Life adds, however, a new feature not found in traditional soft matter: the constant flow of energy and information required to keep living organisms alive. This new feature makes cells of particular interest to physicists because understanding the behavior of *active* living matter requires the development of new theoretical concepts and experimental techniques.

The eukaryotic cell cytoskeleton is a perfect example of this novel type of *active material*. The cytoskeleton allows the cell to carry out coordinated and directed movements such as cell crawling, muscle contraction, and all the changes in cell shape in the developing embryo [1]. The cytoskeleton also supports intracellular movements such as the transport of organelles in the cytoplasm and segregation of chromosomes during cell division [2]. It is highly inhomogeneous, with a large variety of different dynamical *supramolecular* structures. Examples are contractile elements like stress fibres, or the contractile ring in mitosis, or astral objects like the mitotic spindle that forms during cell division [1, 2].

Self-assembled filamentous protein aggregates play an important role in the mechanics and self-organization of the cytoskeleton. In addition, a number of other proteins interact with them and modulate their structure and dynamics. Cross-linking proteins bind to two or more filaments together to form a dynamical gel. Molecular motor proteins bind to filaments and hydrolyze nucleotide Adenosine triphosphate (ATP). This process, coupled to a corresponding conformational change of the protein, turns stored chemical energy into mechanical work. Capping proteins modulate the polymerization and depolymerization of the filaments at their ends.

A key question is how the elements of the cytoskeleton cooperate to achieve its function. To what extent is there a "cellular" brain and how closely does it control cellular mechanisms? How much of a role does spontaneous self-organization driven by general physical principles play?

Much of the recent progress in the understanding of the complex structures and processes that control the behavior and function of the cytoskeleton has been linked to the development of new biophysical probes allowing an unprecedented view of subcellular processes at work. Mechanical probes such as optical and magnetic tweezers [3], atomic force microscopes [4] and micropipettes probe the response of the elements of the cytoskeleton to locally applied forces. Visualization techniques using fluorescence microscopy (e.g., fluorescence imaging with one-nanometer accuracy [5] or single-molecule high-resolution colocalization [6] based on organic dyes) allow one to follow the dynamics of single molecules inside living cells (*in vivo*), giving insights into the microscopic processes underlying cellular dynamics. Many of these experimental developments are reviewed in this book.

Because of the large number of unknown components, it is also of interest to study simplified systems consisting of a smaller number of well-characterized elements *in vitro*. This has led to a number of experimental biophysical studies of purified solutions of cytoskeletal filaments and associated molecular motors that have established that motor-induced activity drives the formation of a variety of spatially inhomogeneous patterns, such as bundles, asters, and vortices [7, 8, 9, 10, 11, 12, 13, 14]. These are reminiscent of some of the supramolecular structures present in the cytoskeleton [15, 16]. The mechanical properties of filament-motor systems have also been studied, showing qualitative differences from passive filament suspension. Because of the controlled nature of their preparation, and the detailed knowledge of their constituents, *in vitro* studies are particularly amenable to a quantitative description using techniques from theoretical physics. In this review, we will be mostly concerned with describing the behavior of such simplified systems on large time scales (times ≥ microsecond) where the atomistic details are not important and a coarse-grained phenomenological description may suffice.

The reductionist viewpoint typified by this approach also has its drawbacks. A simplified system necessarily can provide only a subset of the phenomena observed in living cells because only a small fraction of the components are present. A choice must also be made of *which* simplified system to study because different combinations of components may give different or similar behavior. This choice must, of course, be heavily influenced by previous experiments [7, 8]. A living cell is a highly optimized, complex *system* of interacting agents with the ability to modulate its response to complex changes in its environment. This complexity will be missing from simple mechanical models described here. There is some hope, however, that this complexity can eventually be combined with the physical picture emerging from the approach we present here to give a more complete "biophysical" picture of motility in cells in which the laws of physics provide important constraints on the possible "system" dynamics. Finally, even within our limited frame of reference, we will also make a number of simplifying assumptions in developing the models. Some of the important physical phenomena ignored here, such as active polymerization, treadmilling [19, 20] and filament flexibility [21, 22, 23, 24, 25, 26, 27], will be incorporated in future work.

We first review some recent theoretical approaches to describe active filament suspensions. We then describe some of our current work and give perspectives for the future.

7.2 Theoretical Modeling of Active Systems

There have been a number of recent theoretical studies of the collective dynamics of mixtures of rigid filaments and motor clusters. First and most microscopic, numerical simulations with detailed modeling of the filament-motor coupling have yielded patterns similar to those found in experiments [11, 13], including vortices and asters. These simulations modeled the filaments as elastic rods with motor clusters being parameterized by three binding parameters, the on and off rates, and the off-rate at the plus end of the filament. It was found that the rate of motor unbinding at the polar end of the filaments plays a crucial role in controlling the vortex to aster transitions at high motor densities [12].

A second interesting development has been the proposal of "mesoscopic" mean-field kinetic equations first studied in one dimension [28, 29], where the effect of motors is incorporated via a motor-induced relative velocity of pairs of filaments, with the form of such a velocity inferred from general symmetry considerations. Kruse and collaborators [30, 31, 32] proposed a one-dimensional model of filament dynamics and showed the existence of instabilities from the homogeneous state to contractile states [30] and traveling-wave solutions [32]. We generalized the kinetic model to higher dimensions [33, 34] and used it to classify the nature of the homogeneous states and their stability [35, 36]. Related kinetic models have also been discussed by other authors [37, 38, 39, 40].

Finally, phenomenological continuum theories have been proposed where the mixture is described in terms of a few coarse-grained fields whose dynamics are inferred from symmetry considerations [41, 43, 42, 44, 45, 46, 47, 48, 49, 50, 51].

Lee and Kardar [43] proposed a simple hydrodynamic model for the coupled dynamics of a coarse-grained filament orientation and the motor concentration, ignoring fluctuations in the filament density. These authors argued that filament growth by polymerization provides a mechanism for an instability of the system from an isotropic to an oriented state [43, 42], with large-scale aster and vortex structures. They obtained a phase diagram for the system showing a transition from vortices to asters. This model was subsequently generalized by Sankararaman et al. [44] to include varying populations of bound and free motors, as well as an additional coupling of filament orientation to motor gradients. The effects of boundary conditions on the steady states of the system was also studied numerically.

A phenomenological hydrodynamic description for polar gels and suspensions including momentum conservation has been discussed by several authors [45, 46, 47, 48]. These equations generally consider incompressible suspensions and incorporate momentum conservation in the Stokes approximation, by assuming the form of constitutive equation for the suspension's stress tensor on the basis of symmetry consideration. The coupling of flow and polar order is described via an equation for the local polarization of the suspension. This model has been used to identify the nonequilibrium defect structures that can occur in the polar state [45, 46] and to analyze the behavior of an active polar suspension in specific geometries [52, 47]. In particular, it was shown that the interplay between order and activity can yield a spontaneous flowing state for a solution near a wall [47]. Closely related hydrodynamic models have been used to describe generically the collective dynamics of self-propelled particles in solution, such as swimming bacterial colonies, in both nematic and polar states [49, 50, 51]. This work builds on earlier work by Toner and Tu on hydrodynamic models of flocking, where it was shown that the nonequilibrium nature of internally driven systems allows for novel symmetry breaking phase transitions that are forbidden in equilibrium systems with continuous symmetry in one and two dimensions [53, 54, 55].

The main objective of our work has been to establish the connection between microscopic single-polymer dynamics and the phenomenological hydrodynamic models by deriving the hydrodynamic equations from a mean-field kinetic equation of filament dynamics. In the phenomenological approach, the system is described in terms of a few coarse-grained fields (conserved densities and broken symmetry variables) whose dynamics is inferred from symmetry considerations. The strength of this method is its generality. Its drawback is that for systems that are far from thermal equilibrium, and therefore lack constraints such as those provided by the fluctuation-dissipation theorem or the Onsager relations, all the parameters in the equations are undetermined. We have bridged the gap between microscopic models and continuum theories by deriving the hydrodynamic equations through a systematic coarse-graining of the microscopic dynamics. This derivation provides an estimate of the various parameters in the equations in terms of experimentally accessible quantities. We start with a Smoluchowski equation for filaments in solution, where motor proteins are described as active cross-linkers capable of exchanging forces and torques between filaments. The active currents arising from such motor-mediated exchange of forces and torques are obtained by considering the kinematics of two filaments cross-linked by a single active protein cluster that can rotate and translate at prescribed rates as a rigid object relative to the filaments. The hydrodynamic equations are then obtained by suitable coarse-graining of the Smoluchowski equation. This method yields a general form of the hydrodynamic equations that incorporates all terms allowed by symmetry, yet it provides a connection between the coarse-grained and the microscopic dynamics. In a series of earlier publications, we described in detail the derivation of the hydrodynamic equations for filaments in a quiescent solvent [33, 34, 56, 35, 36]. Here we generalize this work by incorporating the flow of the solvent. This is essential for describing the rheological properties of the solution. A brief account of some of the results presented here have been given elsewhere [57].

7.3 Hydrodynamics of a Solution of Polar Filaments

We consider a collection of rigid polar filaments in a viscous solvent. The solution forms a quasi two-dimensional film, of a thickness much smaller than the length of the filaments. Our goal is to study the interplay of order and flow in controlling the phases and the rheology of the system. The filaments diffuse in the solution and can be cross-linked by both active and stationary protein clusters. Active cross-linkers are small clusters of motor proteins that use chemical fuel as an energy source to generate forces and torques on the filaments, sliding and rotating filaments relative to each other [2, 12]. In addition, other small proteins, such as α-actinins, act as stationary cross-linkers and induce filament alignment [1].

As in passive solutions of rigid filaments, the large scale dynamics can be described in terms of a set of hydrodynamic equations for continuum fields that relax on time scales much longer than microscopic ones. These include the conserved variables of the systems, as well as any field associated with broken symmetries. Various forms of these equations have been written down phenomenologically by other authors. What distinguishes our work from these phenomenological approaches is that we derive the hydrodynamic equations from a mesoscopic model of coupled motor-filament dynamics. This allows us to estimate the various parameters in the hydrodynamic equations (that are undetermined in the phenomenological approach) and relate them to quantities that can be controlled in experiments. To make contact with the existing literature, we first present the equations and then discuss their derivation via coarse-graining of a Smoluchowski equation for rigid rods in a viscous solvent.

The conserved densities in a suspension of interacting filaments (rods) in a solvent are the mass densities of filaments (rods) $\rho_r(\mathbf{r}, t) = \mathrm{m}c(\mathbf{r}, t)$ and solvent $\rho_s(\mathbf{r}, t)$, and the total momentum density $\mathbf{g}(\mathbf{r}, t) = \rho(\mathbf{r}, t)\mathbf{v}(\mathbf{r}, t)$ of the solution (rods+solvent), with $\mathbf{v}(\mathbf{r}, t)$ the flow velocity and $\rho(\mathbf{r}, t) = \rho_s + \rho_r$ the total density. Here, $c(\mathbf{r}, t)$ is the *number* density of rods and m the mass of a rod. The conserved densities satisfy conservation laws, given by

$$\partial_t \rho = -\boldsymbol{\nabla} \cdot \mathbf{g} , \tag{7.1}$$

$$\partial_t c = -\boldsymbol{\nabla} \cdot \mathbf{J} , \tag{7.2}$$

$$\partial_t g_i + \partial_j (g_i g_j / \rho) = \partial_j \sigma_{ij} + \rho F_i^{ext} , \tag{7.3}$$

where $\mathbf{J}(\mathbf{r}, t)$ is the current density of rods and $\mathbf{F}^{ext}$ the external force on the suspension. The stress tensor σ_{ij} is the i-th component of the force exerted by the surrounding fluid on a unit area perpendicular to the j-th direction of a volume element of solution. It includes all forces on a volume of suspension exerted by the surrounding fluid. It can be written as the sum of solvent and filament contributions as

$$\sigma_{ij} = \sigma_{ij}^s + \sigma_{ij}^r . \tag{7.4}$$

The solvent contribution has the usual form appropriate for a viscous fluid,

$$\sigma_{ij}^s = 2\eta_0 u_{ij} + (\eta_b - \eta_0)\delta_{ij} u_{kk} - \delta_{ij} \Pi_s(\rho) , \tag{7.5}$$

where η_0 and η_b are the shear and bulk viscosity of the solvent, $\Pi_s(\rho)$ is the pressure of the solvent, and u_{ij} is the symmetrized rate of strain tensor:

$$u_{ij} = \frac{1}{2}\left(\partial_i v_j + \partial_j v_i\right) . \tag{7.6}$$

In the low Reynolds number limit we can ignore the inertial terms on the left-hand side of the solvent momentum equation, Equation (7.3).

In a solution of long filaments states with liquid crystalline order are possible. Polar rods can form polarized and nematic states, both characterized by orientational order, but with different symmetries for the order parameters. Polar order in a fluid of N rods is characterized by a vector order parameter or polarization, $\mathbf{P}(\mathbf{r}, t)$, defined by

$$\mathbf{P}(\mathbf{r}, t) = \frac{1}{c(\mathbf{r}, t)} \left\langle \sum_{\alpha=1}^{N} \hat{\mathbf{u}}_\alpha(t) \delta(\mathbf{r} - \mathbf{r}_\alpha(t)) \right\rangle , \qquad (7.7)$$

where $\mathbf{r}_\alpha$ is the position of the center of mass of the α-th rod and $\hat{\mathbf{u}}_\alpha$ is a unit vector directed along the polar direction. The angular brackets denote an ensemble average. The polarization vector $\mathbf{P}$ can be written as

$$\mathbf{P}(\mathbf{r}, t) = P(\mathbf{r}, t) \mathbf{p}(\mathbf{r}, t) , \qquad (7.8)$$

where the magnitude of the polarization $P(\mathbf{r}, t)$ is the scalar order parameter and the unit vector $\mathbf{p}(\mathbf{r}, t)$ identifies the direction of broken symmetry in the ordered state.

Nematic order is described by the conventional nematic order parameter or alignment tensor, defined as

$$Q_{ij}(\mathbf{r}, t) = \frac{1}{c(\mathbf{r}, t)} \left\langle \sum_{\alpha=1}^{N} \left(u_{\alpha i}(t) u_{\alpha j}(t) - \frac{1}{d} \delta_{ij} \right) \delta(\mathbf{r} - \mathbf{r}_\alpha(t)) \right\rangle . \qquad (7.9)$$

The subtracted part ensures that the order parameter vanishes in the isotropic state in d dimensions. The alignment tensor Q_{ij} is thus a traceless and symmetric second-order tensor field, with two independent degrees of freedom in $d = 2$. For uniaxial nematics, the alignment tensor takes the form

$$Q_{ij} = S(\mathbf{r}, t) \left[n_i(\mathbf{r}, t) n_j(\mathbf{r}, t) - \frac{1}{d} \delta_{ij} \right] , \qquad (7.10)$$

where $S(\mathbf{r}, t)$ is the scalar nematic amplitude and $\mathbf{n}(\mathbf{r}, t)$ is the familiar nematic director. The nematic state has orientational order ($S \neq 0$) and it is invariant under inversion of the director, i.e., for $\mathbf{n} \to -\mathbf{n}$. The polarized state, in contrast, is not invariant for $\mathbf{p} \to -\mathbf{p}$. In a polarized state, the alignment tensor Q_{ij} is slaved to the polarization and acquires a nonzero value, with $\mathbf{n} = \mathbf{p}$.

The dynamical equations for polarization and alignment tensor have the form (for simplicity, we give the form for $d = 2$ only)

$$D_t(cP_i) = cF_i(\boldsymbol{\kappa}, \mathbf{P}) - \partial_j J_{ij} - R_i , \qquad (7.11)$$

$$D_t(cQ_{ij}) = cF_{ij}(\boldsymbol{\kappa}, \mathbf{Q}) - \partial_k J_{ijk} - R_{ij} , \qquad (7.12)$$

where $D_t = \partial_t + \mathbf{v} \cdot \boldsymbol{\nabla}$, $\boldsymbol{\kappa}$ is the rate of strain tensor, $\kappa_{ij} = \partial_j v_i$, and

$$F_i(\boldsymbol{\kappa}, \mathbf{P}) = -\omega_{ij} P_j + \lambda_P u_{ij} P_j - \frac{5}{4} u_{kk} P_i \quad , \quad \text{or}$$

$$\mathbf{F}(\boldsymbol{\kappa}, \mathbf{P}) = \frac{1}{2} (\boldsymbol{\nabla} \times \mathbf{v}) \times \mathbf{P} + \lambda_P \left[\boldsymbol{\nabla} \mathbf{v} + (\boldsymbol{\nabla} \mathbf{v})^T \right] \cdot \mathbf{P} - \frac{5}{4} (\boldsymbol{\nabla} \cdot \mathbf{v}) \mathbf{P} , \qquad (7.13)$$

$$F_{ij}(\boldsymbol{\kappa}, \mathbf{Q}) = -\left(\omega_{ik}Q_{kj} + \omega_{jk}Q_{ki}\right) + \frac{1}{3}\left(u_{ik}Q_{kj} + u_{jk}Q_{ki} - \delta_{ij}u_{kl}Q_{kl}\right)$$
$$- \frac{4}{3}u_{kk}Q_{ij} + \lambda\left(u_{ij} - \frac{1}{2}\delta_{ij}u_{kk}\right). \tag{7.14}$$

Here, λ_P and λ are the flow alignment parameters in the polarized and nematic states, respectively, and

$$\omega_{ij} = \frac{1}{2}\left(\partial_i v_j - \partial_j v_i\right). \tag{7.15}$$

The low-density derivation based on the Smoluchowski equation described in the next Section gives $\lambda_P = 1/2$ and $\lambda = 1/(2S_0)$, with S_0 the magnitude of the nematic order parameter. Since typically in the nematic state even quite far from the I-N transition, $S_0 \ll 1$, we expect $\lambda > 1$, as required for flow alignment. In addition it is well known that deep in the nematic phase, higher order correlations can further increase the value of λ. In the following, we will treat both λ_P and λ as unknown parameters. The first terms on the right hand side of Equations (7.11) and (7.12) generalize the convective derivative on the left-hand side of the equation to the case of long, thin rods. These are standard terms in nematohydrodynamics that have been derived from a microscopic model before [58]. Different values for some of the numerical coefficients are reported in the literature, depending on the closure scheme used in evaluating various angular averages. The second and third terms on the right-hand side of Equations (7.11) and (7.12) represent translational and rotational currents, including contributions from diffusion, excluded volume, and both stationary and active cross-linkers. The relaxation of the order parameters is controlled by the rotational currents and is non-hydrodynamic. In contrast, the relaxation of the broken symmetry variables $\mathbf{p}$ and $\mathbf{n}$ is controlled by hydrodynamic Goldstone modes, as appropriate in ordered states.

The long wavelength description of the solution is then given by the Equations (7.1-7.3) and (7.11-7.12). To close the hydrodynamic equations we must derive the constitutive equations for the fluxes ($\mathbf{J}$, J_{ij}, J_{ijk}), the rotational currents (R_i and R_{ij}), and the filament contribution to the stress tensor, σ_{ij}^r, as functions of the system's properties (density, filament concentration and order parameters) and of the driving forces (applied mechanical stresses and activity, as measured by the ATP consumption rate). This derivation is carried out below by adapting methods from polymer physics appropriate for a dilute solution of rigid rods. Although the specific expressions obtained by this method for the parameters in the hydrodynamic equations only apply at low concentration of filaments, the structure of the equations is general and remains the same at high density.

7.4 Derivation of Hydrodynamic Constitutive Equations

Our goal is to derive the constitutive equations for the various hydrodynamic currents and stresses starting from a semi-microscopic model of the dynamics of single filaments coupled pairwise by active and stationary cross-linkers. The filaments are modeled as rigid rods of fixed length l and diameter $b \ll l$ immersed in a viscous solvent. They diffuse independently in the solvent and interact via excluded volume. In addition, filaments can be coupled pairwise by both stationary and active cross-linkers that generate additional active currents. Active cross-linkers are described as rigid links that can walk along the filaments towards the polar end at a prescribed rate $u(s)$ proportional to the rate of ATP consumption. Generally, $u(s)$ varies with the point s of attachment along the filament ($0 \leq s \leq l$). Both active and stationary cross-links also mediate the exchange of torques between the filaments by acting as torsional springs of prescribed stiffness, κ. Our goal is to obtain a coarse-grained description of the system, where all the parameters in the hydrodynamic equations are characterized in terms of $u(s)$, κ, and the density of cross-linkers. Collective effects arising from multiple cross-linkers are neglected and the density of cross-linkers is assumed constant for simplicity. We also neglect the dynamics of cross-linkers binding and unbinding, which occurs on faster time scales than those of interest here, so that we can treat a constant fraction of them as *bound*. The dynamics of cross-linkers binding and unbinding was considered, for instance, in [44] and it was found that varying the rates of motor binding and unbinding did not affect the nonequilibrium steady states of the active solution. The derivation of the active contributions to the various fluxes has been presented elsewhere [36] and will be summarized here for completeness. We also present novel results on the evaluation of the filament contribution to the stress tensor up to terms of first order in gradients of the hydrodynamic fields.

To proceed, we also make a series of simplifying assumptions on the dynamics of the solution. First, we assume that the friction between filaments and solvent is large and the filaments move at the flow velocity $\mathbf{v} = \mathbf{g}/\rho$ of the solution. In many fluid mixtures, internal friction mechanisms are so strong that the flow velocities of the two components relax on microscopic time scales to the common value $\mathbf{v}$. There are situations, however, where the relaxation time of the relative momenta of the two species is slow enough to have a significant influence even on hydrodynamic time scales. In this case, a two-fluid description is appropriate and useful. Such a "two-fluid model" of the system (rods and fluid background) will be described elsewhere, where we will show under which conditions one approaches the one-fluid model (which is always the true hydrodynamic limit).

We also limit ourselves to the case of incompressible solutions, with $\rho = \rho_s + \rho_r =$ constant, which requires

$$\nabla \cdot \mathbf{v} = 0 . \tag{7.16}$$

Finally, we neglect fluid inertial effect compared to the frictional forces between the colloidal rods and the solvent. In this limit, the momentum equation (7.3) reduces to the Stokes equation

$$\partial_j \sigma_{ij} = -\rho F_i^{\text{ext}} , \tag{7.17}$$

or, in the absence of external forces,

$$\eta_0 \nabla^2 v_i - \partial_i \Pi_s = -\partial_j \sigma_{ij}^r . \tag{7.18}$$

Equation (7.18) shows that the flow velocity of the solution is determined by the stress introduced by the filaments. In turn, the forces that the filaments exert on each other and on the solvent depend on the flow of the suspension in which they are immersed, and the problem must be solved self-consistently.

The dynamics of a dilute suspension of rods in the presence of a macroscopic flow field $\mathbf{v}(\mathbf{r})$ can de described by the Smoluchowski equation for the probability distribution $\hat{c}(\mathbf{r}, \hat{\mathbf{u}}, t)$ of rods with center of mass at $\mathbf{r}$ and orientation $\hat{\mathbf{u}}$ at time t. The Smoluchowski equation describes the mean-field Brownian dynamics of extended colloidal particles at low Reynolds number, under the assumption that the particles' velocities have equilibrated on microscopic time scales to a local Maxwell-Boltzmann distribution at a temperature T_a [59, 60]. The effective temperature T_a incorporates nonthermal noise sources as may arise from fluctuations in motor concentration and ATP consumption rate. The Smoluchowski equation is given by

$$\partial_t \hat{c} + \nabla \cdot \mathbf{J}_c + \mathcal{R} \cdot \boldsymbol{\mathcal{J}}_c = 0 , \tag{7.19}$$

where $\mathcal{R} = \hat{\mathbf{u}} \times \partial_{\hat{\mathbf{u}}}$ is the rotation operator. The *translational* probability current, $\mathbf{J}_c$, and the *rotational* probability current, $\boldsymbol{\mathcal{J}}_c$, are given by

$$J_{ci} = \hat{c} v_i - D_{ij} \nabla_j \hat{c} - \frac{D_{ij}}{k_B T_a} \hat{c} \, \nabla_j U_{\mathrm{ex}} + J_{ci}^A , \tag{7.20}$$

$$\mathcal{J}_{ci} = \hat{c} \omega_i - D_r \mathcal{R}_i \hat{c} - \frac{D_r}{k_B T_a} \hat{c} \mathcal{R}_i U_{\mathrm{ex}} + \mathcal{J}_{ci}^A , \tag{7.21}$$

where $\omega_i = \epsilon_{ijk} \hat{u}_j \hat{u}_l \partial_l v_k$. Also $D_{ij} = D_{\parallel} \hat{u}_i \hat{u}_j + D_{\perp} (\delta_{ij} - \hat{u}_i \hat{u}_j)$ is the translational diffusion tensor and D_r is the rotational diffusion rate. For a low-density solution of long, thin rods $D_{\perp} = D_{\parallel}/2 \equiv D/2$, where $D = k_B T_a \ln(l/b)/(2\pi \eta_0 l)$, and $D_r = 6D/l^2$. The potential U_{ex} incorporates excluded volume effects that give rise to the nematic transition in a solution of hard rods. It can be written by generalizing the Onsager interaction to inhomogeneous systems as $k_B T_a$ times the probability of finding another rod within the interaction area of a given rod. In two dimensions, this gives

$$U_{\mathrm{ex}}(\mathbf{r}_1, \hat{\mathbf{u}}_1) = k_B T_a \int d\hat{\mathbf{u}}_2 \int_{s_1 s_2} |\hat{\mathbf{u}}_1 \times \hat{\mathbf{u}}_2| \, \hat{c}(\mathbf{r}_1 + \boldsymbol{\xi}, \hat{\mathbf{u}}_2, t) , \tag{7.22}$$

where s_i, with $-l/2 \leq s_i \leq l/2$, parameterizes the position along the length of the i-th filament for $i = 1, 2$, and $\int_{s_i} \ldots \equiv \int_{-l/2}^{l/2} ds_i \ldots \equiv \langle .. \rangle_{s_i}$. The filaments are constrained to be within each other's interaction volume, i.e., in the thin rod limit $b \ll l$ considered here, have a point of contact. The factor $|\hat{\mathbf{u}}_1 \times \hat{\mathbf{u}}_2|$ represents the excluded area of two thin filaments of orientation $\hat{\mathbf{u}}_1$ and $\hat{\mathbf{u}}_2$ touching at one point [61]. Finally, $\boldsymbol{\xi} = \mathbf{r}_2 - \mathbf{r}_1 \simeq \hat{\mathbf{u}}_1 s_1 - \hat{\mathbf{u}}_2 s_2$, is the separation of the centers of mass of the two rods. The translational and rotational active current of filaments with center of mass at $\mathbf{r}_1$ and orientation along $\hat{\mathbf{u}}_1$ are written as

$$\mathbf{J}_c^A(\mathbf{r}_1, \hat{\mathbf{u}}_1) = \hat{c}(\mathbf{r}_1, \hat{\mathbf{u}}_1, t) b^2 m \int_{\hat{\mathbf{u}}_2} \int_{s_1 s_2} \mathbf{v}_a(1; 2) \hat{c}(\mathbf{r}_1 + \boldsymbol{\xi}, \hat{\mathbf{u}}_2, t) , \tag{7.23}$$

$$\boldsymbol{\mathcal{J}}_c^A(\mathbf{r}_1, \hat{\mathbf{u}}_1) = \hat{c}(\mathbf{r}_1, \hat{\mathbf{u}}_1, t) b^2 m \int_{\hat{\mathbf{u}}_2} \int_{s_1 s_2} \boldsymbol{\omega}_a(1; 2) \hat{c}(\mathbf{r}_1 + \boldsymbol{\xi}, \hat{\mathbf{u}}_2, t) , \tag{7.24}$$

where m is the density of bound cross-linkers and $(1; 2) = (s_1, \hat{\mathbf{u}}_1; s_2, \hat{\mathbf{u}}_2)$. Finally, $\mathbf{v}_a(1; 2)$ and $\boldsymbol{\omega}_a(1; 2)$ are the translational and rotational velocities, respectively, that Filament 1 acquires due to the cross-linker-mediated interaction with Filament 2, when the centers of mass of the two filaments are separated by $\boldsymbol{\xi}$ (see Figure 7.1).

The derivation of the form of the active velocities in terms of motor parameters (the stepping rate $u(s)$ and the torsional stiffness κ) has been discussed in detail elsewhere [56, 36]. The angular velocity is

$$\boldsymbol{\omega}_a = 2 \left[\gamma_P + \gamma_{NP} (\hat{\mathbf{u}}_1 \cdot \hat{\mathbf{u}}_2) \right] (\hat{\mathbf{u}}_1 \times \hat{\mathbf{u}}_2) , \tag{7.25}$$

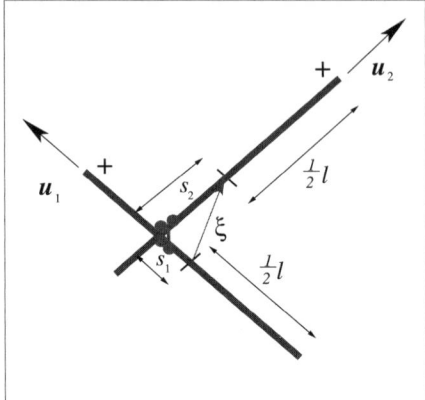

Figure 7.1. The geometry of overlap between two interacting filaments of length l cross-linked by an active cluster. The cross-link is a distance $s_1, (s_2)$ from the center of mass of filament 1(2). The distance between centers is $\boldsymbol{\xi} = \mathbf{r}_2 - \mathbf{r}_1 = s_1 \hat{\mathbf{n}}_1 - s_2 \hat{\mathbf{n}}_2$.

with γ_P and γ_{NP} motor-induced rotation rates due to polar and nonpolar cross-linkers, respectively (see Figure 7.2). The motor-induced translational velocity has the form $\mathbf{v}_a(1;2) = \frac{1}{2}\mathbf{v}_r + \mathbf{V}_m$, with [36]

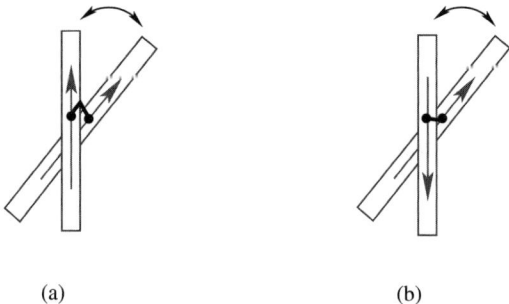

        (a)   (b)

Figure 7.2. Polar and nonpolar clusters interacting with polar filaments. Assuming that clusters always bind to the smallest angle, polar clusters bind only to filaments in configuration (a) while nonpolar clusters bind to both configurations equally.

$$\mathbf{v}_r = \frac{\tilde{\beta}}{2}(\hat{\mathbf{u}}_2 - \hat{\mathbf{u}}_1) + \frac{\tilde{\alpha}}{2l}\boldsymbol{\xi} \, ,$$
$$\mathbf{V}_m = A(\hat{\mathbf{u}}_2 + \hat{\mathbf{u}}_1) + B(\hat{\mathbf{u}}_2 - \hat{\mathbf{u}}_1) \, ,$$

where $\tilde{\alpha} = \alpha(1 + \hat{\mathbf{u}}_1 \cdot \hat{\mathbf{u}}_2)$ and $\tilde{\beta} = \beta(1 + \hat{\mathbf{u}}_1 \cdot \hat{\mathbf{u}}_2)$. The expressions for A and B have been obtained in [36] using momentum conservation. For long thin rods with $\zeta_\perp = 2\zeta_\parallel \equiv 2\zeta$, to leading order in $\hat{\mathbf{u}}_1 \cdot \hat{\mathbf{u}}_2$, we find $A = -[\beta - \alpha(s_1 + s_2)/2]/12$ and $B = \alpha(s_1 - s_2)/24$.

The rotational rates γ_P and γ_{NP}, and the active velocities α and β, can be related to the torsional stiffness κ of the cross-linkers and to the rate $u(s)$ at which a motor cluster attached at position s steps along a filament towards the polar end. This rate will, in general, depend on the point of attachment s, due for instance to crowding or stalling of motors near the polar end. The mean (averaged along the filament) stepping rate $u_0 = \langle u(s) \rangle / l$ is simply proportional to the mean rate R_{ATP} of ATP consumption via the characteristic step length, which we take of order of the thickness b of the filaments, $u_0 \sim b R_{ATP}$. We emphasize, however, that in general the stepping rate $u(s)$ (and other active parameters) may be a *non-linear* and possibly even non-monotonic function of the rate of ATP consumption, R_{ATP}.

In our model, there are three coupled mechanisms for cross-linker-induced filament dynamics, described by the parameters α, β, and the rotational rates γ_P and γ_{NP}. The first is the bundling of filaments of similar polarity at a rate α given by [56]

$$\alpha = \int_{-l/2}^{l/2} \frac{ds}{l} \frac{s}{l} \, u(s) \approx u_0 (b/l) \, , \qquad (7.26)$$

where the last approximate equality applies in situations where $u(s)$ exhibits strong spatial variations on length scales of order b, as may arise for instance from motor stalling at the polar end [56]. It is apparent from Equation (7.26) that $\alpha = 0$ if $u(s)$ is constant. Bundling is driven by the contractile nature of motor clusters, and in our mean field model requires spatial inhomogeneities at the rate at which motors step along the filaments. As we will see, it tends to build up density inhomogeneities and is the main pattern-forming mechanism. The second mechanism of motor-induced dynamics will be referred to as "polarization sorting", although generally it involves coupled filament rotation and spatial separation of filaments of different polarity. It occurs at the rate

$$\beta = \int_{-l/2}^{l/2} \frac{ds}{l} u(s) = u_0 \, , \qquad (7.27)$$

and vanishes for aligned filaments. This mechanism is especially important in the polar state where it allows for terms in the hydrodynamic equations corresponding to convection of filaments along the direction of mean polarization, and it provides the mechanism for the transition to a state with spontaneous flow [47]. Finally, motor-induced filament rotations occur at rates γ_P and γ_{NP} for cross-linkers that preferentially bind to filaments of the same orientation (γ_P) or regardless of their orientation (γ_{NP}). As discussed in [36], we estimate

$$\gamma_P \sim \gamma_{NP} \sim \frac{\kappa}{\zeta_r} \, , \qquad (7.28)$$

with $\zeta_r = k_B T_a / D_r$ (a rotational friction). In general, both active and stationary cross-linkers may induce rotation and be either polar or apolar in nature. In the following, we will restrict ourselves for simplicity to the case where all polar cross-linkers are active motor clusters (of density m_a), while all apolar cross-linkers are stationary (of density m_s). In practice we do expect this to be often the case. The rotational rate γ_P will then depend on ATP consumption, with $\gamma_P \sim R_{ATP}$, while we expect γ_{NP} to be essentially independent of it. As mentioned above, the various active parameters may be nonlinear and even nonmonotonic functions of R_{ATP}. However, these effects will not be considered here.

From the Smoluchowski equation (Equation (7.19)) we obtain the hydrodynamic equations for filament concentration, polarization and alignment tensor by truncating the exact moment expansion of $\hat{c}(\mathbf{r}, \hat{\mathbf{u}}, t)$ as

$$\hat{c}(\mathbf{r}, \hat{\mathbf{u}}, t) = \frac{c(\mathbf{r}, t)}{2\pi}\left\{1 + 2\mathbf{P}(\mathbf{r}, t) \cdot \hat{\mathbf{u}} + 4Q_{ij}(\mathbf{r}, t)\hat{Q}_{ij}(\hat{\mathbf{u}}) + \ldots\right\}, \qquad (7.29)$$

with $\hat{Q}_{ij}(\hat{\mathbf{u}}) = \hat{u}_i\hat{u}_j - \frac{1}{2}\delta_{ij}$ and keeping only the first three moments,

$$\int d\hat{\mathbf{u}}\, \hat{c}(\mathbf{r}, \hat{\mathbf{u}}, t) = c(\mathbf{r}, t) \text{ (density)},$$

$$\int d\hat{\mathbf{u}}\, \hat{\mathbf{u}}\, \hat{c}(\mathbf{r}, \hat{\mathbf{u}}, t) = c(\mathbf{r}, t)\mathbf{P}(\mathbf{r}, t) \text{ (polarization)}, \qquad (7.30)$$

$$\int d\hat{\mathbf{u}}\, \hat{Q}_{ij}(\hat{\mathbf{u}})\, \hat{c}(\mathbf{r}, \hat{\mathbf{u}}, t) = c(\mathbf{r}, t)Q_{ij}(\mathbf{r}, t) \text{ (nematic order)}.$$

The details of the calculation, which involves using a small gradient expansion for the filament probability distribution and evaluating angular averages, are given in [36], where the full expression for the various fluxes and rotational currents are also displayed. The resulting hydrodynamic equations in the isotropic and ordered phases will be given in the next Section.

7.5 Stress Tensor of an Active Solution

In this section, we derive the constitutive equation for the filament contribution to the stress tensor of an active suspension of polar filaments. An important difference as compared to passive solutions is that in active systems stresses can be induced not just by externally applied mechanical deformations (yielding $\kappa_{ij} \neq 0$), but also by motor activity which maintains the system out of equilibrium by supplying energy at a rate R_{ATP}.

In the limit where inertial effects may be ignored (low Reynolds number) and in the absence of external forces, momentum conservation is described by Equation (7.18), with $\nabla \cdot \mathbf{v} = 0$. Using standard methods from polymer physics, the filament contribution to the divergence of the stress tensor of a dilute suspension of hard, thin rods can be written as

$$\nabla \cdot \sigma^r = -\int_\xi \int_{\hat{u}} \hat{c}(\mathbf{r} - \xi, \hat{\mathbf{u}}, t)\left\langle \delta\left(\xi - s\hat{\mathbf{u}}\right)\mathcal{F}^h(s)\right\rangle_s, \qquad (7.31)$$

where $\mathcal{F}^h(s)$ is the hydrodynamic force per unit length exerted by the suspension on a rod at position s along the rod. It arises from interactions with other filaments and proteins, and with solvent molecules. It depends implicitly upon direct interactions between the rods, as well as on hydrodynamic interactions mediated by the solvent.

The hydrodynamic force density on a rigid rod suspended in a viscous solvent can be expressed in terms of the force and torque at its center of mass. A sketch of the derivation is given in the Appendix. Further details of similar calculations can be found in [61, 62]. We find that the stress due to the filaments can be written in the form (to $\mathcal{O}(\nabla^2)$)

$$\nabla \cdot \boldsymbol{\sigma}^r(\mathbf{r}, t) = \int_{\hat{\mathbf{u}}} \hat{c}(\mathbf{r}, \hat{\mathbf{u}}, t) \mathbf{F}^h(\mathbf{r}, \hat{\mathbf{u}}, t)$$

$$- \int_{\hat{\mathbf{u}}} \left\langle \left(\frac{s}{l}\right)^2 \left(\frac{\hat{\mathbf{u}} \cdot \nabla}{l}\right) \hat{c}(\mathbf{r}, \hat{\mathbf{u}}, t) \tau^h(\mathbf{r}, \hat{\mathbf{u}}, t) \right\rangle_s . \tag{7.32}$$

In the absence of inertial effects, the *total* hydrodynamic force, $\mathbf{F}^h(\mathbf{r}, \hat{\mathbf{u}}, t)$, exerted by the suspension *on the center of mass* of a rod, can be found from the condition that all forces acting on the rod must balance. The solvent flow field on a given segment of a rod is calculated using a decoupling approximation where the hydrodynamic coupling to other segments of the same rod are treated explicitly within the Oseen approximation, while the hydrodynamic effects of other rods enter in the determination of a self-consistent value for the flow velocity of the solvent, yielding,

$$\mathbf{F}^h(\mathbf{r}, \hat{\mathbf{u}}, t) = k_B T_a \nabla \ln \hat{c} + \nabla U_{\text{ex}} - \mathbf{F}_a , \tag{7.33}$$

where $-k_B T_a \nabla \ln \hat{c}$ is the Brownian force, $-\nabla U_{\text{ex}}$ is the force due to the direct interaction of the rod with other rods (in this case, via excluded volume), and $\mathbf{F}_a$ is the active force that can be written as

$$F_{ai} = \zeta_{ij}(\hat{\mathbf{u}}) J_{ci}^A / \hat{c} . \tag{7.34}$$

The rod friction tensor $\zeta_{ij}(\hat{\mathbf{u}})$ is proportional to the inverse of the rod diffusion tensor $D_{ij}(\hat{\mathbf{u}})$, with

$$\zeta_{ij}(\hat{\mathbf{u}}) = k_B T_a \left[\mathbf{D}^{-1}(\hat{\mathbf{u}}) \right]_{ij}$$

$$= \zeta_\parallel \hat{u}_i \hat{u}_j + \zeta_\perp (\delta_{ij} - \hat{u}_i \hat{u}_j) , \tag{7.35}$$

with $\zeta_\parallel = 2\pi\eta_0 l / \ln(l/b)$, and $\zeta_\perp = 2\zeta_\parallel$. Similarly, the total hydrodynamic torque is given by

$$\boldsymbol{\tau}^h(\mathbf{r}, \hat{\mathbf{u}}, t) = [k_B T_a \mathcal{R} \ln c + \mathcal{R} U_x - \tau_a] \times \hat{\mathbf{u}}$$

$$- \frac{\zeta_\perp}{2} \hat{\mathbf{u}} \hat{\mathbf{u}} (\hat{\mathbf{u}} \cdot \nabla) \cdot \mathbf{v}(\mathbf{r}) , \tag{7.36}$$

with $\tau_a = \zeta_r \mathcal{J}_c^A / \hat{c}$ the active torque. The last term on the right hand side of Equation (7.36) is a viscous contribution to the stress proportional to the velocity gradient.

The rod contribution to the stress tensor can now be evaluated explicitly using the truncated moment expansion for $\hat{c}(\mathbf{r}, \hat{\mathbf{u}}, t)$, given in Equation (7.29). When evaluating the active contributions to the stress tensor, only terms up to first order in $\hat{\mathbf{u}}_1 \cdot \hat{\mathbf{u}}_2$ are retained in the active force $\boldsymbol{\zeta}(\hat{\mathbf{u}}_1) \cdot \mathbf{v}_a(1; 2)$ exerted by a motor cluster on the filament. This approximation only affects the numerical values of the coefficients in the stress tensor, not its general form.

For simplicity, we consider solutions in the presence of a constant velocity gradient, κ_{ij}, and with a uniform mean rate of ATP consumption. We allow for spatial inhomogeneities in the filament concentration and orientational order parameters and evaluate the stress tensor up to first order in gradients of these hydrodynamic fields. The deviatoric part $\tilde{\sigma}_{ij} = \sigma_{ij} - (1/2)\delta_{ij}\sigma_{kk}$ of the stress tensor of the filaments is

$$\tilde{\sigma}_{ij}^r(\mathbf{r}, t) = \tilde{\sigma}_{ij}^A(\mathbf{r}, t) + \tilde{\sigma}_{ij}^v(\mathbf{r}, t) , \tag{7.37}$$

with

$$\tilde{\sigma}_{ij}^A = 2k_B T_a c\left(1 - \frac{c}{c_{IN}}\right)Q_{ij} - k_B T_a \frac{c^2}{c_{IP}}\left(P_i P_j - \frac{1}{2}\delta_{ij}P^2\right)$$
$$+m_a b^2 \alpha \frac{k_B T_a l^3}{72D}c^2\left(\frac{4}{3}Q_{ij} + P_i P_j - \frac{1}{2}\delta_{ij}P^2\right)$$
$$+m_a b^2 \beta \frac{k_B T_a l^4}{216D}c^2\left[\partial_j P_i - \frac{1}{2}\delta_{ij}\boldsymbol{\nabla}\cdot\mathbf{P} - \frac{1}{4}\left(\partial_i P_j - \partial_j P_i\right)\right], \quad (7.38)$$

where $c_{IP} = D_r/(m_a b^2 \gamma_P l^2)$ and $c_{IN} = c_N/[1 + c_N l^2 m_s b^2 \gamma_{NP}/(4D_r)]$ are the densities for the isotropic-polarized (IP) and isotropic-nematic (IN) transition, respectively, at finite density of active polar motor clusters (m_a) and stationary nonpolar cross-linkers (m_s)[36]. Finally, $c_N = 3\pi/(2l^2)$ is the density of the IN transition in passive systems. There are three types of contributions to the active part of the stress tensor. The first consists of the first two terms on the right-hand side of Equation (7.38). These are equilibrium-like terms, in the sense that they have the same structure one would obtain in a nematic and polar passive fluid, respectively, with the transition densities replaced by their active values. In particular, the first term on the right-hand side of Equation (7.38) should be compared to the corresponding contribution for isotropic ($c < c_N$) passive solutions, $\tilde{\sigma}_{ij}^P = 2k_B T c\left(1 - \frac{c}{c_N}\right)Q_{ij}$. The third term is a homogeneous nonequilibrium contribution that remains nonzero even for $\kappa_{ij} = 0$. This "spontaneous stress" arises from activity and is proportional to the ATP consumption rate that acts as an additional driving force and can build up stresses even in the absence of mechanical deformations. This term is generated by motor-induced filament bundling and is proportional to the bundling rate α. It would therefore vanish in the absence of spatial inhomogeneities in the motor stepping rate. Finally, the fourth term contains active contributions proportional to gradients of the polarization (we have omitted here terms of linear order in the gradients containing both polarization and alignment tensor. The full expression for the stress tensor can be found in the Appendix). These stresses are generated by motor-induced filament sorting and are proportional to β. They are important only in the polarized phase, where we expect they will play an important role in enhancing the relaxation of longitudinal fluctuations of the filaments and the corresponding relaxation of shear via reptation.

Finally, the viscous contribution to the stress is

$$\tilde{\sigma}_{ij}^v = \frac{lc\zeta_\perp}{48}\left[\frac{1}{2}\left(u_{ij} - \frac{1}{2}\kappa_{kk}\delta_{ij}\right) + \frac{1}{3}\left(Q_{ij}\kappa_{kk} - \delta_{ij}\kappa_{kq}Q_{qk}\right)\right.$$
$$\left. + \frac{2}{3}\left(u_{ik}Q_{kj} + u_{jk}Q_{ki}\right)\right], \quad (7.39)$$

7.6 Homogeneous States of a Quiescent Solution

We first examine the case of a quiescent suspension, with $\mathbf{v} = 0$. We consider a system with a concentration m_a of active, polar motor clusters and a concentration m_s of stationary nonpolar cross-linkers. For convenience, we define a dimensionless parameter μ_a measuring activity as

$$\mu_a = m_a b^2 \frac{\gamma_P}{D_r} \sim m_a R_{ATP}, \quad (7.40)$$

where D_r is the rods' rotational diffusion constant and we have assumed that $\gamma_P \sim R_{ATP}$. We also introduce a dimensionless parameter μ_s measuring the effect of stationary cross-linkers as

$$\mu_s = m_s b^2 \frac{\gamma_N P}{D_r} , \qquad (7.41)$$

and assume that μ_s is essentially independent of the ATP consumption rate. The bulk states of the system are determined by the solution of the homogeneous hydrodynamic equations containing only those terms that are of 0th order in the gradients. This is the most coarse-grained description of the system. More detailed descriptions that incorporate slowly varying spatial variations can then be developed by including gradient terms in the hydrodynamics. The possible homogeneous states of the system are obtained as the stationary solution of the homogeneous hydrodynamic equations for filament concentration, polarization, and alignment tensor, setting all gradient terms equal to zero. In this case, the filament concentration is constant, $c = c_0$, and only contributions from rotational currents survive in equations for the polarization and the alignment tensor, which are given by

$$\partial_t P_i = -D_r \left[1 - \mu_a c_0 \right] P_i + D_r \left[4c_0/c_N + (\mu_s - 2\mu_a)c_0 \right] Q_{ij} P_j , \qquad (7.42)$$

$$\partial_t Q_{ij} = -D_r \left[4(1 - c_0/c_N) - \mu_s c_0 \right] Q_{ij} + 2D_r \mu_a c_0 \left(P_i P_j - \frac{1}{2} \delta_{ij} P^2 \right) , \qquad (7.43)$$

where all filament densities are measured in units of l^2 and $c_N = 3\pi/2$.

There are three possible homogeneous stationary states for the system, obtained by solving Equations (7.42) and (7.43) with $\partial_t P_i = 0$ and $\partial_t Q_{ij} = 0$. These are

$$\begin{array}{llll}
\text{isotropic state (I):} & P_i = 0 & Q_{ij} = 0 , \\
\text{nematic state (N):} & P_i = 0 & Q_{ij} \neq 0 , \\
\text{polarized state (P):} & P_i \neq 0 & Q_{ij} \neq 0 .
\end{array}$$

At low density the only solution is $P_i = 0$ and $Q_{ij} = 0$ and the system is isotropic (I). The homogeneous isotropic state can become unstable at high filament and/or motor density, as described below (see Equation (7.51)).

To discuss the instabilities it is convenient to measure time in units of D_r^{-1} and rewrite Equations (7.42) and (7.43) in a more compact form as

$$\partial_t P_i = -a_1 P_i + b_1 c_0 Q_{ij} P_j , \qquad (7.44)$$

$$\partial_t Q_{ij} = -a_2 Q_{ij} + b_2 c_0 \left(P_i P_j - \frac{1}{2} \delta_{ij} P^2 \right) . \qquad (7.45)$$

The coefficients a_1, b_1, a_2, and b_2 are given by

$$a_1 = 1 - \mu_a c_0 , \qquad (7.46)$$
$$a_2 = 4 \left[1 - c_0/c_N - \mu_s c_0/4 \right] , \qquad (7.47)$$
$$b_1 = 4/c_N + \mu_s - 2\mu_a , \qquad (7.48)$$
$$b_2 = 2\mu_a . \qquad (7.49)$$

In the absence of cross-linkers ($\mu_s = \mu_a = 0$), no homogeneous polarized state with a nonzero mean value of **P** is obtained. There is, however, a transition at the density $c_N = 3\pi/2$ from an isotropic state with $Q_{ij} = 0$ for $c_0 < c_N$ to a nematic state with $Q_{ij} = S_0(n_i n_j - \frac{1}{2}\delta_{ij})$, with $\mathbf{n}$ a unit vector along the direction of broken symmetry for $c_0 > c_N$. The transition is identified with the change in sign of the coefficient a_2 of Q_{ij} on the right-hand side of Equation (7.45). A negative value of a_2 that controls the decay rate of Q_{ij} signals an instability of the isotropic homogeneous state. A mean-field description of such a transition, which is continuous in 2d (but first order in 3d), requires that one incorporates cubic terms in Q_{ij} in the equation for the alignment tensor. Adding a term $-a_4 c_0^2 Q_{kl} Q_{kl} Q_{ij}$ to Equation (7.45) we obtain $S_0 = \frac{1}{c_0}\sqrt{-2a_2/a_4} = \frac{1}{c_0}\sqrt{-8(1 - c_0/c_N)/a_4}$.

If $\mu_a = 0$, but $\mu_s \neq 0$, there is again no stable polarized state. The presence of a concentration of nonpolar cross-linkers does, however, renormalize the isotropic-nematic (IN) transition, which occurs at a lower filament density given by

$$c_{IN} = \frac{c_N}{1 + \mu_s c_N/4} . \tag{7.50}$$

The presence on nonpolar cross-links favors filament alignment and shifts c_{IN} downward. It should be noted that this occurs even with a higher effective temperature T_a. A qualitatively similar result has been obtained in numerical simulation of a two-dimensional system of rigid filaments interacting with motor proteins grafted to a substrate [64]. The amount of nematic order S_0 is also enhanced by the cross-linkers, with $S_0 = \sqrt{-2a_2/a_4 c_0^2} = \sqrt{\frac{8}{a_4 c_0^2}(c_0/c_{IN} - 1)}$.

If μ_a is finite, the system can order in both polarized and nematic homogeneous states. The homogeneous isotropic state can become unstable in two ways. As in the case of $\mu_a = 0$, a change in sign of the coefficient a_2 signals the transition to a nematic (N) state at the density c_{IN} given in Equation (7.50). In addition, the isotropic state can become linearly unstable via the growth of polarization fluctuations in any arbitrary direction. This occurs above a second critical filament density,

$$c_{IP} = \frac{1}{\mu_a} , \tag{7.51}$$

defined by the change in sign of the coefficient a_1 controlling the decay of polarization fluctuations in Equation (7.44). For $c_0 > c_{IP}$, the homogeneous state is polarized (P), with $\mathbf{P} \neq 0$. The alignment tensor also has a nonzero mean value in the polarized state as it is slaved to the polarization. The location of the boundaries between the various homogeneous states is controlled by the relative strength and concentration of active polar motor clusters to stationary nonpolar cross-linkers. To simplify the discussion, we fix the value of μ_s that determined the density of the nematic-isotropic transition to $\mu_s = 0$, so that the isotropic-nematic transition takes place at the density c_N of a suspension of rods with no cross-linkers. One can identify two regimes:

I) If $c_{IP} > c_N$, which corresponds to $\mu_a < 1/c_N$, a region of nematic phase exists between the isotropic and the polar state. At sufficiently high filament and motor densities, the nematic state becomes unstable. To see this, we linearize Equations (7.44) and (7.45) by letting $Q_{ij} = Q_{ij}^0 + \delta Q_{ij}$ and $\delta P_i = P_i$. Fluctuations in the alignment tensor are uniformly stable for $a_2 < 0$, but polarization fluctuations along the direction of broken symmetry become unstable for $a_1 \leq c_0 b_1 S_0/2$, i.e., above a critical density

$$c_{NP} = \frac{1}{\mu_a}\left[1 + \frac{b_1^2}{a_4 R}\left(1 - \sqrt{1 + \frac{2a_4 R(1 - R)}{b_1^2}}\right)\right] \qquad (7.52)$$

where $R = c_N/c_{IP}$. The polarized state at $c_0 > c_{NP}$ has

$$P_i^0 = P_0 p_i , \qquad (7.53)$$

$$Q_{ij}^0 = S_P(p_i p_j - \delta_{ij}/2) , \qquad (7.54)$$

with **p** a unit vector in the direction of broken symmetry and

$$P_0^2 = \frac{2a_1 a_2}{c_0^2 b_1 b_2}\left[1 - \left(\frac{2a_1}{b_1 c_0 S_0}\right)^2\right] , \qquad (7.55)$$

$$S_P = S_0\sqrt{1 - \frac{c_0^2 b_1 b_2}{2a_1 a_2}p_0^2} = 2\left|\frac{a_1}{c_0 b_1}\right| . \qquad (7.56)$$

II) For $\mu_a > 1/c_N$, $c_{IP} > c_N$ and the polarity of motor clusters renders the nematic state unstable at all densities and the system goes directly from the I to the P state at c_{IP}, without an intervening N state. The phase diagram has the topology shown in Figure 7.3. At the onset of the polarized state the alignment tensor is again slaved to the polarization field, $Q_{ij} = \frac{b_2}{a_2}c_0\left(P_i P_j - \frac{1}{2}\delta_{ij}P^2\right)$, and $\mathbf{P} = P_0\mathbf{p}$. The value of P_0 is determined by cubic terms in Equation (7.44) not included here.

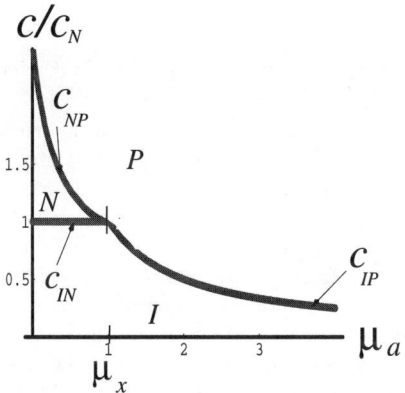

Figure 7.3. (color online) The phase diagram for $\mu_s = 0$. For $\mu_a > 1/c_N$, where c_{IN} and c_{IP} intersect, no N state exists and the system goes directly from the I to the P state ($\gamma_P/D_r = 1$ and $a_4 = 50$).

For a fixed, but nonzero value of μ_s, the phase diagram has the same topology as shown in Figure 7.3, but with c_N replaced by c_{IN} given in Equation (7.50). The value of μ_a, where the three phases coexist, is shifted to a larger value, given by $\mu_a = (1 + \mu_s c_N/4)/c_N$.

Estimates of the various parameters can be obtained using a microscopic model of the motor-filament interaction of the type described in the Appendix. Using parameter values appropriate for kinesin ($\kappa \sim 10^{-22}$nm/rad [65]), we estimate $\gamma_P \sim \gamma_{NP} \sim \kappa/\zeta_r = \kappa D_r/(k_B T_a) \sim 10^{-1}$sec^{-1} where we used the value

$D_r \sim 10^{-2} \text{sec}^{-1}$ appropriate for long thin rods in an aqueous solution [66] and $T_a \sim 300\,\text{K}$. Using $m_a = m_s \equiv m$, $\gamma_P = \gamma_{NP}$, $l \sim 10\mu\text{m}$, $b \sim 10\,\text{nm}$, the value of m above which no nematic state exist is found to correspond to a three-dimensional cross-linker density of about $0.5 - 1\mu\text{M}$ and a sample thickness of order $1\mu\text{m}$. This value is of order of the motor densities used in experiments on purified microtubule-kinesin mixtures such as those of [17], suggesting that the filament solution in this experiment is always in the region where the present mean field model predicts a uniform polarized state.

On the other hand, *in vitro* experiments generally fail to observe states with uniform polarization and report the formation of complex spatial structures. This can be understood in the context of the hydrodynamic theory described here, by examining the dynamics of spatially varying fluctuations of the hydrodynamic fields from their uniform value in each state. It has been shown elsewhere that such fluctuations become unstable in a wide range of parameters. In both isotropic and ordered states, the instability arises from filament bundling (controlled by the rate α) that tends to build up density inhomogeneities, eventually overtaking diffusion and driving the formation of spatially inhomogeneous structures. This instability is described in the next section for the isotropic state. The instability of the nematic and polarized state is driven by the same physical mechanisms, although the details are more subtle as, in this case, one must consider the coupled dynamics of fluctuations in the concentration and in the orientational order parameter. A complete description can be found in [36].

7.7 Hydrodynamics of Flowing Active Suspensions

In this section, we display the explicit form of the hydrodynamic equations for a suspension of active rods obtained by coarse-graining the Smoluchowski equation, as outlined in Sections 7.4 and 7.5. Phenomenological forms of these equations have already been used by other authors to study the interplay of order and flow in active systems in specific geometries [47, 52, 51]. Our work provides a derivation of the continuum theory, starting from the dynamics of single filaments coupled by active cross-linkers and an estimate of the various parameters in the equations in terms of experimentally accessible quantities. As discussed in Section 7.4, we limit ourselves to the case of an incompressible suspension and neglect inertial fluid effects. In this case, the flow velocity $\mathbf{v}$ of the suspension is determined from the solution of Stokes' equation,

$$\eta_0 \nabla^2 \mathbf{v} - \boldsymbol{\nabla} \Pi(c, \mathbf{P}, \mathbf{Q}; \kappa, \mu) = -\boldsymbol{\nabla} \cdot \tilde{\boldsymbol{\sigma}}^r(c, \mathbf{P}, \mathbf{Q}; \kappa, \mu); , \qquad (7.57)$$

with the incompressibility condition

$$\boldsymbol{\nabla} \cdot \mathbf{v} = 0 . \qquad (7.58)$$

The pressure Π is the sum of solvent and filament contributions, $\Pi = \Pi_s(\rho) + \Pi_r$, and we have introduced the deviatoric stress tensor defined by subtracting out the hydrostatic pressure, $\Pi_r = (1/d)\sigma_{kk}^r$, as

$$\tilde{\sigma}_{ij}^r = \sigma_{ij}^r - \delta_{ij}\Pi_r . \qquad (7.59)$$

Both isotropic and ordered (polarized and nematic) suspensions will be considered. The orientational order of the suspension affects the flow through the dependence of the pressure Π and the rods' contributions to the stress tensor $\tilde{\sigma}^r$ on polarization and alignment tensor. The derivation of the constitutive equations for these quantities was described in Section 7.5.

7.7.1 Isotropic State

In an isotropic suspension, the only hydrodynamic variable describing the filaments is the concentration, c. Its dynamics is governed by a nonlinear convection-diffusion equation

$$\partial_t c + \nabla \cdot \left(\mathbf{v}c\right) = \nabla \cdot \mathcal{D}(c)\nabla c \,, \tag{7.60}$$

where $\mathcal{D}(c)$ is an effective (concentration-dependent) diffusion coefficient, softened by active processes. It is given by

$$\mathcal{D}(c) = \frac{3D}{4}(1 + v_0 c) - \alpha \tilde{m}_a c \,, \tag{7.61}$$

with $\tilde{m}_a = m_a b^2$. The first term on the right-hand side of Equation (7.61) is the diffusion coefficient of long thin rods, with $D = D_\parallel = 2D_\perp$, including excluded volume corrections, with $v_0 = 2l^2/\pi$. The second term on the right-hand side of Equation (7.61) arises from filament bundling driven at the rate α given in Equation (7.26) and promotes density inhomogeneities. Equation (7.60) for the concentration couples to the Stokes equation, Equation (7.57), with

$$\Pi_r^I(c,\mu) = k_B T_a c\left(1 + \frac{2c}{\pi}\right) + \tilde{m}_a \alpha \frac{5k_B T_a}{432D}c^2 \,, \tag{7.62}$$

and

$$\tilde{\sigma}_{ij}^{r,I} = \left(2\eta_0 + \frac{k_B T_a}{96D}c\right)u_{ij} \,. \tag{7.63}$$

In an isotropic active suspension there are no active contributions to the deviatoric part of the stress tensor, which has the form usual for passive rods [61]. There is, however, an active contribution to the pressure corresponding to the second term on the right-hand side of Equation (7.62). The first term on the right-hand side of Equation (7.62) is standard for passive rods.

The homogeneous isotropic state in a quiescent suspension is characterized by $\mathbf{v} = 0$ and $c = c_0$. As discussed in the literature [30, 33, 36], the homogeneous state becomes unstable at high filament and motor concentration due to contractile effects generated by motor-induced filament bundling. Bundling is the main mechanism responsible for the instability of both isotropic and ordered homogeneous states in quiescent suspensions. It is therefore instructive to explicitly display the details of this instability for the simple isotropic case. In examining the dynamics of fluctuations in the isotropic state we let $c = c_0 + \delta c$ and $\mathbf{v} = \delta \mathbf{v}$ in Equation (7.60) and only keep terms of first order in the fluctuations. Incompressibility requires $\nabla \cdot \delta \mathbf{v} = 0$, and the linearized equation for δc is simply

$$\partial_t \delta c = \mathcal{D}(c_0)\nabla^2 \delta c \,. \tag{7.64}$$

Expanding δc in Fourier components, $\delta c(\mathbf{r}, t) = \sum_{\mathbf{k}} = c_{\mathbf{k}}(t)e^{i\mathbf{k}\cdot\mathbf{r}}$, one finds immediately that the relaxation of the Fourier amplitudes, $c_{\mathbf{k}}(t) = c_{\mathbf{k}}e^{-z_c(k)t}$, is controlled by a diffusive mode

$$z_c(k) = \mathcal{D}(c_0)k^2 .\tag{7.65}$$

Density fluctuations become unstable when $z_c(k) < 0$, corresponding to $\mathcal{D}(c_0) < 0$ or $c > c_B$, where

$$c_B = \frac{3D}{4\tilde{m}\alpha - 3Dv_0} \sim \frac{3D}{4\tilde{m}\alpha}\tag{7.66}$$

is the concentration above which bundling overtakes diffusion. Using $\alpha \sim (b/l)u_0$, we can express the density c_B in terms of the activity parameter μ_a defined in Equation (7.40) as $c_B = \frac{9}{2\mu_a}(l\gamma_P/\alpha)$, where we have used $D_r = D/(6l^2)$. A possible location of this instability line in the phase diagram is shown in Figure 7.4.

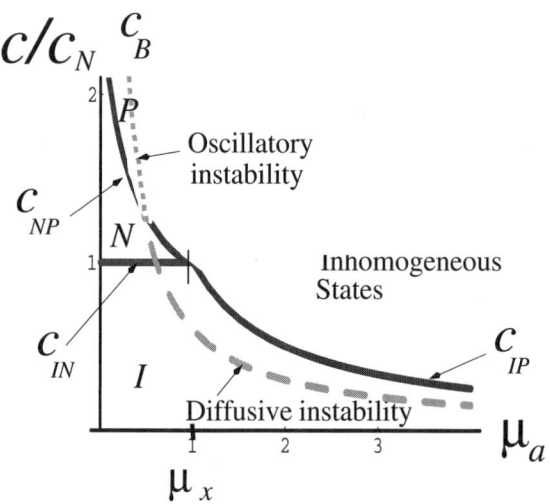

Figure 7.4. (color online) The phase diagram of homogeneous states for $\mu_s = 0$ in the plane of filament density, c_0, and motor activity μ_a, as defined in Equation (7.40), showing the location of the bundling instability at $c_0 = c_B$. The horizontal line at $c_0 = c_N$ for the isotropic-nematic transition crosses c_{IP} at $\mu_a\mu_x = 1/c_N$. The c_B line may lie above the $c_{NP} - c_{IP}$ line or cross through the N and I states, as shown ($l\gamma_P/\alpha = 0.1$, $a_4 = 50$), depending on the value of $l\gamma_P/\alpha$, a numerical parameter to leading order independent of ATP consumption rate. The instability of the I and N states is diffusive (dashed line), while the instability of the P state is oscillatory (dotted line).

7.7.2 Nematic State

The continuum variables describing the large-scale dynamics of an active nematic solution are the density and flow velocity of the solution and the concentration and alignment tensor of the filaments. For simplicity, we consider only the case where there are no stationary apolar cross-linkers, i.e., $m_s = 0$. In this case, the transition from the isotropic to the nematic state occurs at the value c_N of passive suspensions. A finite fraction of stationary apolar cross-linkers lowers the value of the transition density, as discussed in Section 7.6. In addition, it tends to stiffen all the liquid crystal elastic constants [36]. In the absence of external forces, the equations for filament concentration and alignment tensor are given by

$$\left(\partial_t + \mathbf{v} \cdot \nabla\right) c = \partial_i \mathcal{D}_{ij} \partial_j c + \partial_i \mathcal{D}^Q \partial_j (cQ_{ij}) , \tag{7.67}$$

$$\left(\partial_t + \mathbf{v} \cdot \nabla\right)(cQ_{ij}) = cF_{ij}(\kappa, \mathbf{Q}) + H_{ij}(c, \mathbf{Q}) . \tag{7.68}$$

The tensor F_{ij} describes anisotropic convective and flow alignment effects and has the familiar form for passive nematic liquid crystals, as given in Equation (7.14). At low density with the closure approximation described in [36] the alignment parameter has the value $\lambda = 1/(2S_0)$, with S_0 the nematic order parameter defined in Equation (7.10). The tensor H_{ij} plays the role of the equilibrium molecular field for passive nematic liquid crystals, but it contains various active corrections. It is given by

$$H_{ij}(c, \mathbf{Q}) \simeq K\nabla^2(cQ_{ij}) + K'\left[\partial_i\partial_k(cQ_{jk}) + \partial_j\partial_k(cQ_{ik}) - \delta_{ij}\partial_k\partial_l(cQ_{kl})\right]$$
$$+\partial_k\left(K_{ijkl}\partial_l c\right) - 4D_r\left(1 - \frac{c}{c_N}\right)cQ_{ij} - D_r a_4 c^3 Q_{kl}Q_{kl}Q_{ij} , \tag{7.69}$$

where

$$\mathcal{D}_{ij}(c, \mathbf{Q}) = \frac{3D}{4}\left[1 + \left(1 - \frac{2}{3}S^2\right)\frac{3c}{c_N} - \frac{4\alpha\tilde{m}_a c}{3D}\right]\delta_{ij}$$
$$+\left(\frac{Dv_0}{2} - \frac{4}{3}\alpha\tilde{m}_a\right)cQ_{ij} , \tag{7.70}$$

$$\mathcal{D}^Q(c) = \frac{D}{2}\left(1 - \frac{c}{c_N}\right) - \frac{2\alpha\tilde{m}_a c}{3} , \tag{7.71}$$

$$K_{ijkl}(c) = \left[\frac{D}{16}(1 + v_0 c) - \frac{2}{3}\alpha\tilde{m}_a c\right]\left(\delta_{ik}\delta_{jl} + \delta_{jk}\delta_{il} - \delta_{ij}\delta_{kl}\right) , \tag{7.72}$$

$$K(c) = \frac{7D}{12}\left(1 - \frac{c}{c_N}\right) , \tag{7.73}$$

$$K'(c) = \frac{D}{6}\left(1 - \frac{c}{c_N}\right) - \frac{\alpha\tilde{m}_a c}{18} . \tag{7.74}$$

The last term on the right-hand side of Equation (7.69), with $a_4 > 0$ has been introduced by hand. It arises from a quartic term in the free energy of an equilibrium nematic and determines the magnitude of orientational order in passive rod solutions. It is apparent from the form of the various elastic constants in Equations (7.70-7.74) that bundling (described by the parameter α of Equation (7.26)) always decreases the elastic constants of the nematic and therefore ultimately renders the uniform ordered state unstable.

The flow velocity of the suspension is again obtained from Stokes' equation, Equation (7.57), with a rods' contribution to the pressure given by

$$\Pi_r^N(c,\mu) = k_B T_a c \left[1 + \frac{3c}{2c_N}\left(1 - \frac{2}{3}S^2\right)\right] + k_B T_a \frac{c}{36D} u_{kl} Q_{kl}$$
$$+ \tilde{m}_a \alpha \frac{k_B T_a}{432D} c^2 \left(5 - S^2\right), \tag{7.75}$$

where the last term is new and arises from activity. The filament contribution to the deviatoric stress tensor is given by

$$\tilde{\sigma}_{ij}^{r,N} = 2k_B T_a c \left(1 - \frac{c}{c_N}\right) Q_{ij} + \tilde{m}_a \alpha \frac{8 k_B T_a}{432D} c^2 Q_{ij}$$
$$+ k_B T_a \frac{c}{24D}\left[\frac{1}{2}u_{ij} + \frac{2}{3}\left(u_{ik}Q_{kj} + u_{jk}Q_{ki} - \delta_{ij}u_{kl}Q_{kl}\right)\right]. \tag{7.76}$$

Activity modifies the stress tensor of a nematic in two ways. The first term on the right-hand side of Equation (7.76) is equilibrium-like, in the sense that it can be obtained from the corresponding term in the stress tensor of passive rods, $\tilde{\sigma}_{ij}^{r,\text{passive}} = 2k_B T \left(1 - \frac{c}{c_N}\right) Q_{ij}$ by letting $T \to T_a$ (and replacing the transition density c_N by c_{IN}, when $m_s \neq 0$). The second term on the right-hand side of Equation (7.76) is a truly nonequilibrium contribution. It was first proposed phenomenologically by Hatwalne and collaborators [50] who argued that an active element in solution behaves like a force dipole. Correlations among the axis of each dipole build up orientational order and yield active contributions to the stress tensor proportional to the orientational order parameter, Q_{ij}. Our microscopic derivation [57] yields an estimate for the coefficient of this term (undetermined, even in sign, in the phenomenological theory) and shows that the active cross-linkers yield contractile stresses ($\alpha > 0$). Finally, the third term on the right-hand side of Equation (7.76) is the viscous contribution which has the standard form for a solution of rod-like filaments. Finally we note that active contributions proportional to the parameter β given in Equation (7.27) do not appear in the hydrodynamics of the nematic phase. This is expected as terms proportional to β break the inversion symmetry of the ordered state and can only appear in a system with polar order.

7.7.3 Polarized State

The coarse-grained variables describing the dynamics of an active polarized suspension are the density and flow velocity of the solution and the concentration and polarization of the filaments. As shown in Section 7.6, in a polarized state the alignment tensor is slaved to the polarization field and it is not an independent continuum field. On the other hand, because our theory only considers terms that are quadratic in the fields, a nonzero value for $|\mathbf{P}|$ is only obtained by considering the coupled equations for $\mathbf{P}$ to Q_{ij} and eliminating Q_{ij} in favor of $\mathbf{P}$ in the polarization equation to generate a term of order $(\mathbf{P})^3$. To see, consider a filament density well into the polarized state, with $c > c_{IN}$ and $c > c_{IP}$ so that both coefficients $a_1 = 1 - c/c_{IP}$ and $a_2 = 1 - c/c_{IN}$ in Equations (7.44) and (7.45) satisfy $a_1 < 0$ and $a_2 < 0$. Setting the left-hand side of Equations (7.44) and (7.45) to zero, we solve Equation (7.45) for Q_{ij} to obtain

$$Q_{ij} = \frac{b_2 c}{a_2}\left(P_i P_j - \frac{1}{2}\delta_{ij}P^2\right). \tag{7.77}$$

This solution, substituted in Equation (7.44), yields a term $\sim P^2 P_i$ on the right-hand side of Equation (7.44) which has solution $P^2 = (2a_1a_2)/(b_1b_2c^2)$.

The continuum equations for the polarized state are obtained by assuming that the alignment tensor relaxes on microscopic time scales to the form given by Equation (7.77), which is then used to eliminate Q_{ij} in favor of $\mathbf{P}$. With the exception of homogeneous terms such as the $\mathcal{O}((\mathbf{P})^3)$ term just described, this leads to a high density renormalization of the various coefficients in the continuum equations, but does not generate any new terms. For the sake of simplicity in the following, we neglect this renormalization and only keep those terms in the polarization equation generated by the coupling to the alignment tensor that have a qualitatively new structure. We also neglect all excluded volume corrections. The equation for filament concentration is given by

$$\partial_t c = -\boldsymbol{\nabla} \cdot c\Big(\mathbf{v} + \frac{7}{36}\tilde{m}_a \beta c\mathbf{P}\Big) + \partial_i\Big(\mathcal{D}_{ij}^p(c)\partial_j c\Big)$$
$$-\frac{1}{2}\alpha\tilde{m}_a \partial_i\big[c^2 \partial_j(P_i P_j)\big] \,, \tag{7.78}$$

with

$$\mathcal{D}_{ij}^p(c,\mathbf{P}) = \Big(\frac{3D}{4} - \alpha\tilde{m}_a c\Big)\delta_{ij} - \alpha\tilde{m}_a c\Big(P_i P_j + \frac{1}{2}\delta_{ij}P^2\Big)\,. \tag{7.79}$$

The equation for the polarization vector has the form

$$\big(\partial_t + \mathbf{v}\cdot\boldsymbol{\nabla}\big)(c\mathbf{P}) = \frac{1}{2}(\boldsymbol{\nabla}\times\mathbf{v})\times(c\mathbf{P}) + \frac{\lambda_P}{2}\big[\boldsymbol{\nabla}\mathbf{v} + (\boldsymbol{\nabla}\mathbf{v})^T\big]\cdot c\mathbf{P}$$
$$+\mathbf{H}(c,\mathbf{P})\,. \tag{7.80}$$

where $\mathbf{H}(c,\mathbf{P})$ generalizes the molecular field of equilibrium polar fluids [63] by including active contributions. It is given by

$$H_i(c,\mathbf{P}) \simeq -\big[D_r - \gamma_P\tilde{m}_a c + a_3 P^2\big]cP_i + \frac{2}{9}\tilde{m}_a\beta c\partial_i c - \frac{1}{36}\tilde{m}_a\beta\partial_j\Big[c^2\Big(P_i P_j - \frac{5}{2}\delta_{ij}P^2\Big)\Big]$$
$$+\big[\partial_j K_p\partial_i(cP_j) + \partial_i K_p\partial_j(cP_j)\big] + \partial_j K_p\partial_j(cP_i)$$
$$-\partial_j\mathcal{D}_{ijk}^p(c,\mathbf{P})\partial_k c + \gamma_P\frac{\tilde{m}_a c}{24}\nabla^2(cP_i)\,, \tag{7.81}$$

where

$$K_p(c) = \frac{D}{8} - \frac{\alpha\tilde{m}_a}{4}c\,, \tag{7.82}$$

$$K_p'(c) = \frac{5D}{8} - \frac{\alpha\tilde{m}_a}{4}c\,, \tag{7.83}$$

$$\mathcal{D}_{ijk}^p(c,\mathbf{P}) = c\Big[\Big(\frac{Dv_0}{8} + \frac{\alpha\tilde{m}_a}{3}\Big)(P_i\delta_{jk} + \delta_{ij}P_k) + \Big(\frac{17Dv_0}{8} + \frac{2\alpha\tilde{m}_a}{3}\Big)P_j\delta_{ik}\Big] \tag{7.84}$$

The parameter a_3 determines the value P_0 of the magnitude of the polarization in a quiescent (active) suspension, with $P_0^2 = a_3/[D_r(c/c_{IP} - 1)]$.

In contrast to the case of the nematic, all three active mechanisms of motor-induced filament dynamics controlled by α, β, and γ_P appear in the hydrodynamic equations of polarized active suspension. Polarization sorting at a rate $\beta \sim u_0$ yields novel convective contributions in the first term on the right-hand side of the equation for the filament concentration, Equation (7.78). In an equilibrium suspension the filament concentration is convected with the suspension flow velocity $\mathbf{v}$. In an active polar suspension, by contrast, the filament concentration is convected with the effective velocity $\sim \mathbf{v} + \tilde{m}_a \beta \mathbf{P}$. The terms linear in the gradients proportional to β in the polarization equation are of similar origin. These terms were also incorporated in the continuum description of self-propelled particles proposed by Simha and Ramaswamy [49]. Bundling effects controlled by the rate α soften both the diffusion constant $\mathcal{D}_{ij}^p(c, \mathbf{P})$ in the concentration equation, and the effective bend and splay elastic constants K_p and K_p' of the polar fluid. Finally, the rotation rate γ_P builds up polar order and controls the very existence of a polar state.

For an incompressible suspension, the flow field $\mathbf{v}$ is obtained again from the Stokes equation, Equation (7.57). The filament contribution to the pressure is given by

$$\Pi_r^P = k_B T_a c \left(1 + \frac{c}{\pi}\right) + \tilde{m}_a \alpha \frac{k_B T_a}{144D} c^2 \left(\frac{5}{3} + 2P^2\right). \tag{7.85}$$

The filament contribution to the deviatoric stress tensor of a polarized suspension is

$$\begin{aligned}
\tilde{\sigma}_{ij}^{r,P} &= \tilde{m}_a \alpha \frac{k_B T_a}{72D} c^2 \left(P_i P_j - \frac{1}{2}\delta_{ij} P^2\right) + \frac{k_B T_a}{48D} c u_{ij} \\
&\quad + \frac{k_B T_a}{36D} \frac{c^2 b_2}{a_2} \left[u_{ik} P_k P_j + u_{jk} P_k P_i - \delta_{ij} P_k u_{kl} P_l - u_{ij} P^2\right] \\
&\quad + \tilde{m}_a \beta \frac{k_B T_a}{432D} c^2 \left[\partial_j P_i - \frac{1}{2}\delta_{ij} \boldsymbol{\nabla} \cdot \mathbf{P} - \frac{1}{4}\left(\partial_i P_j - \partial_j P_i\right)\right]. \tag{7.86}
\end{aligned}$$

The first term on the right-hand side of Equation (7.86) is the active contribution to the stress tensor first discussed by Hatwalne and collaborators for a nematic suspension [50]. The second and third term arise from the viscous coupling of filaments to the solvent. Finally, the last term contains active contributions proportional to gradients of the polarization. These are controlled by the polarization sorting rate $\beta \sim u_0$. Terms of these type are unique to the polarized state and vanish in a nematic suspension. They are expected to play an important role in renormalizing the rate of stress relaxation via reptation.

Continuum equations for a polarized active suspension have been written down phenomenologically by several authors [49, 48, 51]. It is useful to make contact with this work. The phenomenological description can be recovered from our model by making a few simplifying approximations. An equation for the concentration of filaments of the form given in Equation (7.78) was proposed by Ramaswamy and collaborators [49, 51], although these authors neglected the diffusion terms, which play a crucial role in controlling the bundling instability of quiescent suspensions. The equation for the polarization vector $\mathbf{P}$ reduces to the form used by Voituriez et al. [48] and by Simha et al. [49, 51], if all terms containing higher order gradients of the concentration ($P_i \nabla^2 c$, $P_i(\boldsymbol{\nabla} c)^2$, $\mathbf{P} \cdot \boldsymbol{\nabla} \partial_i c$, $(\mathbf{P} \cdot \boldsymbol{\nabla} c)(\partial_i c)$), as well as terms containing both gradients of concentration and of polarization (($\boldsymbol{\nabla} \cdot \mathbf{P})(\partial_i c)$, $(\partial_j c)(\partial_j P_i)$, $(\partial_j c)(\partial_i P_j)$) are neglected. With this approximation, Equation (7.80) becomes

$$\begin{aligned}
\left(\partial_t + \mathbf{v} \cdot \boldsymbol{\nabla}\right) P_i &= \Gamma\left(1 - \frac{|\mathbf{P}|^2}{P_0^2}\right) P_i + \frac{1 + \lambda_P}{2}(\partial_j v_i) P_j - \frac{1 - \lambda_P}{2}(\partial_i v_j) P_j \\
&\quad - w_1 c(\mathbf{P} \cdot \boldsymbol{\nabla}) P_i - w_2 c P_i(\boldsymbol{\nabla} \cdot \mathbf{P}) + w_3 c \partial_i |\mathbf{P}|^2
\end{aligned}$$

$$+\left[\left(w_4 + w_5 P^2\right)\delta_{ij} - w_6 P_i P_j\right]\partial_j c$$
$$+(K_1 - K_3)\partial_i \boldsymbol{\nabla} \cdot \mathbf{P} + K_3 \nabla^2 P_i , \tag{7.87}$$

where $\Gamma = \gamma_P \tilde{m}_a c - D_r > 0$ and $P_0^2 = \Gamma/a_3$. Here, the coefficients w_i have the form $w_i = c_i \tilde{m}_a \beta$, with c_i numerical coefficients of order one. Note, however, that the terms proportional to w_i other than w_1 are equilibrium-like, in the sense that they could also be obtained from a polar contribution to the free energy of the form $\delta F_p = \int_{\mathbf{r}} \left[B_1 \delta c(\boldsymbol{\nabla}\cdot\mathbf{P}) + B_2 c|\mathbf{P}|^2(\boldsymbol{\nabla}\cdot\mathbf{P}) + B_3|\mathbf{P}|^2\mathbf{P}\cdot\boldsymbol{\nabla}c + ...\right]$. The w_1 term, in contrast, is a true nonequilibrium contribution induced by activity and cannot be obtained from a free energy. All the remainder w_i's generally contain both equilibrium-like contributions determined by the B_i and nonequilibrium ones proportional to $\beta \sim R_{ATP}$. Finally, K_1 and K_3 are the splay and bend elastic moduli, respectively, with

$$K_1(c) = \frac{7D}{8} - \frac{3\tilde{m}_a \alpha c}{4} + \frac{\tilde{m}_a \gamma_P c}{24} , \tag{7.88}$$

$$K_3(c) = \frac{5D}{8} - \frac{\tilde{m}_a \alpha c}{4} + \frac{\tilde{m}_a \gamma_P c}{24} . \tag{7.89}$$

The first term on the right-hand side of Equation (7.87) guarantees the formation of a uniformly polarized state with $|\mathbf{P}| = P_0$. The next two terms are conventional couplings of liquid crystalline order and flow, with λ_P the flow alignment parameter. Our low density calculation yields $\lambda_P = 1/2$. The three terms on the second line are nonequilibrium terms analogous to those first written down by Toner and Tu in models of flocking [53, 54, 55]. The third line describes nonequilibrium changes in polarization driven by concentration gradients. Only the first of these terms ($\sim w_4$) is generally included in phenomenological theories. Equations (7.88) and (7.89) show that motor-induced filament bundling ($\sim \alpha$) softens both the splay and bend elastic constants, while polar cross-linkers ($\sim \gamma_P$) tend to stiffen them. Such effects, i.e. the dependence of elastic constants on the active elements are clearly beyond the scope of phenomenological theories with arbitrary elastic constants. In addition, the microscopic derivation also provides contributions to the stress tensor that are higher order in gradients without the need for new unknown parameters, e.g., the expression for the stress tensor given in Equation (7.86) contains novel contributions proportional to gradients of polarization that were not considered by other authors [49, 50, 46, 47] but whose microscopic origin is the same as those of lower order in the gradients.

Acknowledgements

Some of the work described in this article has been done in collaboration with Aphrodite Ahmadi. MCM acknowledges support from the National Science Foundation through grants DMR-0305407 and DMR-0219292. TBL acknowledges support from the Royal Society. We also thank S. Ramaswamy for many helpful discussions.

A Appendix: Derivation of the Rods' Stress Tensor

We model very long, thin rods as rigid strings of spherical beads of diameter $b \ll l$ suspended in a fluid of viscosity η_0. We assume each rod consists of an odd number $l/b = 2M + 1$ of such beads, as sketched in Figure 7.5. The beads on the α-th rod are indexed by an integer m that runs from $-M$ to $+M$, and the center of the m-th bead is at $\mathbf{r}_\alpha(m) = \mathbf{r}_\alpha + mb\hat{\mathbf{u}}_\alpha$, with $\mathbf{r}_\alpha = \sum_m \mathbf{r}_\alpha(m)$ the center of mass of the rod. Momentum conservation at low Reynolds number is described by the Stokes

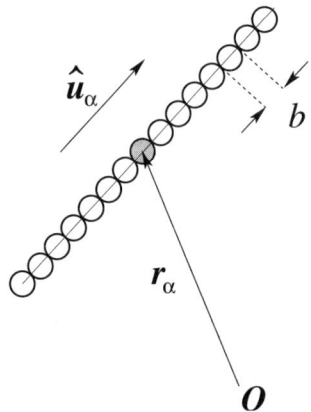

Figure 7.5. The bead model of a rigid filament (color online).

equation

$$\eta_0 \nabla^2 \mathbf{v}(\mathbf{r}) - \nabla \Pi - \sum_\alpha \sum_{m=-M}^{M} \delta\left(\mathbf{r} - \mathbf{r}_\alpha(m)\right) \mathbf{f}_\alpha^h(m) = 0 , \qquad (7.90)$$

where we have modeled each bead as a point-force on the fluid at the position of its center of mass. Faxén's theorem for the hydrodynamic force on a sphere in an inhomogeneous flow relates the force $\mathbf{f}_\alpha^h(m)$ exerted by the fluid on a bead to the flow velocity field $\mathbf{v}_0(\mathbf{r})$ at the bead's position in the absence of that bead, according to

$$\mathbf{f}_\alpha^h(m) = -\zeta_b \left[\mathbf{v}_\alpha(m) - \mathbf{v}_0(\mathbf{r}_\alpha(m)) \right] , \qquad (7.91)$$

where $\mathbf{v}_\alpha(m) = \mathbf{v}_\alpha + mb\boldsymbol{\omega}_\alpha \times \hat{\mathbf{u}}_\alpha$ is the velocity of the bead, with $\mathbf{v}_\alpha$ and $\boldsymbol{\omega}_\alpha$ the center of mass and angular velocity of the rod, and $\zeta_b = 3\pi\eta_0 b$ is the Stokes friction coefficient of a sphere of diameter b in an unbounded fluid of viscosity η_0. Using the linearity of the Stokes equation and the principle of superposition, the velocity of fluid at the position of the bead is given by

$$\mathbf{v}_0(\mathbf{r}_\alpha(m)) = \mathbf{v}(\mathbf{r}_\alpha(m)) - \sum_{n \neq m} \mathbf{H}\left(\mathbf{r}_\alpha(m) - \mathbf{r}_\alpha(n)\right) \cdot \mathbf{f}_\alpha^h(n) , \qquad (7.92)$$

where $\mathbf{v}(\mathbf{r})$ is the velocity of the fluid taking into account the presence of other rods and $H_{ij}(\mathbf{r}) = \dfrac{1}{8\pi\eta_0 r}\left(\delta_{ij} + \hat{r}_i \hat{r}_j\right)$ is the Oseen tensor. Here, the hydrodynamic interactions between beads on the same rod have been included explicitly, while the hydrodynamic coupling to other rods is implicitly taken into account in determining the flow velocity $\mathbf{v}(\mathbf{r})$. The force on bead m on the α-th rod is therefore given by

$$\mathbf{f}_\alpha^h(m) = -\zeta_b \left[\mathbf{v}_\alpha(m) - \mathbf{v}\left(\mathbf{r}_\alpha(m)\right)\right] - \frac{3}{8} \sum_{n \neq m} \frac{1}{|n-m|} (\boldsymbol{\delta} + \hat{\mathbf{u}}_\alpha \hat{\mathbf{u}}_\alpha) \cdot \mathbf{f}_\alpha^h(n) . \quad (7.93)$$

Now we take the limit $l \gg b$ and introduce the continuous variable $s = bm$, where $-l/2 \leq s \leq l/2$, so that $\mathbf{r}_\alpha(s) = \mathbf{r}_\alpha + s\hat{\mathbf{u}}_\alpha$. The hydrodynamic force per unit length, $\mathcal{F}_\alpha^h(s)$, satisfies the equation

$$\mathcal{F}_\alpha^h(s) = -\zeta_s \left(\mathbf{v}_\alpha(s) - \mathbf{v}\left(\mathbf{r}_\alpha(s)\right)\right) - \frac{3}{8} \int_{|s-s'| \geq b} \frac{ds'}{|s-s'|} (\boldsymbol{\delta} + \hat{\mathbf{u}}_\alpha \hat{\mathbf{u}}_\alpha) \cdot \mathcal{F}_\alpha^h(s') , \quad (7.94)$$

where $\zeta_s = 3\pi\eta_0 = \zeta_b/b$. Approximating

$$\frac{1}{|s-s'|} \approx \left\langle \frac{1}{|s-s'|} \right\rangle' \delta(s-s') , \quad (7.95)$$

where

$$\left\langle \frac{1}{|s-s'|} \right\rangle' = \frac{1}{l^2} \int_s \int_{s'} \Theta(|s-s'|-b) \frac{1}{|s-s'|} = 2\ln(l/2b) , \quad (7.96)$$

with $\Theta(x)$ the Heaviside function, we obtain

$$\frac{3\ln(l/2b)}{4} (\boldsymbol{\delta} + \hat{\mathbf{u}}_\alpha \hat{\mathbf{u}}_\alpha) \cdot \mathcal{F}_\alpha^h(s) \simeq -\zeta_s \left[\mathbf{v}_\alpha(s) - \mathbf{v}\left(\mathbf{r}_\alpha(s)\right)\right] . \quad (7.97)$$

Because $\mathbf{v}_\alpha(s) = \mathbf{v}_\alpha + s\boldsymbol{\omega}_\alpha \times \hat{\mathbf{u}}_\alpha$, then integrating Equation (7.97) over s, we can obtain expressions for the hydrodynamic force, $\mathbf{F}_\alpha^h = \langle \mathcal{F}_\alpha^h(s) \rangle_s$ and torque $\boldsymbol{\tau}_\alpha^h = \langle \hat{\mathbf{u}}_\alpha s \times \mathcal{F}_\alpha^h(s) \rangle_s$ at the center of mass of a rod, with $\langle ... \rangle_s = \int_{-l/2}^{l/2} ds...$ as

$$-\boldsymbol{\zeta}^{-1}(\hat{\mathbf{u}}_\alpha) \cdot \mathbf{F}_\alpha^h = \mathbf{v}_\alpha - \frac{1}{l} \langle \mathbf{v}(\mathbf{r}_\alpha(s)) \rangle_s , \quad (7.98)$$

$$-\frac{1}{\zeta_r} \boldsymbol{\tau}_\alpha^h = \boldsymbol{\omega}_\alpha - I^{-1} \frac{1}{l} \langle \hat{\mathbf{u}}_\alpha \times s\mathbf{v}(\mathbf{r}_\alpha(s)) \rangle_s , \quad (7.99)$$

where $\zeta_{ij}(\hat{\mathbf{u}}) = \zeta_\perp(\delta_{ij} - \hat{u}_i\hat{u}_j) + \zeta_\parallel \hat{u}_i\hat{u}_j$, $\zeta_\perp = 2\zeta_\parallel = \frac{4\pi\eta_0 l}{\ln(l/2b)}$, $\zeta_r = \frac{\pi l^3 \eta_0}{3\ln(l/2b)}$ and $I = l^2/12$. Performing a Taylor expansion of the fluid velocity about the center of mass, we obtain to lowest order in gradients

$$-\boldsymbol{\zeta}^{-1}(\hat{\mathbf{u}}_\alpha) \cdot \mathbf{F}_\alpha^h = \mathbf{v}_\alpha - \mathbf{v}(\mathbf{r}_\alpha) - \frac{I}{2} (\hat{\mathbf{u}}_\alpha \cdot \boldsymbol{\nabla})^2 \mathbf{v}(\mathbf{r}_\alpha) + O(\nabla^4) , \quad (7.100)$$

$$-\frac{1}{\zeta_r} \boldsymbol{\tau}_\alpha^h = \boldsymbol{\omega}_\alpha - \hat{\mathbf{u}}_\alpha \times (\hat{\mathbf{u}}_\alpha \cdot \boldsymbol{\nabla})\mathbf{v}(\mathbf{r}_\alpha) + O(\nabla^3) . \quad (7.101)$$

Finally, we require that the hydrodynamic forces and torques be balanced by all other forces and torques on the rod. This gives

$$\mathbf{F}_\alpha^h = \boldsymbol{\nabla}_\alpha U_{ex} + k_B T_a \boldsymbol{\nabla} \ln \hat{c} - \mathbf{f}_\alpha^a , \quad (7.102)$$

$$\boldsymbol{\tau}_\alpha^h = \mathcal{R}_\alpha U_{ex} + k_B T_a \mathcal{R}_\alpha \ln \hat{c} - \boldsymbol{\tau}_\alpha^a , \quad (7.103)$$

where we have included contributions from fluctuations (non-equilibrium osmotic pressure), excluded volume interactions and active driving by the motors. Using Equations (7.97) and (7.100), we can calculate the hydrodynamic force per unit length on the rod as

$$\mathcal{F}_{\alpha i}^{h}(s) = \zeta_{ij}(\hat{\mathbf{u}}_\alpha)\big[v_{\alpha j} - v_j(\mathbf{r}_\alpha) + s\,((\boldsymbol{\omega}_\alpha \times \hat{\mathbf{u}}_\alpha)_j - (\hat{\mathbf{u}} \cdot \boldsymbol{\nabla}_\alpha)v_j(\mathbf{r}_\alpha))$$
$$-\frac{s^2}{2}(\hat{\mathbf{u}} \cdot \boldsymbol{\nabla}_\alpha)^2 v_j(\mathbf{r}_\alpha)\big] \, . \tag{7.104}$$

from which we obtain Equation (7.32). Furthermore, defining the translational and rotational currents as

$$\mathbf{J}_c(\mathbf{r}, t) = \left\langle \sum_\alpha \mathbf{v}_\alpha \delta(\mathbf{r} - \mathbf{r}_\alpha(t)) \right\rangle \tag{7.105}$$

$$\mathcal{J}_c(\mathbf{r}, t) = \left\langle \sum_\alpha \boldsymbol{\omega}_\alpha \delta(\mathbf{r} - \mathbf{r}_\alpha(t)) \right\rangle \, , \tag{7.106}$$

the Smoluchowski equation (7.19) for the dynamics of $\hat{c}(\mathbf{r}, \hat{\mathbf{u}}, t)$ follows.

From equation (7.32), we perform the coarse-graining procedure to obtain the stress tensor. Retaining all terms of first order in gradients of the hydrodynamic fields, the pressure is

$$\Pi_r^P = k_B T_a c\left(1 + \frac{c}{\pi}\right) + \tilde{m}_a \alpha \frac{k_B T_a}{144D} c^2 \left(\frac{5}{3} + 2P^2\right) \, , \tag{7.107}$$

and the deviatoric stress tensor is given by

$$\begin{aligned}
\tilde{\sigma}_{ij}^A = {} & 2k_B T_a c\left(1 - \frac{c}{c_{IN}}\right) Q_{ij} - k_B T_a \frac{c^2}{c_{IP}}\left(P_i P_j - \frac{1}{2}\delta_{ij}P^2\right) \\
& + \tilde{m}_a \alpha \frac{k_B T_a}{72D} c^2 \left(\frac{4}{3}Q_{ij} + P_i P_j - \frac{1}{2}\delta_{ij}P^2\right) \\
& + \tilde{m}_a \beta \frac{2k_B T_a}{432D} c^2 \Big[\partial_j P_i - \frac{1}{2}\delta_{ij}\boldsymbol{\nabla}\cdot\mathbf{P} - \frac{1}{4}\big(\partial_i P_j - \partial_j P_i\big) \\
& + \frac{5}{3}\Big(Q_{jk}\partial_k P_i - P_i\partial_k Q_{jk} - Q_{ik}(\partial_j P_k + \partial_k P_j) + (P_k\partial_j + P_j\partial_k)Q_{ik} \\
& - Q_{ij}\boldsymbol{\nabla}\cdot\mathbf{P} + \mathbf{P}\cdot\boldsymbol{\nabla}Q_{ij}\Big) + \frac{2}{3}\Big(Q_{jk}(\partial_k P_i + \partial_i P_k) - (P_i\partial_k + P_k\partial_i)Q_{jk}\Big) \\
& + \frac{5}{6}\delta_{ij}\Big(Q_{kl}\partial_k P_l - P_l\partial_k Q_{kl}\Big)\Big] \, . \tag{7.108}
\end{aligned}$$

References

1. B. Alberts, A. Johnson, J. Lewis, M. Raff, K. Roberts, and P. Walter, 2002, *Molecular Biology of the Cell* 4th ed. (Garland, New York)
2. J. Howard, 2000, *Mechanics of Motor Proteins and the Cytoskeleton*, (Sinauer, New York).
3. Mehta AD, Pullen KA, Spudich JA. 1998, FEBS Lett. **430**(1-2), 23-7.
4. Dammer, U., Popescu, O., Wagner, P., Anselmetti, D., Güntherodt, H.-J. and Misevic, G.N., 1995, Science 267, 1173-1175.
5. Yildiz A, Forkey JN, McKinney SA, Ha T, Goldman YE, Selvin PR, 2003, *Science* **300**, 2061-2065.
6. Churchman LS, Okten Z, Rock RS, Dawson JF, Spudich JA, 2005, *Proc. Nat. Acad. Sci.* **102**, 1419-1423.
7. A. Szent-Gyorgi, 1951, *Chemistry of Muscle Contraction* (Academic Press, New York).
8. J. Trinick and G. Offer, 1979, J. Mol. Biol. **133**, 549-556.
9. K. Takiguchi, 1991, *J. Biochem.*, **109**, 520.
10. R. Urrutia et al, 1991, *Proc. Natl. Acad. Sci. USA*, **88**, 6701.
11. F. J. Nédélec, T. Surrey, A. C. Maggs and S. Leibler, 1997, Nature **389**, 305.
12. F. J. Nédélec, 1998 Ph.D. thesis (Université Paris XI).
13. T. Surrey, F. J. Nédélec, S. Leibler and E. Karsenti, 2001 Science **292**, 1167.
14. F. Backouche, L. Haviv, D. Groswasser and A. Bernheim-Groswasser, 2006, *Phys. Biol.* **3** 264-273
15. Vallotton P., C. M. Waterman-Storer, and G. Danuser, 2004, Proc. Natl. Acad. Sci. USA. **101** :9660-9665.
16. Verkhovsky, A.B., Svitkina, T.M. and Borisy, G.G., 1999, Current Biology, **9**:11-20.
17. D. Humphrey, C. Duggan, D. Saha, D. Smith and J. Käs, 2002, *Nature* **416**, 413 .
18. L. Le Goff, F. Amblard and E. Furst, *Phys. Rev. Lett.*, **88**, 018101 (2002)
19. M. Dogterom, A. C. Maggs and S. Leibler, 1995, Proc. Nat. Acad. Sci. USA **92**, 6683.
20. Mogilner, A., G. Oster, 2003, Biophys. J. **84**(3):1591-1605.
21. H. Isambert and A.C. Maggs, 1996, *Macromolecules* **29**, 1036.
22. F.C. MacKintosh, J. Käs and P.A. Janmey, 1995, *Phys. Rev. Lett.* **75**, 4425.
23. K. Kroy and E. Frey, 1996, *Phys. Rev. Lett.* **77**, 306.
24. D.C. Morse, 1998, *Macromolecules* **31**, 7030; 1998, *Macromolecules* **31**, 7044.
25. F. Gittes and F.C. MacKintosh, 1998, *Phys. Rev. E.* **58**, R1241.
26. R. Everaers et al, 1999, *Phys. Rev. Lett.* **82**, 3717.
27. Storm C, Pastore JJ, MacKintosh FC, Lubensky TC, Janmey PA, 2005, Nature. 2005; **435**(7039):191-4.
28. H. Nakazawa and K. Sekimoto, 1996, J. Phys. Soc. Jpn. **65** 2404.
29. K. Sekimoto and H. Nakazawa, in *Current Topics in Physics*, Y. M. Cho, J B. Homg and C. N. Yang, eds. (World Scientific, Singapore, 1998).
30. K. Kruse and F. Jülicher, 2000, Phys. Rev. Lett. **85**, 1779.
31. K. Kruse and F. Jülicher, 2003, *Phys. Rev. E* **67**, 051913.
32. K. Kruse, S. Camalet and F. Jülicher, 2001, Phys. Rev. Lett. **87**, 138101.
33. T. B. Liverpool and M. C. Marchetti, 2003, Phys. Rev. Lett. 90, 138102;
34. T. B. Liverpool and M. C. Marchetti, 2004, Phys. Rev. Lett. 93, 159802;

35. A. Ahmadi, T.B. Liverpool and M.C. Marchetti, 2005, Phys. Rev. E 72, 060901 (R).
36. A. Ahmadi, T.B. Liverpool and M.C. Marchetti, 2006, Phys. Rev. E **74** 061913.
37. F. Ziebert et al., 2004, Phys. Rev. Lett. 93, 159801;
38. F. Ziebert and W. Zimmermann, 2004, Phys. Rev. E **70**, 022902.
39. I.S. Aranson and L.S. Tsimring, 2005 Phys. Rev. E **71**, 050901 (R).
40. I.S. Aranson and L.S. Tsimring, 2006 Phys. Rev. E **74**, 031915.
41. B. Bassetti, M. C. Lagomarsino and P. Jona, 2000, Eur. Phys. J. B **15**, 483.
42. J. Kim et al, 2003, *J. Korean Phys. Soc.* **42** 162.
43. H. Y. Lee and M. Kardar, 2001, Phys. Rev. E **64**, 56113.
44. S. Sankararaman, G.I. Menon and P.B. Sunil Kumar, 2004 *Phys. Rev. E* **70**, 031905.
45. K. Kruse, F. Joanny, F. Jülicher, J. Prost, and K. Sekimoto, 2004, Phys. Rev. Lett. **92**, 078101 .
46. K. Kruse, F. Joanny, F. Jülicher, J. Prost and K. Sekimoto, 2006, Eur. Phys. J. E **16**, 5-16.
47. R. Voituriez, J. F. Joanny, and J. Prost, 2005, Europhys. Lett. **70**, 404-410.
48. R. Voituriez, J. F. Joanny, and J. Prost, 2006, Phys. Rev. Lett. **96**, 28102 .
49. R. A. Simha and S. Ramaswamy, 2002, Phys. Rev. Lett. **89**, 058101 .
50. Y. Hatwalne, S. Ramaswamy, M. Rao, and R.A. Simha, 2004, Phys. Rev. Lett. **92**, 118101.
51. S. Muhuri, M. Rao, and S. Ramaswamy, 2006, http://arxiv.org/abs/cond-mat/0610025
52. A. Zumdieck, M. Cosentino Lagomarsino, C. Tanase, K. Kruse, B. Mulder, M. Dogterom, and F. Jülicher, 2005, Phys. Rev. Lett. 95, 258103.
53. J. Toner and Y. Tu, 1995, Phys. Rev. Lett. **75**, 4326-4329.
54. J. Toner and Y. Tu, 1998, Phys. Rev. E **58**, 4828-4858.
55. J. Toner, Y. Tu and S. Ramaswamy, 2005, Ann. Phys. **318**, 170
56. T. B. Liverpool and M. C. Marchetti, 2005, Europhys. Lett. 69, 846.
57. T.B. Liverpool and M.C. Marchetti, 2006, Phys. Rev. Lett., **97**, 268101.
58. N. Kuzuu and M. Doi, J. Phys. Soc. Jpn. **52**, 3486 (1983).
59. S.A. Langer, A.J. Liu, 2000, *Europhys. Lett.* **49**, 68.
60. T.B. Liverpool, A.C. Maggs and A. Ajdari, 2001, *Phys. Rev. Lett.* **86**, 4171.
61. M. Doi and S. F. Edwards, 1986, *The theory of polymer dynamics*, (OUP, Oxford).
62. J.K.G. Dhont and W.J. Briels, 2003, *J. Chem. Phys.* **118**, 1466.
63. W. Kung, M. C. Marchetti, and K. Saunders, 2006, Phys. Rev. E **73**, 031708.
64. P. Kraikivski, R. Lipowsky and J. Kierfeld, 2006, Phys. Rev. Lett. **96**, 258103.
65. J. Howard, *Nature* **389**, 561, (1997).
66. A.J. Hunt and J. Howard, 1993, *Proc. Natl. Acad. Sci. USA*, **90**, 11653.

8

Collective Effects in Arrays of Cilia and Rotational Motors

Peter Lenz

Fachbereich Physik, Philipps-Universität Marburg, D-35032 Marburg, Germany
peter.lenz@physik.uni-marburg.de

Summary. The movement of cilia in arrays is very often coordinated. Neighboring cilia beat cooperatively in a synchronized fashion or they maintain a constant phase difference creating a metachronal wave. This collective ciliar motion is used by many microorganisms for swimming and feeding. There is also strong evidence that ciliar (nodal) flow plays an important role in the establishment of left and right in developing vertebrates. In this short review, we summarize current theoretical efforts in analyzing the influence of hydrodynamic interactions on collective effects in arrays of cilia. Analytical and numerical models for the beating of a single cilium are introduced and used to calculate the motion in the surrounding flow. In arrays, the velocity field induced by a cilium exerts forces on the neighboring cilia, giving rise to hydrodynamic interactions. The importance of these interactions for collective ciliar beating is analyzed in specific models and in a more general framework. We also discuss the influence of boundary conditions and the tilting of monocilia on the nodal flow. Finally, the importance of hydrodynamic interaction for collections of rotational motors are studied. We conclude with some examples of possible applications.

8.1 Introduction

For bacteria and small animals, swimming and generation of fluid flow is a difficult task. On length scales comparable to their size, motion in water is completely overdamped. No net thrust is created by reciprocal motion that is symmetric under time-reversal. To overcome these difficulties nature has come up with several solutions.

Many bacteria move with the aid of flagella. They consist of a rotary motor, driven by proton or sodium ion gradients, and a rigid superhelical filament (with a typical length of $1 - 3\mu$m and a typical radius of 0.02μm) that functions as a propeller that is rotated at high speed (up to several 100Hz) [1, 2, 3, 4].

Bacterial flagella can appear at poles, or all over the cell surface (such as in *E. coli*). Some cells are helical with flagellar rotations at the poles (as in *Spirillum volutans*), or the cell is helical and the flagella are located between the cell membrane and the cell wall leading to internal rotation (e.g., in spirochaetes).

A similar method of propulsion is employed by many eukaryotic cells that also swim by means of flagella. Although prokaryotic and eukaryotic flagella bear the same name they are quite different. The eukaryotic flagella are not only at least ten times larger than bacterial flagella in both diameter and length (they can be up to $70\mu m$ long and $0.8\mu m$ wide), they are completely dissimilar in structure, function, and in the genes that encode their components. The eukaryotic flagellum is built of microtubules, while the bacterial flagellum is composed of the protein flagellin. The eukaryotic flagella also do not rotate but actively propagate bending waves [5]. However, propulsion by both kinds of flagella relies on the fluid mechanical fact that the sideways motion of thin rods encounters greater viscous drag than forward motion [6].

Other organisms, the *ciliates*, are covered with cilia, hair-like projections, which are uniform and aligned in rows [7]. The cilia are used by the cell for swimming and feeding[1]. Cilia also play an important role in the human body. For example, thousands of cilia are attached to the bronchioles creating an air current transporting mucus and small dust particles out of the human lungs. There is also experimental evidence that during development cilia-generated flow contributes to the placement of our organs (see Section 8.4).

Most cilia have (similar to eukaryotic flagella) the "9+2" structure [9], where nine sets of microtubule doublets surround a pair of single microtubules in the center. Motors (dyneins) cause bending deformations, giving rise to characteristic beating patterns typically consisting of a power and a recovery stroke [10]. Somewhat special are monocilia that are primary cilia lacking the central pair of microtubules and thus have a "9+0" microtubule arrangement [11, 12]. They perform a rapid rotational motion [13] and very often are tilted giving rise to non-symmetrical velocity fields (see Section 8.4).

In populations of cilia, beating is very often coordinated. Two types of cooperative motion can be distinguished: i) adjacent cilia beat cooperatively in a synchronized fashion [14]. Here, all cilia oscillate with same frequency and vanishing phase difference. ii) Cilia maintain a constant phase difference, creating a *metachronal wave* [15]. Metachronal waves can propagate in the direction of the effective stroke (symplectic metachronal waves), in the opposite direction (antiplectic), perpendicular direction (laeoplectic or dexioplecit), or oblique direction [16].

Animals, such as the freshwater protozoa *Paramecium*, use this collective ciliar motion to swim. Typically, the cilia beat with a traveling helical wave, where the direction of the ciliary effective stroke is oblique to the long axis of the body [15]. Thus, *Paramecium* swims in a spiral course, rotating around its longitudinal axis. By changing the axis of the helix, a *Paramecium* can steer and reverse its direction of motion [17]. This means of motion is extremely efficient: the $\sim 100\mu m$ long *Paramecium* can swim with a velocity of order $\sim 1mm/s$.

Cilia-generated flow is also important for feeding of cells. Certain sessile species of protozoa (such as *Stentor* and *Vorticella*) use ciliar motion to produce a fluid current into their mouth region [5, 18].

The main purpose of cilia is to propel fluid over the surface of the cell. However, there are also other biological macromolecules whose complex shape changes induce flow in the surrounding fluid. A prominent example is adenosine triphosphate synthase (ATPsynthase), which exhibits a rotational motion [19, 20]. Although the

---

[1] Cilia can also act a sensors [8]. This aspect is not reviewed here.

origin of its rotation is probably of chemical nature (namely the periodic motion in
a "reaction chamber") it nevertheless induces hydrodynamic interactions in arrays
giving rise to collective effects that will be discussed in Section 8.5.

In this review we summarize recent theoretical efforts to understand collective
effects in biological hydrodynamics. We will focus on cilia and rotating motors.
Summaries of flagellar hydrodynamics can be found in [21, 22, 23].

From a theoretical point of view it is one of the key questions what causes the
cooperative behavior in ciliar arrays. Is the cooperative motion triggered biochemically [24] or are hydrodynamic interactions strong enough to couple the beating
pattern of neighboring cilia [25]? In the latter case, the beating of a cilium is influenced by the fluid flow created by the other cilia. This question is not only of
biological relevance. From the statistical physics point of view, motile bacteria and
cells represent beautiful realizations of non-equilibrium systems exhibiting ordering
transitions and pattern formation.

Since the seminal work of Blake [26] there have been considerable efforts in modeling ciliar motion and calculation of the induced velocity fields from first principles
(for an overview see [27]). Generally, one has to distinguish between "discrete cilia
models" [26, 28] and "envelope models" [29, 30]. In the former, each cilium is treated
separately, and the total velocity field in the system is given by the superposition
of the velocity fields induced by each cilium. In contrast, in the envelope models it
is assumed the cilia are closely packed so that they primarily interact with the surrounding water. However, as shown by Blake and Sleigh [10], the envelope approach
is only valid for symplectic metachronal waves.

For this reason, we present here only approaches that are in the spirit of the
discrete ciliar models. Therefore, we start with descriptions of the beating pattern
of a *single* cilium. Then, the collective effects in ciliar arrays can be studied by
analyzing how the beating of a single cilium is influenced by the beating of its
neighbors. This review is outlined correspondingly. We start with a summary of
the efforts to theoretically describe the beating of a single cilium in Section 8.2.
Next, it is shown how these results can be used to study collective effects in ciliar
arrays. In Section 8.4, the role of nodal flow in vertebrate development is discussed.
Finally, in the last section, we analyze the importance of hydrodynamic interactions
in collections of rotatory motors. We conclude with some (possible) technological
applications of these hydrodynamic effects.

8.2 Beating of a Single Cilium

To study collective effects in ciliar arrays, we first have to analyze the forces a
beating cilium exerts on the surrounding fluid. In doing so, we employ an approach
based on the "discrete cilia model" where one is interested in the velocity field
created by a single cilium. Because the length $L \simeq 5 - 10\mu m$ of a cilium is much
larger than its thickness (radius $R \simeq 100nm$ [27]), one can describe the motion
of the cilium as being created by a set of forces localized at the centerline of the
cilium. Furthermore, frequencies are of the order $\omega \simeq 10s^{-1}$ [10, 31] and the ciliar
beating is thus characterized by the Reynolds number, $Re \simeq 10^{-3}$ (see Appendix A
for a short introduction into the relevant hydrodynamical concepts). Therefore, the
inertial terms in the Navier-Stokes equation (8.39) are much smaller than the viscous
term (and the pressure gradient). For sufficiently small systems (typical extension

of a few 100 μm), momentum injection is instantaneous and the velocity field in the surrounding fluid induced by the beating cilium is a solution of the Stokes equation (8.42).

Two different approaches have emerged in theoretically modeling the beating of a single cilium: i) models that directly describe the beating pattern of the cilium; and ii) models in which the forces the motors exert on the microtubuli are prescribed. From these, the bending deformations are calculated, which determine the velocity field in the surrounding fluid. Both approach have some advantages.

8.2.1 Models with Prescribed Beating Patterns

Here, the velocity field is directly calculated from the prescribed motion of the cilium. Generally, this task can only be done numerically, and only with further simplifications can analytical solutions be obtained. One possibility is to model the cilia as a collection of stokeslets (see Appendix A) which move along prescribed trajectories, see Figure 8.1.

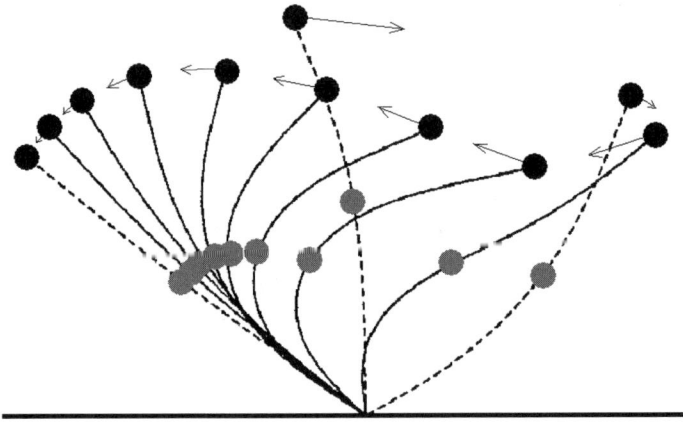

Figure 8.1. Cilia and monocilia can be modeled as a collection of stokeslets moving along fixed trajectories in the fluid. In doing so, it is assumed that each cilium undergoes a prescribed sequence of shape changes that are not altered by the motion of the surrounding fluid. For a single stokeslet, a beating cilium thus corresponds to a phase oscillator. However, more complex beating patterns (as shown) can be described by taking into account several stokeslets. Figure is from [32]. Copyright (2006), with permission from IOP Publishing Ltd.

In the most simple description, the beating cilium is then a phase oscillator corresponding to a stokeslet that moves along a prescribed closed trajectory. This description is certainly over-simplified, but in this case the velocity field induced in the surrounding fluid can be calculated analytically. The advantage of this approach will become evident in Section 8.3.

As an example, we first consider a simple phase-oscillator description of a monocilium. The beating monocilium is here represented by its tip, which is modeled as a spherical bead (of radius a) moving on a circular trajectory $r_t(t)$ of radius R. Here, only rotational vectors normal to the wall, to which the cilium is attached, are considered. Then, the plane of motion is parallel to the wall and $r_t(t) = (R \cos \varphi(t), R \sin \varphi(t), h)$ with constant h. The more general case of tilted motion will be considered in Section 8.4.2.

As mentioned, the ciliar motion is completely overdamped and the equation of motion for a single cilium is simply given by the balance of driving force and drag

$$\zeta R \dot{\varphi} = F_{in}. \tag{8.1}$$

Here, $\dot{\varphi} \equiv d\varphi/dt$, $\zeta = 6\pi\eta a$ is the friction coefficient of the moving bead, and η the viscosity of the surrounding liquid. The driving force is the internal force $\boldsymbol{F}_{in}$ exerted by the motor proteins. Here, we take only the tangential component of the driving force into account by assuming that the normal forces are balanced by the rigid ciliar filament keeping the beat on its circular trajectory. For the circular motion considered here, one has $F_{in} = const$.

The velocity field $\boldsymbol{v}(\boldsymbol{r})$ induced by such a oscillating bead is then in leading order given by the velocity field of a stokeslet (of radius a) moving on the trajectory $\boldsymbol{r}_s = \boldsymbol{r}_s(t)$. In bulk (where no walls are present) one has [33]

$$\boldsymbol{v}(\boldsymbol{r}) = \frac{\boldsymbol{s}}{|\boldsymbol{r} - \boldsymbol{r}_s|} + \frac{(\boldsymbol{r} - \boldsymbol{r}_s)(\boldsymbol{s} \cdot (\boldsymbol{r} - \boldsymbol{r}_s))}{|\boldsymbol{r} - \boldsymbol{r}_s|^3} + \mathcal{O}\left(\frac{a^3}{r^3}\right). \tag{8.2}$$

Here, $\boldsymbol{s} = 3a\dot{\boldsymbol{r}}(t)/4$ is the strength of the stokeslet. A similar expression can be derived for the movement close a (no-slip) wall, see Section 8.3.1.

Obviously, this simple description is only a good approximation if the external forces acting on the cilium (caused for example by the movement of the surrounding fluid) are too weak to significantly alter the beating pattern. This point be will be discussed in detail in Section 8.3. However, the big advantage of this simple model is that it can be easily extended to describe tilted monocilia (see Section 8.4.2 and [34]) and more complex beating patterns.

8.2.2 Models with Prescribed Microscopic Force Fields

Models in which the force fields exerted by the motors on the elastic filaments are prescribed have been studied by several authors (see [35, 36, 37, 38]). To illustrate this approach, we will first discuss the two-dimensional motion of a cilium.

To do so, the shape of the cilium is parameterized by a time-dependent angle $\alpha(s,t)$ between the tangent to the centerline of the cilium and the horizontal x-axis at arclength s. The force generated by the motors inside the cilium has normal and tangential components F_n and F_t, respectively. The normal component is given by

$$F_n(s) = \kappa \frac{d^2 \alpha(s,t)}{ds^2} + S(s,t), \tag{8.3}$$

where κ is the bending rigidity of the filament and $S(s,t)$ is an empirical term describing the active shear force created by the molecular motors. Generally,

$$S(s,t) = f(s)R(\omega t), \tag{8.4}$$

with amplitude $f(s)$ and a periodic function $R(\omega t)$ with frequency ω. To mimic motions with power and recovery stroke, non-symmetrical $R(\omega t)$ have to be considered. A possible choice is [35]

$$S(s,t) = C\zeta_n\omega(s^2 - L^2)\sin(2\pi t)c_1(t)c_2(s,t), \tag{8.5}$$

where C is a constant and ζ_n the normal friction coefficient of the ciliar filament, $c_1(t)$ parameterizes the different beat cycles and $c_2(s,t)$ divides the filament into active and passive portions.

The force balance reads

$$\Phi_n = \frac{dF_n}{ds} + F_t\frac{d\alpha}{ds} \tag{8.6}$$

$$\Phi_t = \frac{dF_t}{ds} - F_n\frac{d\alpha}{ds}, \tag{8.7}$$

where Φ_n and Φ_t are normal and tangential components of the force per unit length exerted on the cilium by the surrounding fluid. The components of the velocity of the cilium are given by

$$\frac{dv_n}{ds} = \frac{d\alpha}{dt} - v_t\frac{d\alpha}{ds} \tag{8.8}$$

$$\frac{dv_t}{ds} = v_n\frac{d\alpha}{ds}. \tag{8.9}$$

The velocity field in the surrounding fluid v created by the moving flexible filament can generally be calculated only numerically. However, in good approximation the filament can be thought of as being built up of a chain of stokeslets. Then, v is simply given by the superposition of the velocity fields created by the stokeslets. In leading order, the cilia do not interact, i.e., the velocity field v_i created by stokeslet i does not fulfill the no-slip boundary conditions on the surfaces of cilia $j \neq i$. The velocity field v induced by this motion is then given by $v = \sum_i v_i$. However, with the appropriate stokeslet field the no-slip boundary condition on the supporting membrane can be taken into account [35].

These equations have to be solved for the boundary conditions

$$\alpha(0,t) = \pi/2, \tag{8.10}$$

$$\left.\frac{d\alpha(s,t)}{ds}\right|_{s=0} = 0, \tag{8.11}$$

that keep the angle at the anchor (at which the cilium is attached to the supporting membrane) fixed. The end at $s = L$ (where L is the cilium's length) is free, implying

$$F_n(L,t) = F_t(L,t) = 0. \tag{8.12}$$

Finally, there is no drag on the anchor yielding

$$\left.\frac{dF_t(s,t)}{ds}\right|_{s=0} = \left.\frac{dF_n(s,t)}{ds}\right|_{s=0} = 0, \tag{8.13}$$

and at $s = L$ the cilium is straight, i.e.,

$$\left.\frac{d\alpha(s,t)}{ds}\right|_{s=L} = 0. \tag{8.14}$$

From these equations, the beating motion of a single cilium can be calculated numerically. The generalization to three dimensions is straightforward [39]. Figure 8.2 shows an example in three dimensions. The prescribed force fields are similar to the one given in Equation (8.5) and the beating pattern consists of a power and a recovery stroke.

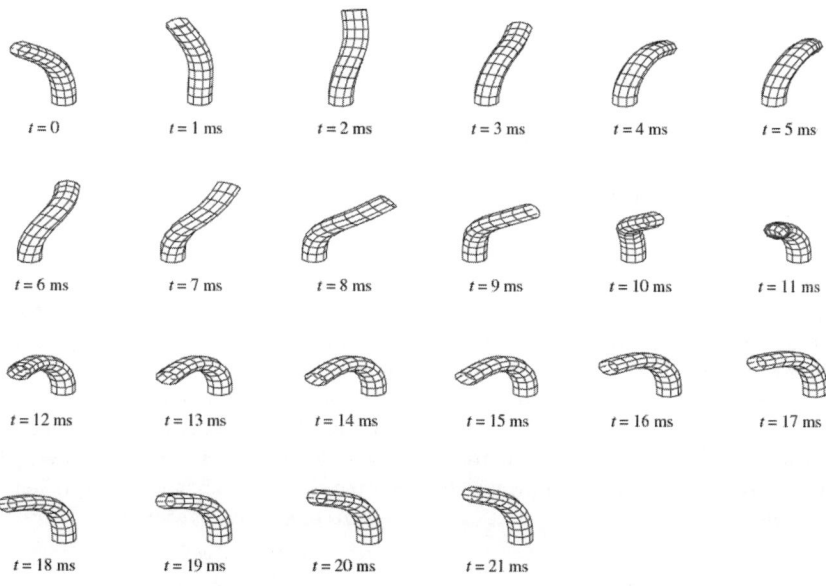

Figure 8.2. Beating of a single cilium in three dimensions consisting of a power stroke ($t = 1-5$ms) and a recovery stroke ($t = 6-21$ms). The details of this beating pattern are determined by the prescribed activity pattern of the molecular motors. Data is for $L = 10\mu$m, $a = 100$nm, $\omega = 25$Hz and $\kappa = \xi_p k_B T$ with $\xi_p = 6$mm. Figure is reprinted from [39]. Copyright (2001), with permission from The Royal Society.

The beating of a cilium can also be implemented numerically in a more direct way as shown in [40]. There, Kim and Netz have studied the pumping efficiency of a periodically beating elastic filament anchored to a solid surface in Brownian dynamics simulations.

In these simulations, the filament is modeled as a chain of N connected spherical beads of radius a. In Brownian dynamics simulations, the equation of motion of bead i at position r_i is given by the position Langevin equation [41]

$$\frac{dr_i}{dt} = -\sum_j \mu_{ij} \nabla_{r_j} U + \mu_{i2} \frac{r_{12} \times \tau(t)}{r_{12}^2} + \xi_i. \qquad (8.15)$$

Here, U is the elastic potential energy, given by a worm-like chain model $r_{ij} \equiv |r_i - r_j|$, and $\tau(t)$ is a prescribed (time-dependent) torque consisting of a phase of small torque (recovery stroke) and one of large torque (power stroke). The torque is applied on the second monomer with the first monomer fixed at the surface (the first bond is thus given by $\boldsymbol{r}_{12}$). The mobility tensor μ_{ij} obeys [41]

$$\mu_{ij}^{\alpha\beta} = \left(1 + \frac{a^2}{6}\nabla_{r_i}^2\right)\left(1 + \frac{a^2}{6}\nabla_{r_j}^2\right)G_{\alpha\beta}(\boldsymbol{r}_i, \boldsymbol{r}_j), \tag{8.16}$$

where $G_{\alpha\beta}(\boldsymbol{r}_i, \boldsymbol{r}_j)$ is the hydrodynamic Green's function, which takes into account the appropriate boundary conditions (here, the presence of a no-slip wall at $z = 0$). The stochastic term $\boldsymbol{\xi}$ mimics the heat bath and obeys the fluctuation-dissipation theorem

$$\langle \xi_i^\alpha \xi_j^\beta \rangle = 2k_B T \mu_{ij}^{\alpha\beta}(\boldsymbol{r}_i, \boldsymbol{r}_j)\delta(t - t'). \tag{8.17}$$

These equations can be directly solved numerically. A typical snapshot of the beating cilium is shown in Figure 8.3

As can be seen from Figure 8.3 during one beating period some net pumping of fluid takes place. In contrast to biological cilia systems, the slow stroke dominates the net pumping. During the slow stroke, momentum transfer to the surrounding liquid is more efficient since the filament is oriented perpendicular to the direction of motion (leading to a large hydrodynamic friction) and the distance to the no-slip wall is largest. In the fast stroke, the filament is bent and closer to the surface.

From these examples, one sees that it is possible to obtain rather specific features of the beating pattern by explicitly prescribing the microscopic forces. Thus, compared with the simple oscillator model, one obtains a more realistic description at the price of more non-universal microscopic details that have to be taken into account.

8.3 Collective Ciliar Motion

Typically, cilia appear in nature not as an isolated entity but belonging to an array. Under these circumstances, a cilium is subject to the velocity field in the surrounding fluid produced by its neighbors. Under suitable conditions, this hydrodynamic interaction can give rise to collective effects such as synchronization of beating and metachronal wave formation. We now first illustrate these phenomena in specific models and then analyze them in a more general approach.

8.3.1 Specific Models

To illustrate the concept of hydrodynamic interactions between beating cilia we will first consider the simple model for monocilia introduced in Section 8.2.1. In a system consisting of two cilia, each cilium is not only subject to internal forces but also to the forces arising from the flow in the surrounding fluid created by the neighboring cilia. In this case, the force balance (8.1) is altered because each cilium follows the flow in the surrounding fluid induced by the other cilium [42, 43]. This hydrodynamic interaction leads to

$$\zeta\left(R\dot{\varphi}_1 - \boldsymbol{t}_1 \cdot \boldsymbol{v}_{12}\right) = F_{in}, \tag{8.18}$$

where v_{12} is the velocity at the position of Cilium 1 induced by Cilium 2. This velocity has to be projected onto the tangential vector $\boldsymbol{t}_1$ of the trajectory of Cilium

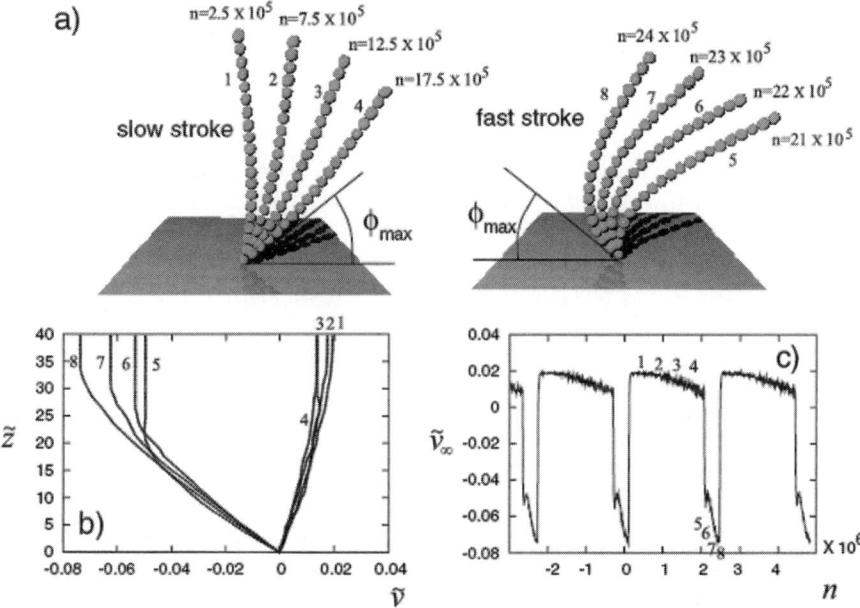

Figure 8.3. Typical snapshot of a Brownian dynamics simulation of an elastic filament (of length L) attached to a no-slip wall. The filaments are modeled as a collection of ($N = 20$) spherical beads with potential energy given by the worm-like chain model (with persistence length $\xi_p/L = 2 \cdot 10^5$). The motion is caused by an applied torque consisting of two phases: a phase of small torque $\tau_s = 2 \cdot 10^5 k_B T$ [corresponding to the sequence of shapes 1-4 shown in Figure (a)] and a phase of large torque $4\tau_s$ [sequence of shapes 5-8 in Figure (a)]. (b) Flow profile as function of distance from the wall shown for the conformations 1-8 of (a). (c) Time-dependence of the induced velocity field. Thus, the main motion of the fluid is caused by the slower stroke. Figure is reprinted from [40]. Copyright (2006), with permission from The American Physical Society.

1. Again, it is assumed that the force arising from the normal component of v_{12} is balanced by the ciliar filament. In deriving the last equation we have implicitly assumed that the hydrodynamic interaction alters only the beating velocity but not the beating pattern of the monocilia. The conditions under which this strong assumption is justified will be discussed in the next section.

If the no-slip boundary condition on the wall is ignored, then the velocity v_{12} is that of a stokeslet (given by Equation (8.2)) with strength $s = 3aR\dot{\varphi}_2 t_2/4$ where t_2 is the tangential vector of the circular trajectory of Cilium 2. Then, the equation of motion of the coupled system becomes

$$\dot{\varphi}_1 + J(\varphi_1, \varphi_2)\dot{\varphi}_2 = 2\pi, \tag{8.19}$$

where time t is measured now in units of the period T of the unperturbed motion, i.e., $\varphi(T) = 2\pi$. It is also assumed that the unperturbed motion of each cilium has the same intrinsic frequency ω. The function $J(\varphi_1, \varphi_2)$ is given by

$$J(\varphi_1, \varphi_2) \equiv -\frac{3}{4}a\frac{t_2 \cdot t_1 + (t_1 \cdot n_{12})(t_2 \cdot n_{12})}{r_{12}}, \tag{8.20}$$

where $n_{12} = r_{12}/|r_{12}|$, and r_{12} is the vector pointing from Cilium 2 to 1. For arrays with low ciliar densities, $R/l \ll 1$ where l is the distance between the centers of the trajectories. One has then

$$J(\varphi_1, \varphi_2) = -\frac{3a}{4l} \left[2\cos(\varphi_1)\cos(\varphi_2) + \sin(\varphi_1)\sin(\varphi_2)\right] + \mathcal{O}\left(\frac{R}{l}\right). \tag{8.21}$$

From Equation (8.20) one immediately sees that two hydrodynamically interacting monocilia do not synchronize. By subtracting the equation of motion for Cilium 1 from that for Cilium 2, one finds

$$(\dot{\varphi}_1 - \dot{\varphi}_2)[1 - J(\varphi_1, \varphi_2)] = 0, \tag{8.22}$$

where it has been used that the drag Cilium 1 exerts on Cilium 2 is given by $\zeta R\dot{\varphi}_1 J(\varphi_2, \varphi_1)$ and $J(\varphi_2, \varphi_1) = J(\varphi_1, \varphi_2)$. Thus, $\varphi_1 - \varphi_2 = const.$ and any initial phase difference persists in time and no synchronization occurs. A similar phenomenon has been found for two rotating helices [44].

The no-slip boundary condition on the wall modifies the velocity field created by the cilia leading to

$$v_{12} = 12h^2\frac{n_{12}(s \cdot n_{12})}{r_{12}^3} + \mathcal{O}(r_{12}^{-5}), \tag{8.23}$$

where higher order terms and the contributions perpendicular to the trajectory have been neglected. In this case, the interaction becomes

$$J(\varphi_1, \varphi_2) = -9a\frac{h^2}{l^3}\left[\cos(\varphi_1)\cos(\varphi_2)\right] + \mathcal{O}\left(\frac{R}{l}\right). \tag{8.24}$$

Thus, even if the boundary conditions on the wall are taken into account, the interaction $J(\varphi_1, \varphi_2)$ is symmetric in φ_1 and φ_2 and again, two cilia do not synchronize. Similar results can be obtained for systems with more than two cilia.

In fact, as will be discussed later in more detail the absence of synchronization is a direct consequence of the basic assumption that hydrodynamic interactions only alter the velocity of the shape changes but not the beating pattern itself. Under this assumption, even more complicated beating patterns (which could be modeled by taking several stokeslets into account, see Figure 8.1) do not lead to synchronization. As we will see in the next section such additional contributions do not alter the general form of the equation of motion (8.19), they only modify the interaction function $J(\varphi_1, \varphi_2)$.

On the other hand, in the specific models of Gueron et al. and Kim and Netz, synchronization of ciliar beating occurs. For example, in [37] the synchronization of the beating of two neighboring cilia was observed numerically (in two dimensions). Synchronization occurred within two beat cycles. Also metachronal waves form spontaneously. Figure 8.4 shows an example for a three-dimensional array with nine cilia in the model of Gueron. In the simulations, such a phase shift between neighboring cilia occurs even though all cilia have the same beating pattern and start with the same initial conditions. This indicates that the synchronized state is not stable (at least for the imposed boundary conditions). One should also be aware that, for example, the wave length of the metachronal wave or the time it takes to synchronize strongly depend on the prescribed microscopic force field driving the ciliar beating.

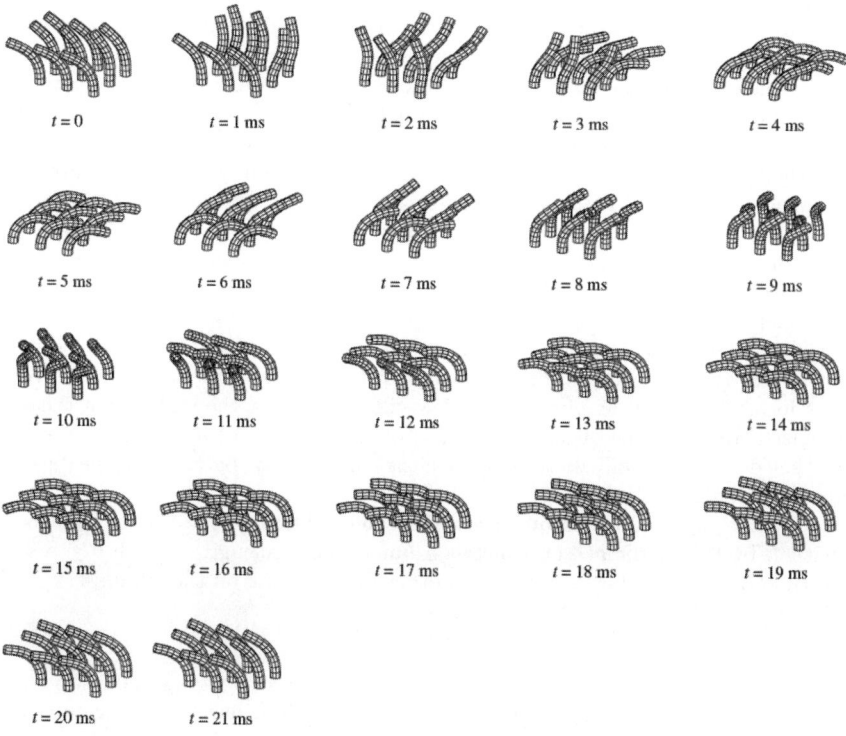

Figure 8.4. In the three-dimensional model of Gueron et al. [35], beating cilia spontaneously form metachronal waves, i.e., they beat with a constant phase lag. Data is for same parameter values as in Figure 8.2. Figure is reprinted from [39]. Copyright (2001), with permission from The Royal Society.

Metachronal wave formation between two neighboring cilia has also been observed numerically in [40]. There, it was also found that the filaments spontaneously beat with a constant phase lag independent of the initial phase difference. The phase lag depends on various parameters such as the persistence length.

8.3.2 A General Approach to Collective Effects in Ciliar Arrays

Although the specific models exhibit synchronization of ciliar beating, there is a certain arbitrariness in prescribing the beating pattern of a cilium. In particular,

Gueron's approach does not reveal which of the obtained results (such as the occurrence of synchronization and metachronal wave formation) are universal (i.e., would also be found with a different description of the ciliar beating) and which are only properties of the underlying model. Generally, all approaches so far have considered only specific beating patterns and not the general aspects of hydrodynamic interactions between cilia.

Only recently, a more general framework has been developed in [32] in which the influence of the beating pattern on the collective effects can be systematically investigated[2]. For this purpose, it is necessary to first focus on arrays with low ciliar density where the hydrodynamic interactions are weak and do not alter the (prescribed) beating pattern of the individual cilia.

We will give a short summary of this approach. Again, we directly prescribe here the beating pattern of the cilium. This has the advantage that our approach is independent of any microscopic details (such as the forces exerted by the molecular motors on the filaments). However, the obtained results do not depend on the explicit form of the beating pattern, they are valid for general motion of the filaments. We now calculate the flow field induced by a moving cilium by (formally) solving the Stokes equation and then using this solution to systematically analyze the influence of hydrodynamic interactions on cooperative beating. In doing so, general criteria can be derived when these interactions lead to synchronization and metachronal wave formation. It turns out that for low ciliar densities, the stability of the synchronized state and the dispersion relation of metachronal waves are non-universal, i.e., they depend crucially on the details of the ciliar beating pattern and the imposed boundary conditions.

For this purpose, we parameterize the conformation of the elastic filament of length L by the vector $\boldsymbol{r}(s,t)$, which is a function of arclength s (with $0 \leq s \leq L$) and time t. The drag forces balance all other forces acting on the cilium

$$\zeta_{ij}\left(\partial_t r_i\left(s,t\right) - v_i\left(\boldsymbol{r}\left(s,t\right)\right)\right) = F_j(\boldsymbol{r}\left(s,t\right),s), \qquad (8.25)$$

where repeated indices have to be summed over and vector $\boldsymbol{v}$ has spatial components v_i. Here, ζ_{ij} is the friction tensor that accounts for the fact that tangential and normal frictions of the filament are different. Furthermore, $\boldsymbol{v}\left(\boldsymbol{r}\left(s,t\right)\right)$ is the velocity in the fluid at position $\boldsymbol{r}$. Here, $\boldsymbol{v} = \boldsymbol{v}_e + \boldsymbol{v}_s + \boldsymbol{v}_n$, where $\boldsymbol{v}_e$ is the external flow field, $\boldsymbol{v}_s$ the contribution arising from the motion of other parts of the filament (at positions $\boldsymbol{r}' \neq \boldsymbol{r}$), and (in ciliar arrays) $\boldsymbol{v}_n$ is the contribution from the motion of neighboring cilia. Finally, $\boldsymbol{F}(\boldsymbol{r}\left(s,t\right),s)$ is the sum of all forces acting on the cilium segment at $\boldsymbol{r}(s,t)$, i.e., elastic and internal driving forces $\boldsymbol{F} = \boldsymbol{F}_{el} + \boldsymbol{F}_{in}$. We assume that these forces are a functional of $\boldsymbol{r}(s,t)$ and only explicitly depend on s but not on t.

As mentioned, we assume that the beating pattern is not altered by the interactions (i.e., for given $\boldsymbol{v}_e$ it is not influenced by $\boldsymbol{v}_n$). We will discuss under which conditions this strong assumption holds and which consequences arise from it. Thus, the sequence of beating shapes is the same for an isolated cilium and for a cilium belonging to an array. However, the velocity of shape changes generally will depend on the hydrodynamic interactions. With this assumption, the sequence of shape changes

---

[2] It should be mentioned that there are also other effective models [45] to describe collective effects in ciliar arrays which, however, do not explicitly take the hydrodynamic interactions into account.

of cilium m is completely described by a single parameter φ_m. The parameterization of the filament then only implicitly depends on time,

$$r(s,t) = r(s, \varphi(t)). \tag{8.26}$$

Upon appropriately rescaling t and measuring time in units of the beating period T, one can set for the unperturbed motion $\dot{\varphi} = 2\pi$. We assume here that all cilia of the array have the same intrinsic beating frequency ω. Arrays of cilia beating with different frequencies will be discussed in Section 8.3.4.

Under this assumption, interacting ciliar filaments undergo the same shape deformations as free cilia but with a different velocity $\dot{\varphi}$. Thus, the beating cilium is a phase oscillator. As shown in Appendix B the equation of motion for two hydrodynamically coupled cilia can again be written as

$$\dot{\varphi}_1 + J(\varphi_1, \varphi_2)\dot{\varphi}_2 = 2\pi, \tag{8.27}$$

where now

$$
\begin{aligned}
&J(\varphi_1, \varphi_2) \\
&= \sum_{n=1}^{\infty} \sum_{m=1}^{\infty} \sum_{\alpha=1}^{\infty} A_{nm\alpha} \frac{(\sin n\varphi_1 + a_{n\alpha} \cos n\varphi_1)(\sin m\varphi_2 + b_{m\alpha} \cos m\varphi_2)}{|r_1^{(0)} - r_2^{(0)}|^{\alpha}} \\
&\equiv \Omega(\varphi_1, \varphi_2) G(r_1^{(0)} - r_2^{(0)}).
\end{aligned}
\tag{8.28}
$$

The Fourier-Taylor coefficients $A_{nm\alpha}$, $a_{n\alpha}$, and $b_{m\alpha}$ characterize the beating pattern of the cilium and can (in principle) be calculated from the solution of the Stokes equation (see Appendix B). Because hydrodynamic interactions are weak, the coefficients $A_{nm\alpha}/l^{\alpha}$, $a_{n\alpha}/l^{\alpha}$ and $b_{m\alpha}/l^{\alpha}$ are of order $\varepsilon \ll 1$ for a typical distance l between neighboring cilia, where the dimensionless parameter ε measures the strength of the hydrodynamic interactions.

This general equation of motion has been studied in detail in [32]. There, general criteria for synchronization of ciliar beating and the dispersion relation of metachronal waves have been derived. Here, we only illustrate these general results in two specific models.

(1) Models with symmetric interactions. An example is given by a cilium described as a stokeslet (of radius a) moving along the line trajectory $r_t(\varphi) = (x, y, z) = (x(\varphi t), 0, h)$ with the power stroke corresponding to the phase interval $(0, \pi)$ and the recovery stroke to $(\pi, 2\pi)$. Thus, in leading order [cf. Equation (8.28)] one has $(x, y, z) = (R \sin(\varphi t), 0, h)$. For two cilia at positions r_1 and r_2 one has $r_m(s, \varphi_m) = r_m(s = 0) + r_t(\varphi_m)\delta(s - L)$ for $m = 1, 2$, where $s = L$ is the position of the tip. By denoting the distance between the cilia with $l \equiv |r_1(s = 0) - r_2(s = 0)|$, the interaction for parallel motion close to a wall is given by

$$J(\varphi_1, \varphi_2) = -\varepsilon a h^{\alpha-1} \frac{\sin^{\kappa}(\varphi_1) \sin^{\kappa}(\varphi_2)}{l^{\alpha}} + \mathcal{O}\left(\frac{R}{l}\right), \tag{8.29}$$

where in leading order $\alpha = 3$ if the no-slip boundary condition on the supporting wall is taken into account. In fact, the obtained results depend only weakly on α. Furthermore, $\kappa = 1, 3, 5.....$ The latter parameter characterizes the movement of the cilium. For large κ, the power and recovery stroke are localized around $\varphi_i = \pi/2$ and $\varphi_i = 3\pi/2$, respectively.

For $\kappa = 1$, J given by Equation (8.29) corresponds to the simplest choice where in-phase motion of two neighboring cilia leads to an increase in phase speed. In fact, monocilia close to a wall interact precisely in this way, cf. Equation (8.24).

It turns out that in this model, the synchronized state and metachronal waves are only marginally stable in linear order. Nonlinear terms destabilize the synchronized state but stabilize the metachronal wave (under periodic boundary conditions) [32].

(2) **Non-symmetric models with a power stroke.** In a somewhat more realistic description of the ciliar beating one can take into account that during the power stroke $0 < \varphi < \pi$ interactions are stronger. A possible choice is

$$ J(\varphi_1, \varphi_2) = -\varepsilon a h^{\alpha-1}(1 + \beta \sin^\nu(\varphi_2)) \frac{\sin^\kappa(\varphi_1)\sin^\kappa(\varphi_2)}{l^\alpha} + \mathcal{O}\left(\frac{R}{l}\right), \qquad (8.30) $$

where $0 < \beta < 1$, $\nu, \kappa = 1, 3, 5, \ldots$. The more complex beating pattern does not alter the stability of the dynamical states. Again, in linear order the synchronized state and metachronal waves are only marginally stable. Numerically, one sees that in this case the synchronized state is also unstable, while metachronal waves are stable [32].

Finally, we have to show that our basic assumption (namely that the hydrodynamic interactions do not alter the sequence of shape changes the filament undergoes) is justified in arrays of low ciliar densities. Indeed, for a cilium consisting of N stokeslets, one has in leading order $v(r) \sim Nr^{-\alpha}$ (and $\alpha \geq 3$). Thus the beating cilium exerts a drag force $F_d \sim N\zeta l^{-\alpha}$ on its neighbor. However, because $F_{in} \sim \zeta N a^{-\alpha}$ one has $\frac{F_d}{F_{in}} \sim \left(\frac{a}{l}\right)^\alpha \ll 1$ justifying the analysis just given.

8.3.3 Discussion

This general approach shows that no synchronization of ciliar motion occurs if hydrodynamic interactions do not alter the (prescribed) beating pattern of the individual cilia (which is the case in low-density arrays).

These general results, however, do not contradict the numerical findings presented in Section 8.3.1. For example, Gueron et al., who consider arrays of high ciliar density, explicitly take into account the dependence of the ciliar beating pattern on the flow in the surrounding liquid, see Equations (8.3), (8.8), and (8.9). In fact, the comparison of the general results with the numerical findings of Gueron et al. suggests that the latter effect is a requirement for the occurrence of synchronization (and metachronal wave formation under non-periodic boundary conditions). This is supported by the findings of [44] where it was shown that two rotating helices synchronize only if their motion is not too much confined.

The phase-oscillator approach can be generalized to arrays with high ciliar densities where hydrodynamic interaction alters the beating pattern [46]. In this case, hydrodynamic interactions generally lead to synchronization. For example, if the trajectories of the monocilia introduced in Section 8.3.1 have a variable radius R, this additional degree of freedom already leads to synchronization of the beating of two neighboring cilia [46]. Within our approach, it will be possible to develop general analytical criteria for the occurrence of these phenomena and to classify their stability for various boundary conditions.

Finally, the phase-oscillator description shows that hydrodynamic interactions generally lead to the formation of metachronal waves. However, for weak hydrodynamic interactions, this dynamical state is only stable for periodic boundary conditions and its dispersion relationship depends on the details of the angular dependence of the interciliar interactions. This non-universal nature of our results shows that one has to be very careful in drawing general conclusions from studies of specific ciliar beating patterns.

8.3.4 Synchronization of Beating Frequencies

If two neighboring cilia have different frequencies ω_1 and ω_2, hydrodynamic interactions can, in principle, lead to synchronization of the frequencies, i.e., to a synchronized movement of the two cilia with the same frequency ω. For example, for cilia with a specific beating pattern [47], this phenomenon has been shown to occur under suitable conditions.

However, this is not an universal feature. For example, for the general hydrodynamic interaction introduced above, the two cilia obey

$$\dot{\varphi}_1 + \dot{\varphi}_2 J(\varphi_1, \varphi_2) = \omega_1, \tag{8.31}$$

$$\dot{\varphi}_2 + \dot{\varphi}_1 J(\varphi_2, \varphi_1) = \omega_2. \tag{8.32}$$

Here, synchronization only occurs for non-symmetric interactions $J(x,y)$, i.e., $J(x,y) \neq J(y,x)$. This can be seen by making the ansatz $\varphi_1 = \phi(t)$, $\varphi_2 = \phi(t) + \delta$, where δ is a time-independent constant. Then, Equations (8.31) and (8.32) yield

$$\omega_1(1 + J(\phi + \delta, \phi)) = \omega_2(1 + J(\phi, \phi + \delta)). \tag{8.33}$$

For symmetric interactions $J(x,y) = J(y,x)$ (as one has, e.g., in the case of the monociliar beating introduced in Section 8.3.1) there is no synchronized solution.

There is an interesting connection to general synchronization phenomena occurring, for example, in arrays of coupled oscillators. One of the best studied models in this context is the Kuramoto model (for a review see [48]). The Kuramoto model consists of N coupled oscillators with natural frequencies ω_i whose motion is described by the dynamical variable θ_i. The equation of motion is given by

$$\frac{d\theta_i}{dt} = \omega_i + \sum_{j=1}^{N} K_{ij} \sin(\theta_j - \theta_i), \tag{8.34}$$

where additionally the coupling-matrix K_{ij} has to be specified. The simplest choice is an identical coupling between all oscillators:

$$K_{ij} \equiv \frac{K}{N}. \tag{8.35}$$

The basic properties of the Kuramoto model can be summarized as follows: For $N \gg 1$ there is a critical coupling strength K^* above which (i.e., $K > K^*$) the oscillator become synchronized to their average phase $\theta_i \approx \psi$. For $K < K^*$ the oscillators behave incoherently. One can also easily see that in the Kuramoto model, two oscillators synchronize (i.e., $\varphi_1 = \varphi_2$) for $\omega_1 = \omega_2$. However, for two cilia with general interaction $J(\varphi_1, \varphi_2)$, this is not the case. Thus, it remains a challenge to develop an understanding of what the essential difference between Kuramoto's and ciliar systems is.

8.4 Cilia-Generated Nodal Flow

Cilia are not only responsible for the motion of cells but they also play an important role in the development of vertebrates. There are strong experimental indications that cilia-generated fluid flow is responsible (or at least a necessary requirement) for the establishment of left and right in developing embryos. We first summarize the experimental findings supporting this view. In Section 8.4.2 we then give a short overview of the current theoretical efforts in modeling these observations.

8.4.1 Experimental Facts

Vertebrate development involves three levels of broken symmetry induced by the specification of three body axes. These events occur at different growth stages partitioning the embryo into anterior and posterior (top-bottom), dorsal and ventral (front-back), and left and right domains [49].

First, the anterior-posterior (A-P) axis of the initially cylindrically symmetric embryos is determined by newly emerging signaling centers along the future midline of the embryo [50]. Assuming that the anterior side is randomly selected (as hypothesized in [51]) then L-R symmetry breaking must occur afterwards. This is consistent with findings that gene expression is transiently symmetric before this stage [52]. The L-R axis must be reliably oriented with respect to A-P and dorso-ventral (D-V) axes [53].

There is now strong experimental evidence that cilia-generated fluid flow plays an important (if not essential) role in breaking of left-right symmetry [11, 54, 55, 56]. More precisely, the flow created by monocilia over a small region of the embryo (node) seems to trigger the establishment of the right and left sides. This node is a major organizing center regulating pattern formation (see Figure 8.5). During gastrulation, several genes essential for formation of L-R axis are expressed at or around the node [57, 58, 59].

As mentioned in the introduction, the monocilia of the ventral node lack the central pair of microtubules and thus have a "9 + 0" microtubule arrangement different from the "9 + 2" arrangements of conventional cilia. They perform a rotation-like motion, creating a leftward flow over the node [11, 54, 61]. The presence of this directed flow is necessary for breaking of the L-R symmetry. Experiments by Nonaka et al. have shown that the absence of functional (motile) monocilia leads to randomization of the left-right placement of organs [11] (see Figure 8.6). However, by subjecting the surface of the mouse embryos to an artificial flow (created by a mechanical pump) L-R patterning was re-established in mice with only non-motile cilia [62].

The ciliar movement has been analyzed at high temporal resolution [56]. The nodal monocilia perform a clockwise rotation (if seen from above) around an axis tilted about $\psi = 40° \pm 10°$ to the posterior (where a tilting angle $\psi = 0°$ corresponds to an axis normal to the surface), see Figure 8.7. The trajectory described by the tip of the monocilium is slightly elliptical. Because the velocity field has to vanish on the surface of the node (i.e., the no-slip boundary condition has to be fulfilled on the nodal membrane), the cilia induce a larger velocity in the surrounding fluid during their leftward swing. The rightward swing close to the surface is less effective. Thus, averaged over one beating pattern, the monocilia effectively push fluid from the right to the left (see Figure 8.7).

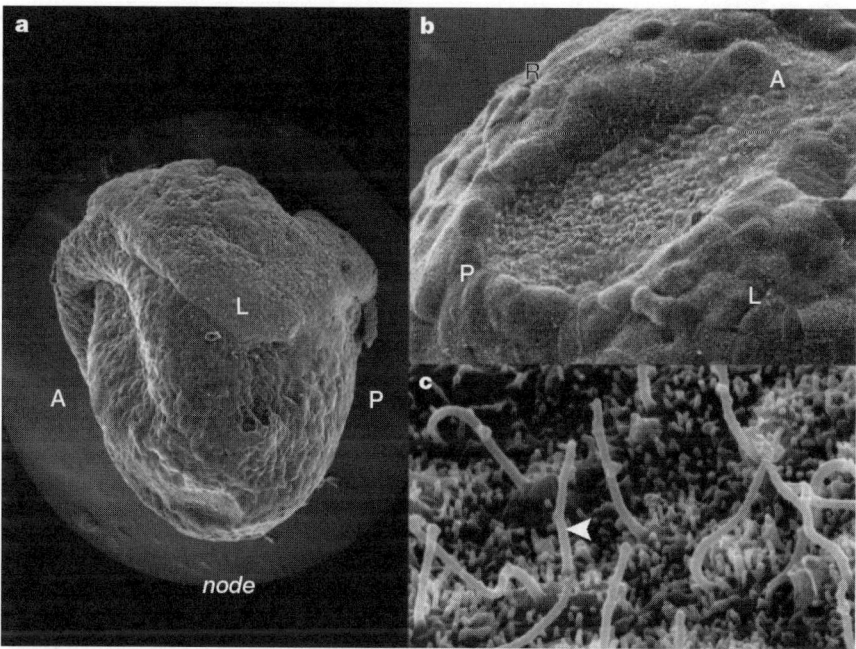

Figure 8.5. Scanning electron micrographs of the node (a and b) and monocilia (arrowhead in c) of a mouse embryo. A, P, R, L indicate the anterior, posterior, right, left side, respectively. Figure is reprinted from [60]. Copyright (2002), with permission from Macmillan Publishers Ltd.

The beating pattern has also been characterized quantitatively. Monocilia have a typical length of $L \simeq 5\mu m$ (for mice) and they beat with typical frequencies $\omega \simeq 10 Hz$ [56].

The flow over the nodal surface induced by the beating of all monocilia is height-dependent. With the help of exogenously introduced latex beads following the nodal flow the velocity can be measured [13]. Okada et al. [56] find at a height of $\simeq 5\mu m$ a typical fluid velocity $v_f \simeq 4\mu m/s$ to the left (earlier measurements of the same group yielded larger velocities $v_f \simeq 20 - 50\mu m/s$ [54]). The counter-flow to the right (which is necessary due to conservation of mass) takes place at a height $\simeq 20\mu m$, with typically $v_f \simeq 2\mu m/s$.

From a theoretical point of view, the rotational movement of the cilia is represented by an axial vector. This chiral structure aligned with respect to the A-P and A-V axes then generates the L-R laterality as proposed by Brown and Wolpert [53]. The direction of the flow is solely determined by the tilting of the axis of rotation. This mechanism of establishing L-R asymmetry is also conserved in rabbits and fish.

The analysis of the nodal flow in mice mutants confirms the role monocilia play in L-R symmetry breaking. *iv* is a mouse mutant that results in randomization of L-R determination [63]. The mutant *inv* shows a complete inversion of the L-R body axis [64]. The experimental analysis of the properties of the monocilia of *inv* mutants gives further support that posteriorly tilted, clockwise rotation of cilia produces the

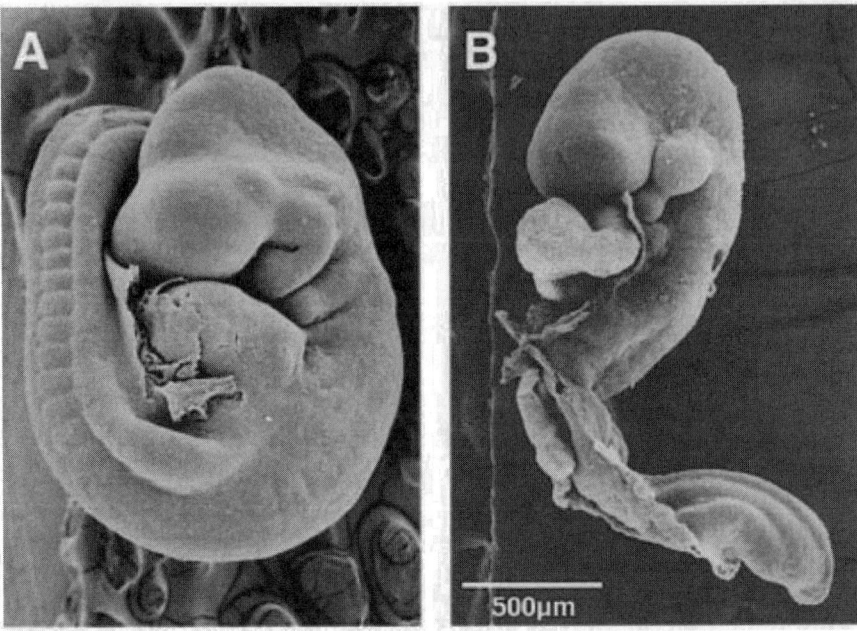

Figure 8.6. In KIF3b⁻-mutants (which lack KIF3B, kinesin superfamily proteins) nodal monocilia are absent and left-right asymmetry is not established. While in the wildtype (left side A) the cilia-generated flow over the node leads to a directed placement of organs, their location is completely random in the mutant (right side B). If mutants are subjected to an artificial external flow (having the same direction as the nodal flow) left-right patterning is, however, re-established [62]. Figure is reprinted from [11]. Copyright (1998), with permission from Elsevier.

leftward flow. As shown by Okada et al. in [56] nearly 20% of all cilia are anteriorly tilted in *inv* mutants. For these monocilia the "power stroke" occurs during their rightward motion, while the leftward occurs close to the wall. Also around 20% of the cilia exhibit a counterclockwise rotation. This disordered arrangement of rotational vector gives rise to the typical "meandering" flow observed in [54]. In *iv* mutants, the monocilia appeared to be very rigid. They hardly moved and apparently were frozen [54].

   Thus, there is strong experimental evidence that monocilia generate the nodal flow from right to left. However, so far it is not clear by which mechanism the ciliar flow influences the vertebrate development. More specifically, how is the movement of fluid across the node translated into asymmetric patterns of gene expression? Currently, several models are being discussed that speculate on the physiological role of the nodal flow: i) models in which the nodal flow leads to the formation of concentration gradients of secreted morphogens (such as Fgf8, GDF1, and Nodal) [11, 56]; and ii) stimulation models with two different kind of cilia. Here it is assumed that only some of the nodal cilia are motile. The remaining cilia act as sensors of

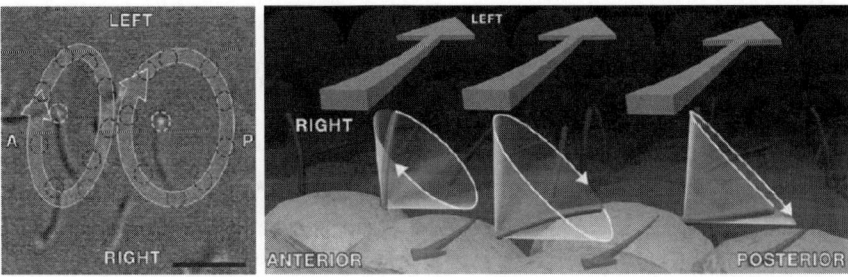

Figure 8.7. The trajectory described by the tip of nodal monocilia is elliptic with the axis of rotation tilted toward the posterior side. The larger the distance between tip and nodal surface, the stronger the fluid layer is pushed into the direction of ciliar motion. The ciliar motion thus gives rise to a fast flow just above the tip from right to left, and to a backward flow close to the nodal surface from left to right. The backward flow is much slower due to the viscous resistance of the wall. Figure is reprinted from [13]. Copyright (2006), with permission from Elsevier. (See color insert.)

the flow (by initiating a calcium-dependent signal transduction event) [65, 66]. In both approaches the cilia-generated flow plays an essential role which triggers the symmetry-breaking event.

Stimulation models were introduced to explain the observed differences in the lateral expression patterns between mutant embryos lacking nodal cilia and mutants with only immotile cilia [11]. Here, we will only discuss the chemical-gradient models. The cilia-generated flow transports morphogens secreted into the cavity of the ventral node leading to an accumulation on the left side. For sufficiently large morphogenic proteins, the nodal flow is fast enough to generate a concentration gradient with a higher concentration on the left side. However, for smaller proteins this gradient might be destabilized by the rightward backflow discussed earlier. To circumvent these difficulties, Hirokawa and coworkers have suggested that vesicles might carry the morphogens which release them by rupturing on the left side [67]. However, the precise rupturing mechanism remains unclear.

Generally, it is believed that morphogens are detected by receptors placed on both the left and right side of the node. However, the concentration gradient then leads to asymmetric signaling transduction events. The precise mechanisms are not known. An overview of currently discussed reaction-diffusion models can be found in [54, 68].

Finally, it should be emphasized that despite considerable progress in the understanding of this system, some key questions remain unanswered: What makes the flow leftward, i.e., what is the origin of the asymmetry in cilia? Is the establishment of the nodal flow the initial L-R symmetry-breaking event in vertebrates? And, how universal is this mechanism in vertebrates?

8.4.2 Theoretical Modeling

One of the main theoretical challenges is to show that the nodal flow can stabilize a morphogen concentration gradient. This question has been addressed by Cartwright et al. in [31]. In this study, the monocilia are modeled as simple rotating spheres (rotlets). For simplicity, boundary effects are not taken into account, the cilia are assumed to beat in bulk. The rotational motion of each monocilium then gives rise to a velocity field v at position r

$$v = \frac{\tau \times r}{8\pi r^3},\tag{8.36}$$

where τ is the applied torque (i.e., the torque driving the rotation of the cilium). Of course, solution (8.36) does not fulfill the no-slip boundary condition on the nodal membrane, i.e., v does not vanish on the nodal membrane as it would be required by hydrodynamics. This point will become important later. Because the Stokes equation is linear, the hydrodynamical velocity field generated by an array of cilia is simply given by the superposition of the rotlet solution (8.36).

If the rotational axis of the cilia are vertical, the velocity field on the node is vortical. The cilia generate no net motion over the center of the node, only a circular flow occurs at the lateral boundaries of the node (which are also no-slip walls and thus also modify the velocity field). This finding is pretty much independent of the actual shape of the node, indicating that the shape of node is not the main reason for breaking of L-R symmetry as suggested in [11, 62].

However, if the cilia are tilted, i.e., if the angle between the axis of rotation and the vertical is smaller than $90°$, then a net current between right and left occurs. More precisely, if the rotlets are tilted toward the posterior side, a net current from right to left is established. However, this current from right to left occurs only *above* the cilia. Below them the flow is from left to right. Because Cartwright et al. neglect the no-slip boundary condition on the nodal membrane, their model yields a too large backward flow below the cilium. In the nodal cavity, this flow is much slower due to the presence of the nodal surface where the velocity field vanishes.

It is important to note that such a backward current has to occur in a finite geometry due to mass conservation. With a correct treatment of the no-slip boundary condition on the nodal surface, it might be even possible to explain the observations of Okada et al. that the backward flow occurs about 20μm above the surface [56].

Nevertheless, the approach of Cartwright et al. is useful to get first estimates of the order of magnitude of the important quantities. As shown in [31], one finds from the observed ciliar velocity field (here taken to be of the order $v_f \simeq 20 - 50\mu$m/s) that the tilt angle $\psi \simeq 15° \pm 10°$ from the vertical[3]. A larger tilting angle leads to a weaker right-left velocity, field which might be an important factor in explaining the observations in the mutants introduced earlier.

The velocity field created by nodal cilia close to the tip is strong enough to stabilize a morphogen concentration gradient. This can be seen by considering the Péclet number Pe $\equiv t_d/t_a = v_f L/D \gg 1$ which compares typical diffusion times with advection times. Here, $L \simeq 50\mu$m is the extension of the node, and $D \simeq 10^{-11} - 10^{-12}\text{m}^2/\text{s}$ is a typical diffusion constant of proteins yielding $25 \leq$ Pe ≤ 250. Such high Péclet numbers imply that transport by the fluid is much faster than diffusion, stabilizing thus a morphogen concentration gradient over the node [31]. Obviously, the concentration of morphogens is only larger on the left side close to

---

[3] Thus, the value of v_f taken here is larger than the experimentally measured value of $v_f \simeq 5\mu$m [56], explaining the deviation in the value of ψ between theory and experiment.

the ciliar tip. Further away from the nodal surface, the opposite is true, i.e., the morphogen concentration is larger on the right. One way to avoid consequence from this uneven distribution is to assign a window of activity to the morphogens [31]. However, such a mechanism is currently purely speculative. All these results are not conclusive yet. It should be emphasized that further theoretical progress can only be achieved if, in the calculation of the cilia-induced velocity field, all the boundary conditions are taken into account. Only in this case can the stability of morphogen gradients over the node be calculated reliably.

A more complex beating pattern than the simple rotational motion of the rotlets has been considered by Buceta et al. [69]. In their model, each cilium is a string of N moving spheres of radius a connected by elastic rods and attached to a nonmoving sphere at the nodal surface. Each sphere i of cilium k describes a trajectory $\boldsymbol{R}_{i,k} = (x_{i,k}, y_{i,k}, z_{i,k})$ and during this motion generates a velocity field corresponding to that of a stokeslet (see Equation (8.2)). Thus, Buceta et al. also do not take into account the no-slip boundary conditions on the wall. However, the influence of the wall is mimicked by a time-dependent radius $a(t)$, where $a(t)$ is large (small) for small (large) z. The spheres are also connected by elastic springs to take bending of the ciliar filament into account.

Each cilium (i.e., each collection of spheres) is tilted towards the vertical and undergoes a two-phase motion consisting of a power and a recovery stroke. The velocity field created by a single cilium is simply given by the superposition of the stokeslet fields created by the sphere $i = 1, .., N$. However, this description is only correct in the far field (i.e., for $r \gg a$), because the no-slip boundary conditions on the surfaces of the spheres are not taken into account.

By appropriately prescribing the bending of each cilium, Buceta el al. are able to reduce the backward flow induced by the recovery stroke by reducing the drag the cilium exerts on the fluid. However, because the recovery stroke takes place close to the surface, it would be essential to take into account the no-slip boundary conditions on the node. In fact, the two-phase beating might be a natural consequence of the no-slip boundary condition on the nodal surface. Thus, by appropriately taking this condition into account it might not even be necessary to prescribe the different beating phases.

Within their approach Buceta et al. obtain a somewhat more realistic characterization of the nodal flow. However, again, the results depend crucially on the prescribed microscopic force fields of the beating pattern and such studies do not reveal any universal properties of these systems. Thus, there is still quite a number of open issues that need to be addressed in future theoretical investigations. First of all, the no-slip boundary condition on the confining walls of the node is essential for the analysis because it not only influences the beating pattern of the cilium but also the backward flow due to mass conservation. Another question is how important synchronization of ciliar beating is. Does the flow from right to left depend on whether the cilia beat in a synchronized fashion or not? Similarly, the beating of the individual cilia and thus the nodal flow might be influenced by membrane fluctuations. The framework developed in [70, 71] provides the appropriate tools to investigate such phenomena. Finally, the influence of disorder on the nodal flow has to be analyzed. In particular, it has to be investigated whether a small number of cilia with a negative tilting angle ψ or opposite rotational direction is already sufficient to significantly alter the nodal flow as experimental studies on mutants seem to indicate.

8.5 Hydrodynamic Interactions in Other Biological Systems

Hydrodynamic interactions also play an important role in other biological systems. As an example, we discuss the properties of a collection of molecular motors exhibiting rotational motion such as ATPsynthase. The major difference from the ciliar arrays is that the motors are mobile. Thus, here hydrodynamic interactions also lead to collective motion and an effective repulsion (or attraction) between rotating motors. The analysis of this system will be presented in Section 8.5.1. In Section 8.5.2 the implications for possible technological applications will be discussed.

8.5.1 Collective Effects in Arrays of Rotatory Motors

Some molecular motors exhibit rotational motion. One of them is ATPsynthase, which is a truly astonishing molecular machine. Its function is to synthesize ATP. In doing so, a large enzymatic protruding portion F_1 is performing a rotation in a "reaction chamber" consisting of α and β subunits (see Figure 8.8). The F_1-part is attached to a membrane-embedded, proton-conducting portion F_0 [19]. It is believed that protons passing through the transmembrane carrier generate the rotation of the F_1 part [72, 73, 74]. When protons flow through F_0, ATP is synthesized in F_1. The motor is reversible and an excess of ATP provokes a rotation in the opposite direction and a reverse flux of protons.

In an ingenious experiment, Kinosita et al. have visualized the rotational motion of ATPsynthase. By attaching a fluorescently labeled actin filament to the F_1-part (see Figure 8.8) the rotation can be directly observed under an optical microscope [20]. With this technique, it is also possible to directly measure the torque τ transfered by the motor to the surrounding liquid. It turns out that $2\pi\tau \simeq 3\Delta G$, where $\Delta G = 20kT$ is the energy difference between ATP and ADP. Because every rotation consumes three ATP this little machine is astonishingly efficient.

In the chloroplasts of plants and in human mitochondria these motors are densely packed. Typically, a membrane of area $1\mu m^2$ contains around 1,000 motors. Thus, the mean distance between them is roughly 30nm. All motors rotate and the induced movement in the cell liquid influences the neighboring motors. Because the motors are mobile in the cell membrane, this induced velocity field gives rise to a repulsive hydrodynamic interaction as shown below (see Figure 8.9). We will now analyze the consequences of this interaction for the collective behavior in an array of motors and the associated order phenomena [42, 43].

To discuss these effects, we focus here on the most simple description of the rotating motor where it is modeled as a rotating sphere of radius R. Furthermore, we neglect the surface viscosity of the embedding membrane. For a more general analysis, see [42, 43]. Then, the rotation of motor i at position $\boldsymbol{x}_i$ induces a velocity field

$$\boldsymbol{v}_i = \frac{R^3}{r^3}\boldsymbol{\omega} \times \boldsymbol{r} = \frac{\tau}{8\pi\eta r^2}\boldsymbol{e}_\varphi, \qquad (8.37)$$

where $\tau = 8\pi\eta R^3\omega$ is the torque transfered by the rotating sphere on the surrounding liquid. The velocity field is in the direction of $\boldsymbol{e}_\varphi \equiv \boldsymbol{e}_z \times \boldsymbol{r}/(r\sin\theta)$, where in spherical coordinates $\boldsymbol{r} = r(\sin\theta\cos\varphi, \sin\theta\sin\varphi, \cos\theta)$ and $\boldsymbol{e}_z = (0,0,1)$.

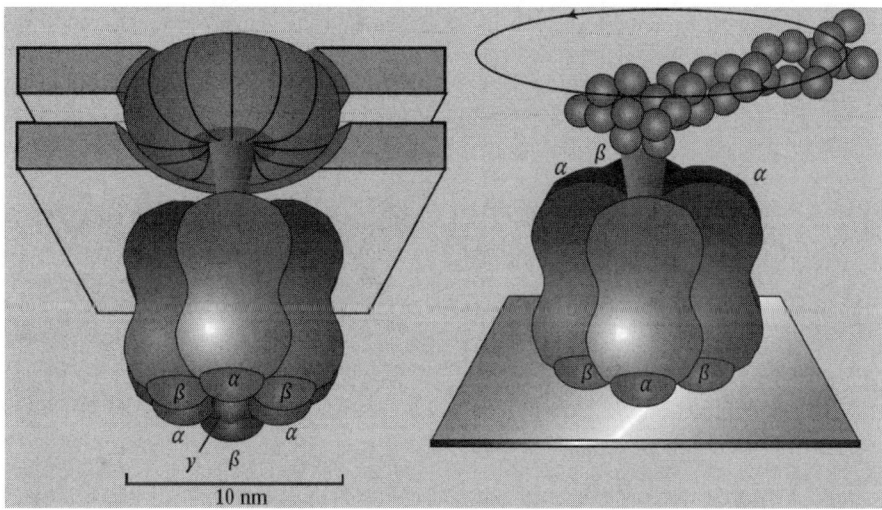

Figure 8.8. (Left) In eukaryotic cells, ATP is synthesized by the molecular motor ATPsynthase. It is believed that protons passing through the transmembrane carrier F_0 of ATPsynthase generate the rotation of F_1 in a "reaction chamber" consisting of α and β subunits. Per rotation three ADP molecules from the surrounding liquid are adsorbed at F_1 and converted to ATP. The motor is completely reversible and an excess of ATP causes rotation of the F_1 part in opposite direction and a reverse flux of protons. (Right) The rotation of the motor part has been visualized with fluorescently labeled actin filaments that were attached to the F_1-part of ATPsynthase (in the presence of ATP).

The velocity field $\boldsymbol{v}_i(\boldsymbol{x})$ puts the other motors $j \neq i$ into translational motion. At the same time, however, motor i follows the velocity field created by the other motors. Because rotational and translational motion of motor i are decoupled, the velocity (of the center of mass of motor i) is given by

$$\boldsymbol{v}_s = \frac{d\boldsymbol{x}_i}{dt} \equiv \boldsymbol{v}_t(\boldsymbol{x}_i) = R^3 \sum_{\substack{j \\ j \neq i}} \frac{\boldsymbol{\omega} \times (\boldsymbol{x}_i - \boldsymbol{x}_j)}{|\boldsymbol{x}_i - \boldsymbol{x}_j|^3} + \mathcal{O}(\frac{R^6 \omega}{|\boldsymbol{x}_i - \boldsymbol{x}_j|^5}). \tag{8.38}$$

The neglected terms arise from the no-slip boundary condition on the surface of motor i under the influence of $\boldsymbol{v}(\boldsymbol{x}_j)$. For small motor densities, these corrections are small.

Next, we consider a system of two motors with identical rotational vectors $\boldsymbol{\omega}$ which for simplicity are assumed to be perpendicular to the supporting membrane. Thus, both motors rotate with the same frequency in the same direction. Here, the hydrodynamic interaction leads to a circular motion around their center of mass. Because the axis of the rotation is perpendicular to the direction of translational motion an additional force acts on each motor: the Magnus force $\boldsymbol{F}_M \sim \boldsymbol{v} \times \boldsymbol{\omega}$. In Newtonian fluids, the Magnus force is a direct consequence of energy conservation and is thus an inertial effect (i.e., depends on the density of the surrounding liquid). In the overdamped systems considered here, this inertial Magnus force F_M is very small. However, in non-Newtonian fluids there is also a viscous Magnus force that depends on the microscopic relaxation time t_m of the additional components (such as polymers) in the surrounding liquid [75]. For the two motors that are moving around their center of mass, the Magnus force leads to an effective repulsion (see Figure 8.9).

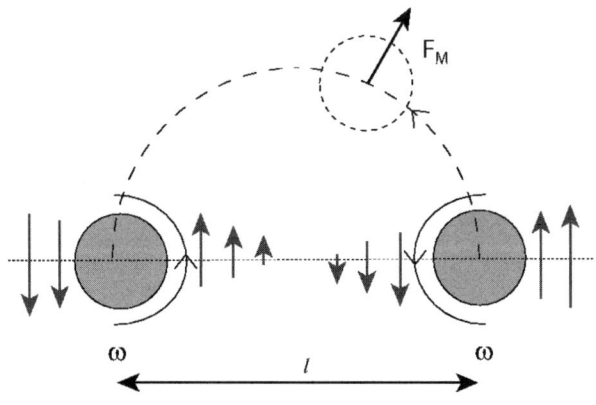

Figure 8.9. In a system with two rotating mobile motors, each motor is forced (by the velocity field created by the other motor) on a circular trajectory around the center of mass. Because the motors perform a rotational and a translational motion they are repelled from each other by the Magnus-force F_M. This repulsion minimizes the total kinetic energy in the surrounding liquid because as the distance l between the motors increases, the region of destructive interference (of the velocity fields created by the motors) becomes larger.

In systems with N rotating mobile motors, this repulsive interaction leads to the formation of a (Wigner) crystal. More precisely, if hydrodynamic interactions are strong enough (i.e., $\omega > \omega_c$), then the motors arrange on a lattice. For weak interactions (i.e., $\omega < \omega_c$) the lattice melts and the motors form a disordered phase. The melting point $\omega = \omega_c$ can be estimated by comparing two time scales. Namely, t_{fluct} (the typical time scale on which a fluctuation occurs which drives a motor away from its equilibrium lattice position) and t_{rel} (the time it takes the displaced motor to reach its initial position again). At the melting point, these two time scales become comparable (i.e., $t_{rel} = t_{fluct}$) [42, 43].

In summary, hydrodynamic interactions can also induce repulsions (or attractions) between mobile motors giving rise to ordering transitions such as the crystallization of rotating motors. Similar effects have been observed in bacterial layers [76]. Here, the competition between advection by the bacteria-generated flow and diffusion can lead to an instability driving bacterial aggregation.

8.5.2 Applications

Collective effects mediated by hydrodynamic interactions might also become important for the production and design of artificial micro-/nano-devices and actuators. For this purpose, it is advantageous to use other than membrane-embedded rotational motors. For example, the group of Whitesides has demonstrated that magnetic disks can be confined to a liquid surface and then brought to rotation by an external magnetic field [77, 78]. Because the position of these motors can be easily controlled, it is possible to realize prescribed velocity fields. In first applications, simple sorting devices and "carousel" systems where built with several motors [79]. Similar approaches use artificial rotors stirred by Laser tweezers [80] or light-torque operated optically trapped nanorods [81].

The next step toward more complex systems is the production of artificial rotational molecular motors exhibiting motion similar to ATPSynthase. An example has been given in [82], where a small (molecular) ring rotates in a synthetic molecular structure. The positional displacements are caused by biased Brownian motion. The sense of rotation is determined by the order in which a series of orthogonal chemical transformations is performed. It might even be possible to make use of red blood cells as motors. When optically trapped, they undergo rotational motion under the influence of circularly polarized light [83].

There has also been considerable experimental efforts to fabricate hybrid micro- or nanoactuators driven by biological molecules [84, 85, 86]. Recently, Hiratsuka et al. were able to use the gliding bacterium *Mycoplasma mobile* to power the rotational motion of an artificial silicon dioxide rotor (of 20μm diameter) [87]. Bacteria were put onto a substrate (with adsorbed fetuin to promote bacterial gliding), which ended in a circular groove track. Onto this track the rotor is docked, to which the bateria bind by biotin-streptavidin interactions. The crawling (bound) bacteria put the rotor into motion. Typical rotation rates were in the range of 2 rpm. It was also found that the torque generated by an individual cell is in the range $2-5\cdot10^{-16}$Nm, which is about four orders of magnitude smaller than that of typical electrostatic microactuators.

Finally, artificial motor-generated flow over surfaces and membranes have been experimentally realized by using "bacterial carpets" [88]. Flagellated swarmer cells of *Serratia mercescens* were adsorbed to a solid surface where they formed a densely packed monolayer with freely rotating flagella. The induced flow can be visualized by tracer beads. The rotation of the flagella gives rise to complex flow patterns consisting of regions with linear and rotational flow. So far, these effects are relatively small. They give rise to enhanced diffusion characterized by a diffusion coefficient of order $D \simeq 20\mu$m^2/s. However, coordination of rotational motion should lead to larger effects with longer range.

Acknowledgements

This work has been supported in part by the Fonds der Chemischen Industrie.

A Short Summary of Overdamped Hydrodynamics

The velocity field $v(r)$ of incompressible fluid flow obeys the Navier-Stokes equation

$$\rho \frac{\partial v}{\partial t} + \rho(v \cdot \nabla)v = -\nabla p + \eta \nabla^2 v. \qquad (8.39)$$

Here, $v = v(r)$ is the velocity in the fluid at position r, $p = p(r)$ the distribution of pressure, and ρ the density of the fluid with viscosity η. The incompressibility of the fluid leads to

$$\text{div } v = 0. \qquad (8.40)$$

The ratio between convective and viscous terms is given by the Reynolds number

$$\text{Re} \equiv \frac{\rho V L}{\eta}, \qquad (8.41)$$

where V and L are a typical velocity and length scale associated with the flow. Here, we are studying small objects moving with small velocities implying $\text{Re} \ll 1$. In this case, the velocity field is determined by Stokes equation[4]

$$\nabla p = \eta \nabla^2 v. \qquad (8.42)$$

The flow field generated by a moving sphere of radius a at large distances $r \gg a$ can be approximated by the flow due to a point force acting within the fluid. The corresponding velocity field is that of a stokeslet moving on a trajectory r_t and obeys

$$\nabla p = \eta \nabla^2 v - f \delta(r - r_t), \qquad (8.43)$$

where $\delta(r)$ is Dirac's delta function. The Green's function (or Oseen tensor) of the last equation is given by

$$G_{\alpha\beta}^S(r) = \frac{1}{8\pi\nu} \left(\frac{\delta_{\alpha\beta}}{r} + \frac{r_\alpha r_\beta}{r^3} \right), \qquad (8.44)$$

where $\delta_{\alpha\beta}$ is Kronecker's delta (i.e., $\delta_{\alpha\beta} = 1$ if $\alpha = \beta$ and $\delta_{\alpha\beta} = 0$ if $\alpha \neq \beta$). The corresponding velocity field is given by $v(r) = G^S(r)f(r)$, which reduces to Equation (8.2).

In the presence of a no-slip wall, the velocity field must vanish on the surface of the wall. As shown by Blake [89], mirror images can be used to fulfill this boundary condition. This is a generalization of a method known from electrostatics where a charge image is introduced to cancel a charge distribution's field on a surface. However, in hydrodynamics it is not sufficient to simply take into account a mirror stokeslet. Additional contributions, denoted as doublets (D) and source doublets (SD) are required. Thus, for a stokeslet at position $r_i = (x_i, y_i, z_i)$ in the presence of a no-slip wall at $z = 0$ the Green's function becomes

$$G_{wall}(r_i, r) = G^S(r - r_i) - G^S(r - r_i') + G^D(r - r_i') - G^{SD}(r - r_i'), \qquad (8.45)$$

where $r_i' = r_i - 2z_i e_z$ is the position of the image stokeslet and G^D and G^{SD} are the contributions of the double and source doublet [89].

---

[4] However, one should note that in large systems even for slow motion momentum injection is not instantaneous everywhere, which has to be taken into account by considering time-dependent velocity fields.

B Derivation of the Phase-Oscillator Equation of Motion

Here, we derive the equation of motion (8.27). The velocity field v_s is (as a solution of the linear Stokes equation) a linear functional of $\dot{r}(s,t)$, i.e., $v_s = \widehat{V}^s\,[\partial_t r(s,t)]$. A more explicit expression for the solution can be found in [35]. To demonstrate the influence of the interaction term v_n, we will now consider two neighboring ciliar Filaments 1 and 2 at positions $r_1 = r_1(s,\varphi_1)$ and $r_2 = r_2(s,\varphi_2)$. Then, v_n at position r_1 is a linear functional of $\dot{r}_2$, i.e., $v_n = \widehat{V}^n\,[\partial_t r_2(s,t)]$. Equation (8.26) implies $v_s = \dot{\varphi}_1(t)\widehat{V}^s\,[\partial_{\varphi_1} r_1(s,\varphi_1)]$ and $v_n = \dot{\varphi}_2(t)\widehat{V}^n\,[\partial_{\varphi_2} r_2(s,\varphi_2(t))]$. Thus, Equation (8.25) becomes

$$\zeta_{ij}\left(\dot{\varphi}_1 \partial_{\varphi_1} r_{1i}(s,\varphi_1) - \dot{\varphi}_1 \widehat{V}_i^s\,[\partial_{\varphi_1} r_1(s,\varphi_1)] - \dot{\varphi}_2 \widehat{V}_i^n\,[\partial_{\varphi_2} r_2(s,\varphi_2)]\right)$$
$$= F_j\left(r_1(s,\varphi_1),s\right), \tag{8.46}$$

where r_{1i} and $\widehat{V}_i^s\,[\partial_{\varphi_1} r(s,\varphi_1)]$ denote the i-th component of r_1 and v_s, respectively. By projecting the forces Filament 2 exerts on Filament 1 onto the tangential vector of the trajectory of Filament 1 and upon summing over all filament pieces (i.e., stokeslets) Equation (8.46) becomes

$$\dot{\varphi}_1 \mathcal{F}_d(\varphi_1) - \dot{\varphi}_2 \mathcal{I}(\varphi_1,\varphi_2) = \mathcal{F}(\varphi_1). \tag{8.47}$$

Here,

$$\mathcal{F}_d(\varphi_1) = \int_0^L ds\,\partial_{\varphi_1} r_{1j}(s,\varphi_1)\zeta_{ij}\left(\partial_{\varphi_1} r_{1i}(s,\varphi_1) - \widehat{V}_i^s\,[\partial_{\varphi_1} r_1(s,\varphi_1)]\right), \tag{8.48}$$

$$\mathcal{I}(\varphi_1,\varphi_2) = \int_0^L ds\,\partial_{\varphi_1} r_{1j}(s,\varphi_1)\zeta_{ij}\widehat{V}_i^n\,[\partial_{\varphi_2} r(s,\varphi_2)], \tag{8.49}$$

$$\mathcal{F}(\varphi_1) = \int_0^L ds\,\partial_{\varphi_1} r_{1j}(s,\varphi_1)F_j\left(r_1(s,\varphi_1),s\right), \tag{8.50}$$

where $r_1(s,\varphi_1)$ is the solution of the Equation (8.25) without interactions [where $\varphi = 2\pi t$]. Because then $2\pi\mathcal{F}_d(\varphi) = \mathcal{F}(\varphi)$, one finds Equation (8.27), where now

$$J(\varphi_1,\varphi_2) \equiv -2\pi\mathcal{I}(\varphi_1,\varphi_2)/\mathcal{F}(\varphi_1). \tag{8.51}$$

Equations (8.19), and (8.48)-(8.50) are valid for arbitrary beating patterns even for arrays of cilia with different (individual) beating patterns. This description is a generalization of the model for interacting monocilia introduced in Section 8.2.1. In this case, $r(s,\varphi(t)) = r_t = (R\cos\varphi(t), R\sin\varphi(t), h)\delta(s - L)$, with $\delta(s)$ denoting Dirac's delta function and $s = L$ is the position of the tip. Furthermore, $\widehat{V}_i^s = 0$, $\widehat{V}_i^n\,[\partial_{\varphi_2} r(s,\varphi_2)] = v_{12}/\dot{\varphi}_2$, where v_{12} is given by Equations (8.2) or (8.23). Then, $\mathcal{I}(\varphi_1,\varphi_2) = R\zeta t_1 \cdot v_{12}/\dot{\varphi}_2$, $\mathcal{F}(\varphi) = Rt_1 \cdot F = RF_{in}$, and $\mathcal{F}_d(\varphi) = R^2\zeta$.

In principle, a continuous description of the bending deformation of the filament could be used to calculate $J(\varphi_1,\varphi_2)$. However, here we will use a different approach. For a system of cilia with identical beating patterns, one has $r_m = r_m(s = 0) + r_t(s,\varphi_m)$, where and r_t parameterizes the time-dependent beating pattern. Because all cilia have the same beating pattern, r_t depends only on φ_m. Because $0 \le \varphi_m \le 2\pi$ the function $J(\varphi_1,\varphi_2)$ is periodic in both arguments and is given by Equation (8.28).

References

1. DePamphilis, M & Adler, J. (1971) *J. Bacteriol.* **105**, 384–95.
2. DePamphilis, M & Adler, J. (1971) *J. Bacteriol.* **105**, 396–407.
3. Berg, H & Anderson, R. (1973) *Nature* **245**, 380–2.
4. Silverman, M & Simon, M. (1974) *Nature* **249**, 73–4.
5. Bray, D. (2001) *Cell Motility.* (Garland, New York).
6. Berg, H. (1993) *Random Walks in Biology.* (Princeton University Press, Princeton).
7. Vogel, S. (1996) *Life in Moving Fluids: The Physical Biology of Flow.* (Princeton University Press, Princeton).
8. Marshall, W & Nonaka, S. (2006) *Curr. Biol.* **16**, R604–14.
9. Alberts, B & et al. (2002) *Molecular Biology of the Cell, 4th ed.* (Garland, New York).
10. Blake, J & Sleigh, M. (1974) *Biol. Rev. Camb. Philos. Soc.* **49**, 85–125.
11. Nonaka, S, Tanaka, Y, Okada, Y, Takeda, S, Harada, A, Kanai, Y, Kido, M, & Hirokawa, N. (1998) *Cell* **95**, 829–37.
12. Wheatley, D. (2004) *Cell. Biol. Int.* **28**, 75–7.
13. Hirokawa, N, Tanaka, Y, Okada, Y, & Takeda, S. (2006) *Cell* **125**, 33–45.
14. Tamm, S. (1984) *J. Exp. Biol.* **113**, 401–408.
15. Tamm, S, Sonneborn, T, & Dippell, R. (1975) *J. Cell Biol.* **64**, 98–112.
16. Knight-Jones, E. (1954) *Q. J. Microsc. Sci.* **95**, 503–21.
17. Jahn, T & Votta, J. (1972) *Ann. Rev. Fluid Mech.* **4**, 93–116.
18. Aiello, E & Sleigh, M. (1972) *J. Cell. Biol.* **54**, 493–506.
19. Yoshida, M, Muneyuki, E, & Hisabori, T. (2001) *Nat. Rev. Mol. Cell. Biol.* **2**, 669–77.
20. Kinosita, K, Yasuda, R, Noji, H, Ishiwata, S, & Yoshida, M. (1998) *Cell* **93**, 21–4.
21. Berg, H. (2000) *Physics Today, January* pp. 25–29.
22. Purcell, E. (1997) *Proc. Natl. Acad. Sci. USA* **94**, 11307–11.
23. Berg, H. (2003) *Annu. Rev. Biochem.* **72**, 19–54.
24. Zagoory, O, Braiman, A, Gheber, L, & Priel, Z. (2001) *Am. J. Physiol. Cell. Physiol.* **280**, C100–9.
25. Okamoto, K & Nakaoka, Y. (1994) *J. Exp. Biol.* **192**, 61–72.
26. Blake, J. (1972) *J. Fluid Mech.* **55**, 1–23.
27. Brennen, C & Winet, H. (1977) *Ann. Rev. Fluid Mech.* **9**, 339–398.
28. Liron, N & Mochon, S. (1976) *J. Fluid Mech.* **75**, 593–607.
29. Lighthill, M. (1952) *Comm. Pure Appl. Math.* **5**, 109–118.
30. Blake, J. (1971) *J. Fluid Mech.* **46**, 199–208.
31. Cartwright, J, Piro, O, & Tuval, I. (2004) *Proc. Natl. Acad. Sci. USA* **101**, 7234–9.
32. Lenz, P & Ryskin, A. (2006) *Phys. Biol.* **3**, 285–94.
33. Landau, L & Lifshitz, E. (1959) *Hydrodynamics.* (Pergamon, New York).
34. Lenz, P. (2007) *to be published.*
35. Gueron, S & Liron, N. (1992) *Biophys. J* **63**, 1045–1058.
36. Camalet, S & Jülicher, F. (2000) *New J. Phys.* **2**, 24.1–24.23.
37. Gueron, S, Levit-Gurevich, K, Liron, N, & Blum, J. (1997) *Proc. Natl. Acad. Sci. USA* **94**, 6001–6.
38. Gueron, S & Levit-Gurevich, K. (1998) *Biophys. J.* **74**, 1658–76.

39. Gueron, S & Levit-Gurevich, K. (2001) *Proc. Biol. Sci.* **268**, 599–607.
40. Kim, Y & Netz, R. (2006) *Phys. Rev. Lett.* **96**, 158101.
41. Ermak, D & McCammon, J. (1978) *J. Chem. Phys.* **69**, 1352–1360.
42. Lenz, P, Joanny, J, Julicher, F, & Prost, J. (2003) *Phys. Rev. Lett.* **91**, 108104.
43. Lenz, P, Joanny, J, Julicher, F, & Prost, J. (2004) *Eur. Phys. J. E* **13**, 379–90.
44. Reichert, M & Stark, H. (2005) *Eur. Phys. J. E* **17**, 493–500.
45. Lagomarsino, M, Jona, P, & Bassetti, B. (2003) *Phys. Rev. E* **68**, 021908.
46. Lenz, P & Eckhardt, B. (2007) *to be published.*
47. Vilfan, A & Julicher, F. (2006) *Phys. Rev. Lett.* **96**, 058102.
48. Acebrón, J. A, Bonilla, L. L, Pérez Vicente, C. J, Ritort, F, & Spigler, R. (2005) *Rev. Mod. Phys.* **77**, 137–185.
49. Gilbert, S. (2003) *Developmental Biology (7th ed.).* (Sinauer, Sunderland, Ma).
50. Beddington, R & Robertson, E. (1999) *Cell* **96**, 195–209.
51. Alarcon, V & Marikawa, Y. (2003) *Biol. Reprod.* **69**, 1208–12.
52. Tsang, T, Kinder, S, & Tam, P. (1999) *Cell. Mol. Biol.* **45**, 493–503.
53. Brown, N & Wolpert, L. (1990) *Development* **109**, 1–9.
54. Okada, Y, Nonaka, S, Tanaka, Y, Saijoh, Y, Hamada, H, & Hirokawa, N. (1999) *Mol. Cell* **4**, 459–68.
55. Essner, J, Vogan, K, Wagner, M, Tabin, C, Yost, H, & Brueckner, M. (2002) *Nature* **418**, 37–8.
56. Okada, Y, Takeda, S, Tanaka, Y, Belmonte, J, & Hirokawa, N. (2005) *Cell* **121**, 633–44.
57. Zhou, X, Sasaki, H, Lowe, L, Hogan, B, & Kuehn, M. (1993) *Nature* **361**, 543–7.
58. Collignon, J, Varlet, I, & Robertson, E. (1996) *Nature* **381**, 155–8.
59. Meyers, E & Martin, G. (1999) *Science* **285**, 403–6.
60. Hamada, H, Meno, C, Watanabe, D, & Saijoh, Y. (2002) *Nat. Rev. Genet.* **3**, 103–13.
61. Takeda, S, Yonekawa, Y, Tanaka, Y, Okada, Y, Nonaka, S, & Hirokawa, N. (1999) *J. Cell. Biol.* **145**, 825–36.
62. Nonaka, S, Shiratori, H, Saijoh, Y, & Hamada, H. (2002) *Nature* **418**, 96–9.
63. Hummel, K. P & Chapman, D. B. (1959) *J. Hered.* **50**, 8–13.
64. Yokoyama, T, Copeland, N, Jenkins, N, Montgomery, C, Elder, F, & Overbeek, P. (1993) *Science* **260**, 679–82.
65. Tabin, C & Vogan, K. (2003) *Genes Dev.* **17**, 1–6.
66. McGrath, J, Somlo, S, Makova, S, Tian, X, & Brueckner, M. (2003) *Cell* **114**, 61–73.
67. Tanaka, Y, Okada, Y, & Hirokawa, N. (2005) *Nature* **435**, 172–7.
68. Tabin, C. (2006) *Cell* **127**, 27–32.
69. Buceta, J, Ibanes, M, Rasskin-Gutman, D, Okada, Y, Hirokawa, N, & Izpisua-Belmonte, J. (2005) *Biophys. J.* **89**, 2199–209.
70. Lenz, P & Nelson, D. (2001) *Phys. Rev. Lett.* **87**, 125703.
71. Lenz, P & Nelson, D. (2003) *Phys. Rev. E* **67**, 031502.
72. Boyer, P. (1993) *Biochim. Biophys. Acta* **1140**, 215.
73. Abrahams, J, Leslie, A, Lutter, R, & Walker, J. (1994) *Nature* **370**, 621–8.
74. Noji, H, Yasuda, R, Yoshida, M, & Kinosita, Jr, K. (1997) *Nature* **386**, 299–302.
75. Brunn, P. (1976) *Rheol. Acta* **15**, 163.
76. Cogan, N & Wolgemuth, C. (2005) *Biophys. J.* **88**, 2525–9.
77. Grzybowski, B & Stone, H.A.and Whitesides, G. (2000) *Nature* **405**, 1033.
78. Grzybowski, B, Jiang, X, Stone, H, & Whitesides, G. (2001) *Phys. Rev. E* **64**, 011603.

79. Grzybowski, B, Radkowski, M, Campbell, C, Lee, J, & Whitesides, G. (2004) *Appl. Phys. Lett.* **84**, 1798–1800.
80. Galajda, P & Ormos, P. (2002) *Appl. Phys. Lett.* **80**, 4653.
81. Bonin, K, Kourmanov, B, & Walker, T. (2002) *Optics Exp.* **10**, 984–989.
82. Hernandez, J, Kay, E, & Leigh, D. (2004) *Science* **306**, 1532–7.
83. Dharmadhikari, J, Roy, S, Dharmadhikari, A, Sharma, S, & Mathur, D. (2004) *Appl. Phys. Lett.* **85**, 6048–6050.
84. Hess, H & Vogel, V. (2001) *J. Biotechnol.* **82**, 67–85.
85. Jia, L, Moorjani, S, Jackson, T, & Hancock, W. (2004) *Biomed. Microdev.* **6**, 67–74.
86. Soong, R, Bachand, G, Neves, H, Olkhovets, A, Craighead, H, & Montemagno, C. (2000) *Science* **290**, 1555–8.
87. Hiratsuka, Y, Miyata, M, Tada, T, & Uyeda, T. (2006) *Proc. Natl. Acad. Sci. USA* **103**, 13618–23.
88. Darnton, N, Turner, L, Breuer, K, & Berg, H. (2004) *Biophys. J.* **86**, 1863–70.
89. Blake, J. (1971) *Proc. Camb. Phil. Soc.* **70**, 303–310.

Index

Printed in the United States of America